POWER AFTER CARBON

POWER AFTER CARBON

BUILDING A CLEAN, RESILIENT GRID

PETER FOX-PENNER

Harvard University Press
Cambridge, Massachusetts, and London, England 2020

Printed in the United States of America

First printing

Library of Congress Cataloging-in-Publication Data

Names: Fox-Penner, Peter S., 1955– author.
Title: Power after carbon : building a clean, resilient grid / Peter Fox-Penner.
Description: Cambridge, Massachusetts : Harvard University Press, 2020. | Includes bibliographical references and index.
Identifiers: LCCN 2019045200 (print) | LCCN 2019045201 (ebook) | ISBN 9780674241077 (cloth) | ISBN 9780674245624 (epub) | ISBN 9780674245631 (mobi) | ISBN 9780674245648 (pdf)
Subjects: LCSH: Electric power systems—United States. | Renewable energy sources—United States.
Classification: LCC TK3001 .F69 2020 (print) | LCC TK3001 (ebook) | DDC 621.31028/6—dc23
LC record available at https://lccn.loc.gov/2019045200
LC ebook record available at https://lccn.loc.gov/2019045201

For Susan, Emmy, Jake, and Plum

No longer can we tear the world apart to make our fire.
The Gates of Prayer

When you have exhausted all possibilities, remember this—
you haven't.
Thomas Edison

Contents

Preface

After *Smart Power* (*SP*) was published in 2010, I spent the next few years pretending that writing more books about electric power was no longer my destiny. I was encouraged to see the debate over whether the electric industry needed fundamental change settled much more quickly and decisively than I feared when I wrote *SP*. Utility managements, outside stakeholders, and most importantly industry regulators seemed to embrace the idea; writings, conferences, and regulatory activity on the "utility of the future" seemed to be sprouting everywhere. We were now into tactics and execution: how-to manuals for industry specialists, not conceptual arguments that might be interesting to both generalists and insiders.

Yet, I gradually became nagged by the notion that the important policy issues were changing. One review of *SP* on Amazon began by saying, "it's now a little bit dated, but . . ." Once I began noodling on what an update would look like, I quickly realized that I had far underestimated how much the frame had shifted. Climate change had become the industry's (and one of civilization's) most urgent public policy challenges, and electrification had become a critical pathway to decarbonization. The latter made the sheer pace of industry expansion, simultaneous with a massive shift in generation resources and industry structure, a global imperative. Speed to the ball and sound execution were now paramount.

Rather than becoming an *SP* update, this book became a more current, wider, and sometimes deeper companion piece. If *SP* is the text for Future Utilities 100, I hope that *Power after Carbon* supports the next course in the sequence. I have tried to make it nearly as accessible to non-specialists as *SP* has proven to be, and I hope it sparks an even greater pace of constructive change.

From Facebook posts to refereed academic papers, transparency regarding sources and support is becoming recognized as an essential ethical dimension of any published work. Toward that end, I close by noting that during the time I wrote this book at the Boston University Institute for Sustainable Energy, the Institute was supported by the Hewlett Foundation, the Energy Foundation, and Bloomberg Philanthropies, among others. The book itself did not receive any direct support from any funder. In addition, I am fortunate to serve as Chief Strategy Officer at Energy Impact Partners, which invests in many clean technology companies (listed at www.energyimpactpartners.com) and where I hold an equity interest. I also hold a small equity-like interest in, and serve as academic advisor to, the Brattle Group, as well as options in EOS Energy Storage.

I am grateful for another chance to contribute to the electricity industry's transition, and to all the people and organizations listed in the Acknowledgments. All errors, of fact and judgment, are my own.

Abbreviations

AC	alternating current
ACEEE	American Council for an Energy Efficient Economy
AI	artificial intelligence
AV	autonomous vehicle
BA	balancing area
BAA	balancing area authority
CCA	Community Choice Aggregation
CCS	carbon capture and storage
CCUS	carbon capture, utilization, and storage
CSAT	customer satisfaction
DC	direct current
DDPP	U.S. Deep Decarbonization Pathways Project
DER	distributed energy resources
DG	distributed generation
DLMP	distribution locational marginal pricing
DOE	U.S. Department of Energy
DR	demand response
DSO	Distribution System Operator
EAAS	energy as a service
ED	energy democracy
EE	energy efficiency
EIA	U.S. Energy Information Administration
EIRP	Energy Innovation Reform Project
ESCO	energy service company
ESU	Energy Service Utility

EV	electric vehicle
FERC	Federal Energy Regulatory Commission
GDP	gross domestic product
GHG	greenhouse gas
GMP	Green Mountain Power
GND	Green New Deal
HVDC	high-voltage direct current
ICT	information and communications technology
IEA	International Energy Agency
IOT	Internet of Things
IOU	investor-owned electric distribution utilities
IPCC	Intergovernmental Panel on Climate Change
IPP	independent power producer
ISO	independent system operator
KPI	key performance indicator
kWh	kilowatt hour
LBNL	Lawrence Berkeley National Labs
MCS	Mid-Century Strategy
NCA	U.S. National Climate Assessment
NEEP	Northeast Energy Efficiency Partnership
NOAA	National Oceanic and Atmospheric Administration
NREL	National Renewable Energy Lab
NWA	non-wires alternative
PBR	performance based regulation
POLR	Provider of Last Resort
PPA	power purchase agreement
PQR	power quality and reliability
PSH	pumped storage hydro
PTG	power-to-gas
P2G2P	power-to-gas-to-power
PV	photovoltaic solar
PwC	PricewaterhouseCoopers
R&D	research and development
RIIO	Revenue = Incentives + Innovation + Outputs
RMI	Rocky Mountain Institute
RPS	renewable portfolio standards
RTO	regional transmission operator

SECC	Smart Energy Consumer Collaborative
SI	Smart Integrator
SMR	small modular nuclear reactor
STEM	science, technology, engineering, and mathematics
TO	transmission operator
TOD	time-of-day pricing
TSO	transmission system operator
UNFCCC	United Nations Framework Convention on Climate Change
VRE	variable renewable energy

POWER AFTER CARBON

PART I

The Need for Power and the Grids That Deliver It

1 Les Jeux Sont Faits

Walk over to the roulette wheel in a French casino and you will hear the croupier say "Les jeux sont faits"—literally *the plays are made*—as she or he launches the wheel. The colloquial translation, *the chips are down,* is now a widely used English metaphor.

When Yale economist William Nordhaus searched for a way to describe the risks of climate change to human civilization, he settled on this metaphor, likening our greenhouse gas (GHG) emissions to betting our future in a casino. "By this I mean that economic growth is producing unintended but perilous changes in the climate and earth systems," he wrote in *The Climate Casino,* a book that would land him a Nobel Prize. "These changes will lead to unforeseeable and probably dangerous consequences. We are rolling the climatic dice."[1]

Writing in 2013, Nordhaus saw hope for climate progress: "But we have just entered the Climate Casino, and there is time to turn around and walk back out." Since then, scientific understanding of climate change has continued to advance, and an ever-stronger scientific consensus is warning us that the dangers of climate change are coming on faster than we anticipated.[2] Contrary to Nordhaus's exhortation, it may be that we have already placed most of the bets on our climate future. Rather than leave the casino, we are called to do everything we can to reduce climate change as much as possible before the roulette wheel stops.

The threats from human-induced global warming are of epic proportions: superstorms that kill, displace millions, and damage billions of dollars of property; droughts that turn fertile farm areas into dust-bowls, triggering food price spikes and shortages; loss of biodiversity and natural habitat unseen in human history; dangerous, fast-moving disease vectors triggered by weather shifts.[3] The last U.S. National Security Strategy before Trump, issued in February 2015, says in its official Defense Department summary that "[it] is clear that climate change is an urgent and growing threat to our national security, contributing to increased natural disasters, refugee flows, and conflicts over basic resources such as food and water."[4] The late physicist Stephen Hawking, often referred to as the smartest person of his generation, ranked global climate change and nuclear war as the two greatest threats facing humankind over the next thousand years.[5] In *The Uninhabitable Earth,* David Wallace-Wells likens it to the five prior mass extinctions, each of which wiped out 75 percent or more of life on earth.[6]

As the world labors to implement the Paris climate accord, all world leaders—save one—agree that climate change must be limited to an average of two degrees Celsius or less. Scientists tell us this requires reductions of GHG emissions by at least 80 percent by 2050—"80 by 50" in climate shorthand. Yet, the energy systems of the world now rely on carbon-emitting fuels for more than 81 percent of all energy used—more than 83 percent in the United States.[7] Fossil fuels provide nearly all transport energy in the world (92 percent in the United States) and just over two-thirds of global electricity.[8] The latest report from the International Energy Agency on global CO_2 emissions from energy production contains astonishingly bad news. Global carbon dioxide emissions *grew* 1.7 percent to an all-time high of 33.1 billion metric tons. U.S. energy-related CO_2 grew by an even larger amount, 3.1 percent, in spite of the fact that coal use for electric power is at a forty-year low in the United States. Europe was the only region of the world where CO_2 from energy declined—by about 1.3 percent.[9]

This calls for rapid change in the power industry's generation stage at a pace far faster than the typical pace—faster than any peacetime industry has ever changed out its production stock.[10] The span and space of change poses titanic challenges for the existing power systems in the developed world—the focus of this volume. Many of these challenges also apply to utilities in emerging economies, but these nations have many unique, added challenges that are beyond our own scope.

Leapfrogging and Euthanasia

In developed countries, almost all power is made in large plants and delivered by the giant grid that laces the countryside. It is still true today that nearly 100 percent of power in most developed countries comes from large sources.

But what about tomorrow? There are really three visions of the power industry of the future. The first is an overwhelmingly local series of energy-independent cities, community grids and microgrids, all powered by clean energy. The second is a predominantly deregulated, transaction-centric marketplace of "prosumers" who make some of their own power, buy the rest, and rely on a large regulated grid as the essential trading platform. The third vision is a transformation of our current utilities into advanced service providers, still regulated and still relying on large and small sources.

In developing countries, the visions are nearly reversed. In the poorest nations of Africa, the grid supplies perhaps three in ten families, often with very unreliable power.[11] It has proven very difficult to expand the grid so as to achieve universal coverage. As a result, a new group of government units, companies, and nongovernmental organizations are increasingly turning to isolated micro- or community grids to supply power. Small, isolated systems that power minimal basic energy service needs, such as lights, water pumps, and cell phones, are increasingly common. It is logical to ask here whether we will simply leapfrog the large power grid and go straight to a preponderance of small- and no-grid regions, much as cell service has leapfrogged land lines in many of these same countries.

Let's put it bluntly. If the inevitable trend is toward the first vision of a fully distributed grid, we should set policies to accelerate this trend and stop building the large power plants and transmission lines. Isn't this how nations and economies win in an increasingly global world—exit rapidly from obsolete and dirty technologies and deploy the best, cleanest, smartest alternatives as fast as they can?

Or should we be talking about the second or third vision, where large power plants, our aging grid, and an explosion of distributed sources must learn to coexist? If so, how do we sort out the size and scale of the plants on the power grid, the amount of transmission and distribution we need, the role of regulation and markets, and the financing and operation of the entire system? What is the path to get from here to there? How easy lies the truce between these vastly different visions for a sector that is the lifeblood of our economy and our only real hope for stopping climate change?

Beyond Decarbonization

While the industry rapidly decarbonizes, it must continue to deliver on its other key performance objectives. In modern economies, electricity must be universally accessible and affordable, highly reliable and abundant, and secure from physical and cyberattacks. As smart energy control and use technologies multiply, we want an industry that makes these innovations widely available. The environmental impacts of the system, including GHG emissions, must remain acceptably low and equitably distributed. With very little electricity storage available today, even small interruptions are costly and dangerous; large-scale blackouts can be catastrophic.

It is easy to describe future visions of the sector as a set of most-desired or lowest-cost technologies. In the fully decentralized future vision, large distant power plants are replaced by distributed and community generators, strong and widely deployed efficiency technologies, and a smart grid that effectively integrates many small sources, storage technologies, and intelligent apps. Another technological future is anchored by large wind and solar plants built where the wind and sun are strong and an expanded grid that sends clean power into the cities of the world. These visions are not choices in overall climate policy; in both visions, we all but eliminate

GHG emissions from the power sector one way or another. One technological pathway might do this a little faster or cheaper, but failing to decarbonize is not an option.

But it is not nearly enough to describe, or advocate for, one or more of these technological futures. As we saw when Hurricane Maria destroyed Puerto Rico's grid in 2017, new power systems do not drop out of the sky. Every piece of technology must be purchased, financed, paid for, and operated by some combination of for-profit or nonprofit energy firms; utilities owned by customers, investors, or local governments; and ultimate consumers. Reliable and economical grids require extensive design engineering, operating rules and standards, and active control on a second-by-second basis. It is almost a reverse image of the Internet, where we need almost no controls whatsoever on the number or type of devices, users, and apps. No other industry can match the degree to which power systems are technologically and economically interdependent.

The core of the system, the power grid itself, is a fully regulated or publicly owned business unit with a long, complex change horizon. The grid, a single integrated machine of continental size, is owned by dozens of disparate firms and run by operators balancing thousands (soon millions) of power sources in real time. Its finances and operations are 100 percent controlled by public agencies operating under aging statutes and economic models. One can design and build a building or a new appliance in a year or two and a new efficient appliance or electrified automobile in well under a decade. Power grids take decades to plan, site, and build, often inspiring monumental opposition and conflict along the way.

Beyond these technological limits, the collection of firms and utilities that make up the power sector must be financially sustainable under acceptable political terms. The unregulated firms must see enough profit to raise and spend capital, operating consistently with the rules set by the grid. The regulated or publicly owned parts of the industry must work well enough to deliver reliable universal service while adhering to due process and other legal requirements. They must also be able to raise the capital they need from regulated or market revenues.

None of these conditions can be guaranteed simply because elementary calculations demonstrate that the total generating capacity of a power system is as large as total customer demand—the customary measure of a power grid's adequacy. When the degree of interdependence in design and operation is added in, along with the need for fiscal soundness, many variations on the primary visions of the industry's technological structure simply won't deliver the key outcomes we need. In a nutshell, our challenge is to steer clear of the technical and institutional pathways that together yield poor service, expensive power, or a failure to decarbonize quickly.

And there is more to it than simply protecting against downsides that jeopardize fast progress. Technological change is enabling a power network that is far more customer friendly, controllable, decentralized, and optimized for greater efficiency. Low cost, universal service, and zero carbon continue to be indispensable, but they are no longer the sole criteria by which industry arrangements and individual suppliers will be judged. Looking toward other sectors, customers in advanced countries will demand a growing range of products and services to choose and control and a modern customer experience. Network operators must run a digital, distributed power system not simply to keep reliability high, but to manage the system to squeeze out higher levels of efficiency and service and provide for other goals.

In this book, I provide an overview of the main choices we face along the way and directions we should head off in to prevent an economic and climate meltdown. In Part I, I describe the need for electricity as a climate solution and the role of distributed resources and local power. This begins with the long-term drivers of power demand and the resulting need for power, which turns out to be at least half again what we use today, if not more than double (Chapter 2). Chapter 3 examines the amount of this power we can expect to generate locally by 2050, and save through energy efficiency, without the need for a large-scale, distant power grid. The answer turns out to be quite a lot—but in almost all cases, we will continue to need large external supplies for our increasingly urbanized world. This leads to Chapter 4's look at the functions of the grid and how they vary with geographic scale, from community-sized grids to supergrids that span nations and continents. Finally, Chapter 5 describes how new technology will make the grid more flexible and resilient, but may weaken its role as a universal and equitable reliability backstop.

In Part II, we examine the large-scale power grid—how it is planned, built, and paid for—and how these processes must change to secure and accelerate decarbonization. Chapter 6 explains the technologies that will be used to supply carbon-free power and how they must fit together to make a reliable system. Chapter 7 examines the unique features of transmission networks—networks that are as essential to low-cost decarbonization as they are difficult to plan and build. Chapter 8 looks beyond technologies and system planning to look at the markets and revenue models that will be needed to induce the massive, rapid investment needed.

In Part III, we return to the downstream world of utilities and local markets that will ultimately organize and optimize the entire system. Chapter 9 introduces the main dimensions along which utilities will choose their business futures: the regulatory model, product mix, and geographic span. Chapters 10 and 11 look at the platform utility I call the Smart Integrator, first from the standpoint of technology and business arrangements and then from the regulators' vantage point. Chapter 12 does the same for the contrasting business model, the Energy Service Utility. Chapter 13 considers the role of political, technical, and social forces outside the utility industry, such as the possible effects of the Energy Democracy movement and future privacy regulations. And no discussion of this capital-hungry industry would be complete without hearing the financial community's views on the industry's business future (Chapter 14). Chapter 15 offers broad concluding thoughts. Recommendations for industry leaders and policy makers, laced throughout each chapter, are collected for ease of use in Appendix A.

We start this journey by asking two simple questions. First, how much electricity will we need as we decarbonize developed-world energy systems? Second, how much of this power can we realistically expect to make from local power sources integrated into our communities? For developed countries such as the United States, the answers aren't as obvious as you might think.

2 The Future Is Electric

To catch a glimpse of our urban energy future, head south from the narrow streets and gleaming glass towers of downtown Boston into the blue-collar neighborhood of Dorchester. There, across from a community playground and just off the Orange Line subway, you'll find a modern new condo complex. The angular shape of the building gives it a bit of a modern flavor, but otherwise it seems to be the same size and shape as the two-flats and condos that fill the neighborhood. The interior of the condo looks even more typical—filled with the same flat screens, digital devices, and stainless-steel kitchen appliances as millions of other similar condos in countless American suburbs.

But this condo is different in one important way: it produces more energy than it consumes. Although it cost the same to build as other same-sized condos in the neighborhood, its walls and windows are thicker and better insulated. The one clue that this condo is different is the small electric heat pump system mounted high on the wall in the open-plan first floor. John Dalzell, the Boston city architect who helped plan the project, looked up at the system and said to me, "You'll see that there's another one of these systems on the second floor. We really didn't need it, but we thought buyers would be so skeptical that a single heater of this size would warm the whole house that we stuck an extra one in on the second floor for show."

The unit hosts the obligatory photovoltaic solar panels on the roof and additional solar absorbers for the hot-water system, which are the source of all the net energy. There is no pretense of this home being "off the grid"—it sends excess solar electricity into the grid most days and buys the electricity it needs at night from the local utility. However, it makes more power than the sum of what it uses and buys. The owner of the unit hasn't paid an energy bill since the month he moved in, three and half years ago, and gives his extra energy credits to his brother-in-law across town.

I marveled at just how normal the building looked and felt. Down the block, another builder was building another net-positive apartment complex that had a traditional New England look. Once again, the only dead giveaway was a little electric heat pump unit mounted near the ceiling.

I asked John how long he thought it would take before every new home in Boston would be net zero or net positive in electricity, acknowledging that changing the techniques and materials used by housebuilders and demanded by home buyers was a slow proposition. I expected his response would be 2030 or 2040. It was evident that he had thought about this quite a bit. It only took him a few moments to look at me and calmly say, with surprising precision, "2022."

Forecasting anything thirty years from now is surely a fool's errand. Starting with McKinsey's infamous prediction that fewer than one million cell phones would ever be sold (there are now 7.6 billion in use), it is easy to find many gleeful celebrations of humans' inability to forecast almost anything.[1]

Long-term energy forecasts are very much a part of this tradition. Vaclav Smil, one of the world's leading energy analysts, writes:

> I recognize the value of exploratory forecasts to stimulate critical thinking and as tools for formulating and criticizing new ideas. But most energy forecasts fall into a much less exalted category of plain failures.[2]

One particularly well-known episode of persistent mis-forecasting occurred during the late 1970s. Two researchers analyzed the long-term forecasts by electric utilities looking back ten years from 1985. Year after year, the utility forecasts were found to overestimate demand and had to be lowered the following year. When graphed on a single page, the series of annual forecast recalibrations had the shape of an unfolded Chinese fan, leading the two researchers who studied it to call it "the NERC fan." The name stuck, and to this day, insiders use the term as code for persistently inaccurate forecasts.

We've now had thirty more years to watch electricity use grow around the world, alongside the population and the economy, and to deepen our understanding of the determinants of electricity demand. Perhaps we can now rise to the task of determining roughly what sort of long-term electricity demand growth we can expect through the next few decades, clearing the rather low bar set by Professor Smil. It is certainly worth a try. What could be more important to the industry's future than its long-run prospects for sales?

Deconstructing Electricity Growth

Electricity demand increases are most often modeled as functions of gross national product growth and other statistical drivers. Long-term forecasts are created by statistical regressions that forecast based on dedicated economic growth and other "explanatory variables." Predictions using this approach typically fall victim to two main sources of error. First, the relationship between the explanatory variables and power use embedded in the equation isn't stable over many decades. Second, forecasts of these variables are themselves inevitably incorrect.

In any event, our purpose here is not to forecast the precise level of future power use, but rather to get a good idea of its range. As a result, it is better to think of this future range as the product of the main processes that create demand: basic access to electricity, the affordable electrification of traditional end uses, and newly invented electricity applications.

Since the birth of commercial electricity, the fundamental process driving demand growth has followed a recognizable pattern. Electricity service requires access to the grid, so the spread of the electric grid is always a critical precondition to demand growth. And it isn't just physical access that counts—the costs of connection and the necessary hardware must be affordable at the current stage of economic development. Affordable access is nearly universal in developed countries, but it is still an important constraint on demand in developing Asia and Africa.

Affordable access increases as the grid spreads out and economic growth expands disposable incomes. During this period, homes and businesses do two essential things. First, they gradually convert activities that use other forms of energy to activities that use power. During the first phase of electrification (1890–1950 in the United States), lighting, clothes washing, mechanical shaft power, and many other *end uses* were completely converted from human or steam power to electricity.

The second essential phenomenon is the invention of new electric-only end uses. I am sure there is a wood-fired television set somewhere in a Jetsons cartoon, but in the real world, electronic communication and recreation end uses are more a new use of electricity than a substitution away from similar processes that use nonelectric fuels. Thousands of business applications of electricity, from assembly robots to laser-guided surgical knives, are unthinkable without electric power.

It is easy to see this pattern unfold in the United States during the twentieth century. In 1907, the first year federal records were kept, electricity was a luxury installed in 8 percent of U.S. dwellings, effectively all of them in cities (it would take twenty more years before rural access reached this level). Access increased rapidly to 35 percent by 1920, 60 percent by 1925, 80 percent by 1940, and 98 percent by 1955. At the same time, average household use also rose from 264 kWh per year in 1912 to 547 kWh in 1930 to 1,845 kWh in 1950—a sevenfold increase over 1912. This was the period in which all the older end uses, such as lighting and dishwashing, transitioned to electricity. Roughly the same thing happened with industrial use.

In the 1960s, we started to see entirely new electricity end uses. Technically, we could call electronic computing the electrification of mechanical calculators, but computing technology is so large a leap forward that it is almost a new application. Air conditioning was another large new use for power, along with a steady stream of new gadgets such as fax machines and videocassette recorders. These new uses had a huge impact; average household use doubled between 1950 and 1960, then doubled again (to 7,066 kWh per home) by 1970, and increased by another 50 percent by the late 1990s.

But there is one more part of this story that is also critically important. In the same era that new end uses were born, both new and existing uses became more efficient. U.S. refrigerators today are 36 percent more efficient than they were in 1990.[3] Computers have doubled their efficiency every one-and-a-half years since their inception, according to the oft-cited Moore's Law.[4] Light-emitting diode light bulbs use about one-sixth of the energy of incandescent bulbs and three quarters of the energy of compact fluorescents.[5] Today's average air conditioner is 250 percent more efficient than its counterpart twenty years ago.[6]

All of these changes in electrical efficiency act to reduce power use, counteracting the effects of increased population, income growth, and the invention of new applications. You can think of it as a balancing act: if all existing plus new uses of electricity, including those prompted by population and economic growth, grow only as quickly as electric technology gets more efficient, then electricity demand will flatten out. The drivers of growth will be perfectly offset by greater end use efficiency. If efficiency gains are even larger than population growth, economic growth, and new inventions together, then electricity use will actually trend downward, at least for a while.

This takes us to about where we are today in the highly developed economies of the world, including the United States. Per capita electricity use in the United States has climbed steadily since 1907 with two exceptions: the ten years of the Great Depression (1930–1940), when it remained flat, and the period since 2007, when it has *declined* 8 percent. As of 2016, *total* U.S. electricity generation has not returned to the peak it set in 2007 (about 4,000 billion kWh), and the latest U.S. government forecasts for the next few years, during which the population and the economy are both expected to grow healthfully, are also

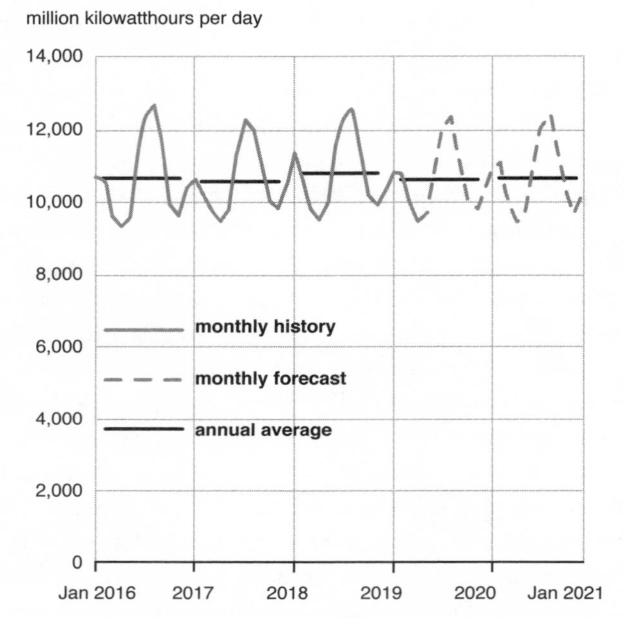

Figure 2-1 U.S. Electricity Consumption, 2016–2021.
U.S. Energy Information Administration, Short-Term Energy Outlook (September 2019).

expected to be pretty much flat (see Figure 2-1). In New England, where energy efficiency (EE) policies are some of the best and population growth is modest, the official ten-year forecast for power sales is *minus* 0.6 percent per year.[7]

The recent trend is even more remarkable because it has occurred during a time when real electricity prices have been going down, not up. As we learn in freshman economics, the demand for almost everything goes up as its price declines, and electricity has historically never been an exception. Yet, U.S. inflation-adjusted retail prices declined between 2006 and 2017 by about 2 percent, which should have boosted power demand slightly—and probably did.[8] So, efficiency improvements not only surmounted economic and population growth—they also overcame the effects of lower prices.[9]

There is no doubt that the recent trend to lower per capita use and overall flat lining is partly attributable to the Great Recession, which hit hard in 2008 and lingered until recently. Even without this, it would be vastly premature to declare a permanent inflection point in per capita power use (indeed wrong, as we'll see in a moment). Nevertheless, it is worth pausing to absorb the importance of what has occurred during the last ten years. Despite the addition of about *twenty-two million* Americans since 2007 and real gross domestic product (GDP) growth of almost $2 trillion, the United States hasn't used one additional kilowatt-hour of electricity.[10] The thirty-six largely mature economies of the Organization for Economic Cooperation and Development performed even better; their energy use is no higher than it was in 2000, while total economic growth increased $8.5 trillion or 26 percent.[11] Some of this might be attributable to shifts away from energy-intensive industries such as manufacturing, but those shifts haven't been large lately, nor do they affect population-driven electricity increases. In addition, there were no major shifts in the applications for power during this period. For at least the last decade, the pull of greater electrical efficiency has outmatched the push from population and economic growth.

Impressive as they have been, electrical efficiency improvements share a unique, extremely inconvenient feature: consumers and industries must be willing to adopt the better technologies. These new technologies often have higher initial costs than their less efficient rivals, often require a capital outlay earlier than originally planned, and require customers to experience some inconvenience or change a process or habit. We must put up with the time and transaction costs of changing out our old hardware, learning how to operate something new, financing a capital improvement, and so on. Many studies confirm that most consumers don't adopt all the efficient technologies that save them money in the long run, and economists have long argued about why this is the case.[12]

This economic debate has not stopped policy makers from recognizing that they could take actions to accelerate the adoption of efficient technologies. These policies include building energy codes, programs and rebates offered by utilities, tax incentives, low-cost financing options, and appliance efficiency standards, to name just a few. To be sure, the normal forces of markets, technological change, stewardship, and environmental values induce quite a lot of efficiency regardless of what governments do. Some research suggests that these "fundamental drivers" are responsible for the great majority of efficiency progress since the 1980s.[13] However, most researchers also think that government policies substantially affect efficiency improvements, and therefore electricity demand. To cite one example, the International Energy Agency (IEA) recently concluded that:

> Energy efficiency investments are set to keep growing, driven by more assertive and more comprehensive policies. Several factors indicate that the energy efficiency market will remain robust in the medium term. Principal among these is the existence of strong and increasingly stringent policies, which recognize energy efficiency measures as being among the most cost effective means of helping to tackle energy security, productivity, local air pollution and climate change challenges.[14]

For forecasters, the strong dependence of efficiency outcomes on policies is no blessing. If power forecasts depend on the battle between population, economy, and efficiency, and the latter depends strongly on policies as well as technology, we have no choice but to predict both the course of technology improvements and the future strength of efficiency policies. Trends in technical change are often relatively stable, and when there are surprises, they tend to boost the trend, not change its direction. But forecasting government policies over the long-term? In a world where pollsters have trouble predicting the winners of elections a few weeks or months away, one can almost hear Professor Smil chuckling.

Long-Term Efficiency Trends

The last few decades of efficiency improvements have been unquestionably impressive. Nonetheless, assuming that the future will look like the past is forecasting's cardinal sin. After thirty years of improvements in the current end uses of power, how much potential is left to tap? And how much can we count on our political leaders to adopt policies that realize gains that don't emerge on their own from market forces? The answers to these questions probably spell the difference between flat or even negative kilowatt-hour sales growth and a robust, growing sales path.

As to the technical potential, we appear to be quite far from exhausting further efficiency gains. Scientists and manufacturers foresee several generations of efficiency gains in most end uses from technologies now in laboratories and prototypes.

As far back as 2010, the official U.S. forecast predicted 53 and 45 percent increases in heating and water heating efficiency by mid-century, respectively, 70 percent increases for lighting, 60 percent increases for washers, and 39 percent increases for personal computers.[15] Because these forecasts explicitly assume *no* changes in EE policies, they confirm that technological efficiency improvements still have lots of runway and that market processes will have very large impacts. In addition, big data and the smart grid are enabling new forms of EE that were not previously possible.

Existing policies *are* also likely to change through 2050. We've already met a good example of these oncoming policy changes: net-zero buildings. A building code that requires all new homes to use zero net energy not only reduces residential energy sales growth—it ends it. Ironically, when net-zero homes become the required standard, population and housing growth will reduce net utility residential sales, since homes will routinely make more energy than they use and send the surplus back. Existing homes will become more efficient too, though they'll achieve net zero on a much longer timetable.[16] John Dalzell's 2022 prediction for this policy shift may have been optimistic, but there are many signs that net-zero construction is moving toward becoming the norm. California has already started to adopt a building code requiring new homes to be net zero by 2020 and wants to add commercial buildings by 2030.[17] The City of Vancouver has set 2030 as a net-zero requirement for all new buildings, and expects to convert even existing buildings to net zero by 2050.[18]

The timing and geographic breadth of these policy shifts is the hardest aspect of all this to predict. Electricity forecasters typically skirt this challenge by using policy scenarios—the efficiency gains that *could* occur if policy makers adopted weak, moderate, or strong policies. These studies begin by examining the largest savings that are technologically achievable, regardless of likely customer behavior, and then adjust them down into cases, with labels such as "maximum achievable" for the strongest policies forecasters think are realistic or "business as usual" for continued current policies.

These scenarios give us a range of technology-plus-policy forecasts for efficiency gains and power demand—mainly for current uses—by 2050. The official U.S. government forecast, which freezes policies but assumes continued population and economic growth, projects a 33 percent sales increase by 2050.[19] In this world, population and economic growth outweigh efficiency improvements, considerable as they are. At the other end of the spectrum, Amory Lovins recently estimated vast new untapped EE potential through integrated design.[20] The American Council for an Energy Efficient Economy (ACEEE) examined electricity growth if the United States was to adopt thirteen groups of bold new efficiency policies. For example, ACEEE assumed that 80 percent of all U.S. new construction would be net zero starting in 2031, and that 80 percent of all homes would have artificial intelligence (AI)-enabled thermostats by 2040.[21] It also assumed many increased product efficiency standards, recognizing:

> Achievement of the full savings potential for new standards will require various steps, including improved test procedures on some products (so that tests approximate performance in the field); market introduction of an increased number of models at today's highest efficiency levels; efforts by manufacturers, distributors, utilities, governments, and large customers to promote these most-efficient products; and, ultimately, a rulemaking by DOE to adopt new standards that require increased but cost-effective levels of efficiency.[22]

The impacts of ACEEE's strong EE policy scenario are extraordinary. By 2040, electricity use declines from the U.S. Department of Energy (DOE)'s business-as-usual projection of 4,805 TWh (800 TWh above

2017 levels) to ACEEE's estimated use of 3,100 TWh—900 TWh *below current* power use.[23]

It is highly unlikely that the United States as a nation will suddenly increase its will to adopt all these very bold policies. The real point is that the ACEEE scenario and many others like it suggest that there is enough technical *and* policy potential to eliminate long-term electricity growth for traditional uses.[24] The barrier is our willingness to adopt these policies and their ability to propagate across the United States and other countries.

Most long-term U.S. forecasts don't hew to either of the extremes found in DOE's zero-change scenario or ACEEE's ambitious future. Instead, they assume that EE policies gradually get stronger and thus move somewhat steadily toward "achievable potential." Under these conditions, you get forecasts much like the IEA's 2050 estimate for U.S. power use, which is only about 8 percent higher than 2018, or their estimate for the EU, which is practically flat. The State of California—a longtime leader in EE—creates official estimates with and without just one (albeit major) policy, its utility EE programs. Through the next ten years, these programs cut total electricity growth in half, from 14 to 7 percent—still not fully outweighing population and economic drivers, but yielding a hardly perceptible growth rate of 0.67 percent a year.

In 2008, when I was writing *Smart Power,* I didn't have the benefit of seeing the last ten years of flat electricity sales. Nonetheless, I thought the signs of an end to kilowatt-hour sales growth were strong enough to predict "roughly flat" grid power use for the next several decades or so. This prediction was based on a judgment that EE policies were going to get a little stronger a little faster than in the past.[25] Up until that time, EE policies were primarily motivated by life-cycle savings to consumers, with environmental gains as a lesser driver. I knew that EE policies were almost always a major ingredient in climate-change policies, and believed that the political importance of fighting climate change would increase. The likelihood that EE policies would become weaker was a lot less than the chance they'd become stronger as part of climate policy packages. Stronger EE policies would motivate greater effort by technologists, and the overall effect would be lower sales growth than was now forecast—certainly less than DOE's 33 percent increase.

I think this view remains accurate. If we limit ourselves to applications that already use electricity, such as air conditioning and lighting, my bet is that total power use for all this will continue to be "flat to down" for the next few decades. Population and economic growth will be offset by greater efficiency, spurred by city, state, and national climate-change policies as well as the marketplace.

But when I suggested a long-term end to *total* electricity sales growth, I missed something enormous and thereby might just have earned myself a place in Smil's forecasting hall of shame.

Enter Carbon

We are well into the era in which policy makers and planners have committed to reducing greenhouse gases (GHGs) so as to limit global warming to less than two degrees Celsius. Dozens of studies have examined how the major economies of the world, and sometimes also subnational areas such as states and cities, can adopt policies that reduce emissions to these levels. These studies—typically called climate plans or *pathway studies*—are intended to guide policy makers to action portfolios strong enough to meet reduction goals without causing other undesirable effects, such as job losses or more energy poverty.

Nearly all these plans and studies are based on three major elements. First, save as much energy as possible by maximizing cost-effective EE. The less energy is demanded, the less zero-carbon energy you need to reduce emissions and the lower your energy bills. Second, decarbonize the electric power system. Every plan or study assumes something a little different about the future mix of solar, wind, geothermal, nuclear, and coal or gas plants with carbon capture and storage (CCS), but they all agree that the future power system must emit little or no GHGs. Finally, every fossil-fueled end use that can be converted to use electricity with equal performance and reasonable cost should be electrified, making more use of carbon-free power. In short, decarbonize the grid and electrify every feasible fossil-fueled end use.

Because most experts are confident that power grids can be deeply decarbonized by 2050, the support for all three of these elements, including electrification, is quite broad. Before leaving office, the Obama Administration issued a Mid-Century Strategy (MCS) for decarbonizing the entire U.S. economy by 2050. After surveying the literature, the MCS declared that "Nearly all deep decarbonization scenarios show large increase in the deployment of certain technologies and strategies, including energy efficiency, electrification, wind, solar, and biomass."[26] An extensive global review by the Energy Innovation Reform Project (EIRP) concluded that "each of the economy-wide studies reviewed envisions electricity supplying greater shares of heating, industry, and transportation energy demand by 2050." A review by the IEA called electrification "a backbone of the clean energy transformation."[27] Or, as one leading Chinese power official puts it, "the world has stepped into a re-electrification era."[28]

The reason these studies and plans turn to electrification is straightforward. Without it, there appears to be no reasonably priced way to ensure an adequate, reliable energy supply with near-zero emissions. The leading U.S. pathway study, Pathways to Deep Decarbonization in the United States, explains the reasoning crisply:

> Meeting the 2050 target requires almost fully decarbonizing electricity supply and switching a large share of end uses from direct combustion of fossil fuels to electricity (e.g., electric vehicles), or fuels produced from electricity (e.g., hydrogen from electrolysis). In our four decarbonization cases, the use of electricity and fuels produced from electricity increases from around 20% at present to more than 50% by 2050. As a result, electricity generation would need to approximately double (an increase of 60–110% across scenarios) by 2050 while its carbon intensity is reduced to 3–10% of its current level.
>
> Concretely, this would require the deployment of roughly 2,500 gigawatts (GW) of wind and solar generation (30 times present capacity) in a high

renewables scenario, 700 GW of fossil generation with Carbon Capture and Storage (CCS) (nearly the present capacity of non-CCS fossil generation) in a high CCS scenario, or more than 400 GW of nuclear (4 times present capacity) in a high nuclear scenario.[29]

To fill in the details of these long-term plans, researchers examine the prospects for electrifying each major energy end use that burns fossil fuels. About half of all buildings are heated by burning fossil fuels,[30] and many buildings also use gas or oil for water heating and sometimes cooling. With the aid of supportive policies, researchers believe that most of this heating and cooling can be converted to electric heat pumps, water heaters, and chillers. In transportation, the electrification strategy is obvious: shift to electric vehicles (EVs) of all types and sizes, including electric trains and mass transit. In industry, electrify processes such as steelmaking or fertilizer manufacturing.[31] As an example, the Obama MCS calls for about 60 percent electric auto travel, industrial electricity use expanded from 20 percent today to 50 percent, and almost all electric space and water heating by the year 2050.[32]

Most of the plans and studies are somewhat vague as to exactly how policy makers are going to cause electrification, but the toolkit of policy options is quite familiar by now. Carbon taxes, cap-and-trade schemes, codes and standards, tax credits, utility and customer incentives, low-interest financing, and direct government funding have all been widely proposed or used. Ideally, policy makers will use economic incentives as much as is feasible and protect against regressive or otherwise unbalanced impacts.

Recognizing that we don't know the specific policies that will make them happen, the implications of these pathways for electricity demand are profound. Appropriately, every one of these climate plans assumes that policy makers first take relatively strong measures to increase EE and only then electrify. Yet, in almost every case, the net effect is to increase electricity demand substantially by 2050. The Obama MCS is a good example; despite extensive measures to improve efficiency, electricity generation grows by 60 percent. Net generation growth is even larger in the Deep Decarbonization Pathways studies. Jenkins and Thernstrom, the two EIRP researchers, summarize several other prominent findings:

> The global decarbonization scenarios summarized by Krey et al. (2014) envision global electricity demand rising roughly 35–150% by 2050, with electricity supplying 20–50% of energy demand by midcentury . . . In eight of nine models reviewed by Morrison et al. (2015), electricity demand in California increases 8–226% by 2050.[33]

A study for New England by the Northeast Energy Efficiency Partnership (NEEP) using a detailed, sector-by-sector computation predicted 2050 increases of 30–55 percent versus DOE business-as-usual estimates.[34] This is consistent with my own look at transport electrification, which found that EVs alone are likely to increase U.S. demand by 18–24 percent. Add on demand from electrifying some industrial processes and other forms of transport and you are well into positive growth territory unless you really push efficiency to its political (if not technical) limits.[35]

To give an idea of the wide range of results in this area, Figure 2-2 compares forecasted 2050 power use in twenty-six climate plans from a wide group of countries. Forecasted 2050 power demand is shown as a multiple of current use in the left-hand margin. A mark on the chart at the level of 1 means 2050 power use is unchanged from today; a mark at 2 shows forecasted power use doubling. As the chart shows, only a handful of studies predict flat-to-down growth (1 or below), while the vast majority of forecasts cluster between 25 percent (1.25 on the chart) and 125 percent (2.25) growth. On the chart, the United Kingdom shows especially high rates of growth because it expects high electrification of its proportionately large fossil-fueled industries.

As with all modeling exercises, these results are burdened by layer upon layer of assumptions. The models they use often underestimate long-term efficiency gains and can overestimate the speed of electrification. There's some chance that breakthroughs in hydrogen or biofuels will displace electrification as a strategy, though there is little sign of this today.[36] The

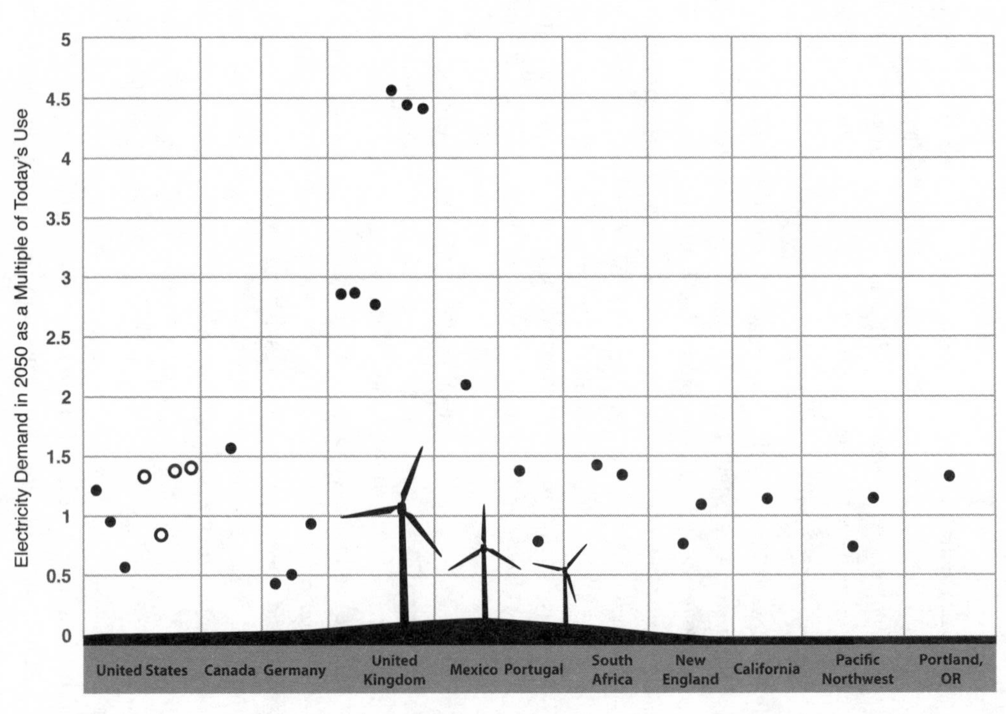

Figure 2-2 2050 Electricity Demand in Selected Regions: Forecasts from Deep Decarbonization Studies.

See Appendix C.

approaches that I think are most realistic, such as the NEEP sectoral estimates, tend to have lower sales growth than the others. Nonetheless, the inaccurate assumptions and political miscalculations (including for EE) will have to be awfully large to bring power 2050 growth in the wake of sound climate policies down to the single digits.[37]

Climate-based electrification policies blow up the third term in our electricity demand decomposition, new uses for electricity. Electricity demand in 2050 is pushed up by increases in population and economic growth and then down by greater efficiency for current uses. It is pushed back up again, quite substantially, by the electrification needed to meet climate goals. Energy policy makers can influence (though not dictate) the size of the downward push of efficiency and upward push of electrification by the strength of their policies, but they are unlikely to change the overall dynamic. And there's one more wild card before we finish.

The AI Wild Card

One day, around 2009, my old, wired desk telephone rang. It was my good friend, climate blogger Joe Romm. He called to ask, rather incredulously, whether it was possible the Internet was using something like 8 percent of all electricity. He had read a report dating back to 1999 that made this claim. I told him that I had no idea, but it sounded high.

Joe went on to research this thoroughly with colleagues at Lawrence Berkeley National Labs (LBNL) and the Rocky Mountain Institute. They concluded that the Internet used about 1 percent of all U.S. electricity in 2000; all computers used about 3 percent. They also argued strongly that this relatively small use of energy enabled vastly larger savings from information-enabled efficiency, greater industrial productivity, telecommuting rather than traveling, and many other efficiency gains. A year later, Joe blogged that

> the internet is a net energy saver—and a big one—since it increases efficiency (especially in things like the supply chain) and dematerialization (it uses less energy to research online than in person). The fact that U.S.

energy intensity (energy consumed per dollar of GDP) began dropping sharply in the mid-1990s is but one piece of evidence that internet- and IT-driven growth is less energy intensive.[38]

The fact that total electricity use has remained flat to down over the past ten years as Internet traffic, data centers, and the entire information and communications technology (ICT) complex has mushroomed certainly supports this thesis. Indeed, the very same LBNL re-examined electricity use by data centers (servers, storage, network equipment, and infrastructure—but not end user equipment) and found that it had increased only to 1.8 percent of U.S. use by 2014. They predicted that data center use would grow about 4 percent a year, faster than almost every other electricity use but nowhere near enough to offset economy-wide efficiency gains.[39] Once again, we are reminded of the tremendous potential of EE improvements, including improvements in the ICT infrastructure itself.

Nonetheless, there is some chance that we are entering an era in which the spread of IP-accessible devices, AI, edge computing, and the coming "data tsunami" will lead to a noticeable uptick in power demand. It is not that I doubt the Romm-Koomey-Rosenfeld-Lovins thesis has held so far, but it seems we are entering a new era in which computing is becoming exponentially more common in all devices. Machine learning is changing the nature of computing into one where you throw as much computing power and bandwidth at as much data as you can collect and let the machines churn until they settle on an answer.

The first sign of this sort of phenomenon is bitcoin and blockchain. These products use an approach where computers are pitted against other computers to see which ones can solve a complex problem as fast as possible. This sets off a competition to see who can install the largest bank of fastest computers to do so-called bitcoin mining and thereby process transactions. Writing in the distinguished journal *Joule,* PwC researcher Alex de Vries estimated that bitcoin computers around the world were performing 26 quintillion calculations every second, using at least 22.3 TWh—roughly equal to the electricity use of Ireland—and quite possibly three times this much when the energy to cool all these computers is included.[40] The website Digiconomist, which now operates an index of worldwide bitcoin power use, claimed it has reached 54.9 TWh (1.7 percent of U.S. use) by March 2019.[41] It is hard to see how any of this, alone equal to the rest of the power used by data centers, is contributing to dematerialization or operating efficiency.

One could say bitcoin and blockchain are anomalies, and perhaps they are. Many blockchain developers are now exploring versions of software that use the so-called proof-of-stake algorithm, which requires much less energy. Regardless, there is an inexorable trend toward outfitting every possible device with as much computing power as possible and relying on the continuous massive transfer of unprocessed data to devices, the processed results to other devices, and huge systems to attempt to manage and optimize using this intelligence.

Another ICT-driven innovation that has been the subject of significant energy research, self-driving cars, also raises the specter of increased energy use. Autonomous vehicles (AVs) will transform our transport infrastructure and urban landscapes more dramatically than anything since the automobile itself, but the nature and timing of the effects on electricity demand are highly uncertain. Many features of AVs reduce demand, but mainly only once they are almost ubiquitous and connected to each other through an effective communications and optimization infrastructure. At this point, vehicle crashes and congestion will (we think) disappear, and vehicles can be downsized dramatically and run as a vastly more efficient system.[42]

However, in the critical next few decades when we must decarbonize, AVs will also induce much more driving and energy use. When vehicles are fully autonomous, we can send them to fetch junior at soccer practice, sleep overnight in them while they drive us to another city, and summon them for trips we'd otherwise take by foot, bike, or transit. Nearly all AV energy experts agree that the net effect of these changes is highly uncertain, and that in many scenarios, energy and electricity use could increase.[43]

There is no doubt that this will lead to greater opportunities to discover and unlock greater efficiency in energy use throughout the economy. I recently attended a presentation at Google X where it was claimed that the use of AI applied to the design and management of a data center led to an astonishing 40 percent reduction in power use. A 2015 report for the World Economic Forum written by McKinsey generalizes this claim, arguing that the Internet of Things (IOT), machine-to-machine communications, and AI "are likely to represent the next major step change in asset productivity" and will be the key to creating a low-impact circular economy.[44]

In short, the Romm et al. thesis may apply even more strongly to a world where tens of billions of vehicles and other devices are connected and 5G networks provide ten times the current mobile bandwidth. However, I think there is also some chance that the bitcoin phenomenon may recur—that is, the massive use of computing power and bandwidth may start to create apps that consume more energy than ICT squeezes out of the rest of the economy.

Some evidence supporting this thesis has been advanced by Anders Andrae, a researcher at Huawei, the sometimes-controversial manufacturer of 5G equipment.[45] In 2017, Andrae estimated that world data centers were growing at a pace that would lead to the use of 20 percent of global electricity by 2025—a claim as remarkable as the one that prompted Joe's phone call to me ten years ago. Because many of the tech giants building global data centers are moving to carbon-free energy, Andrae estimated that this 20 percent of use would create only 3.5 percent of global carbon emissions. If all this ICT lowered the rest of the globe's emissions by more than 3.5 percent, a figure that is certainly plausible, we'd still be better off. But still, I worry.

Whatever the ultimate net effect on electricity demand, ICT will itself increase electricity use within its sector significantly. It is fortunate that many of the tech giants building data centers have committed to 100 percent carbon-free energy. Nonetheless—as if we needed it—ICT power use is one more reason why decarbonizing the entire grid as fast as possible is so important.

Electricity's Third Act

It would be far better for the planet and our economy if, in the coming decades, the effects of efficiency policy measures and secular technology changes outweighed the demand pull from electrification and AI. However, realism tempers this hope. Across many countries and eras, the political economy of energy has generally favored expanded supply over reduced demand. The industries that benefit from electrification, including electric utilities, are relatively concentrated and easier to organize politically than the efficiency industry, which is spread across many verticals and firms. Policy makers can see the economic benefits of supply-side measures more easily than they can see the diffuse, though often larger and more equally distributed, gains from greater efficiency. The sun is reportedly setting on Moore's Law just as the IOT and AI disruptions are ramping up. When all is said and done, it's hard to believe that advanced-economy power sectors will double in the next thirty years, but it's harder to believe that they won't grow (net of efficiency gains) by 25–60 percent. Qualitatively, the overall situation is illustrated in Figure 2-3.

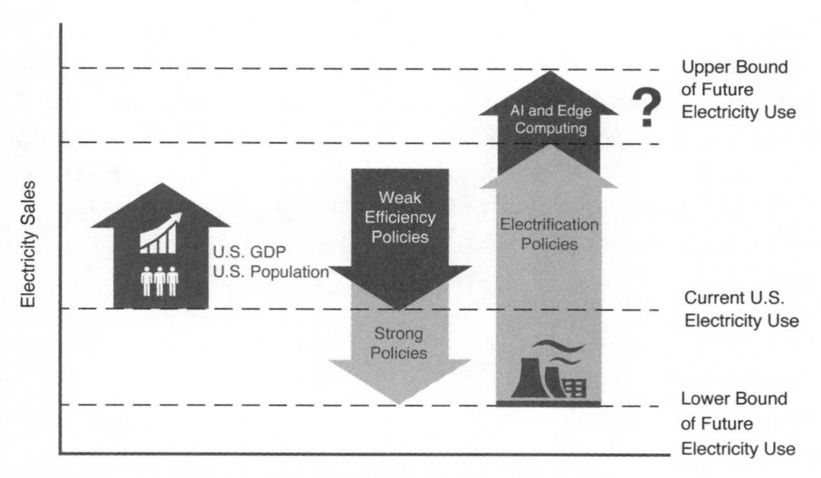

Figure 2-3 Long-Term Drivers of Demand Change for Electricity.

About thirty years ago, I was visiting the Florida Power Corporation, which has since been acquired by Duke Energy. I happened into a room where the company kept its customer swag and quickly spied something I liked very much. It was an auto license plate with a slogan written in a hippie-inspired typeface that told me it was produced in the late 1960s, when psychedelia was in fashion. The license plate read *The Future is Electric.*

Florida Power undoubtedly gave these plates to customers who were moving into homes that were advertised as "all electric" because they had no gas heat or cooking. These same customers were experiencing the spread of electric air-conditioning in homes, offices, and theaters, separating them from the omnipresent Southern Florida heat and humidity. Policy makers were not thinking about EE, and neither utilities nor anyone else had policies to encourage it. In fact, regulatory policies encouraged power sales, but ever-lower prices and economic growth did most of the work fueling demand growth. Sales were doubling every ten years—a rate faster than China or India experiences today.

We're about to enter a new, smarter era of electrification. This time around, all policy makers absolutely must recognize the enormous importance of EE and fight for it. At the same time, it is appropriate to promote new markets for electricity that Florida Power could not have dreamed of. My future-is-electric license plate may end up on an all-EV made of steel from an electric arc furnace, driven to a home that uses no net energy at all for heating or cooling. If climate policies are successful—and they had better be—power sales growth from 2030 to 2050 is going to be quite healthy by recent standards—easily 1–2 percent a year. The question is, whose sales will they be?

3 La Vida Local

Silicon Valley product launches are rock-star affairs, and the introduction of Tesla's new Powerwall battery was no exception. Elon Musk—CEO of Tesla Motors, chairman of Solar City, founder of SpaceX, and easily the most famous entrepreneur of his era—entered a stage lit with blue and green floods to a pounding electronic music fanfare. An adoring crowd and three walls filled with video cameras and reporters recorded every moment.

On this evening, April 30, 2015, Elon was here to reveal the Powerwall, the first battery designed to store solar energy in millions of American homes. He began by making a statement so sweeping that one might be tempted to dismiss it immediately as Silicon Valley hype. "What I'm going to talk about tonight," he began, "is a fundamental transformation of how the world works, about how energy is delivered across the Earth."

The wall behind him immediately lit up with pictures of enormous clouds of black smoke billowing from the stacks of coal-fired power plants. He told the audience, "This is actually how most power in the world is generated . . . The way the grid works today is you've got coal, natural gas, nuclear, hydro and then wind and solar, but not enough wind and solar obviously." He showed a graph of actual and projected concentrations of greenhouse gases in the atmosphere and said, "I think we collectively should do something about this and not try to win the Darwin award"—a clever way of urging stronger action to combat climate change.

But the evening's purpose was not to dwell on the problem. "What we're here to talk about is the solution. I actually think it's a fairly obvious solution, but it's something that we need to do, and the solution is in two parts."

The first part, he explained, was solar power. No big surprises here: the solar energy falling on the earth in an hour and a half equals all energy used on earth in 2001.[1] Technologists as diverse as Thomas Edison, Henry Ford, and Raymond Kurzweil have predicted that solar energy will ultimately be our major source of electric power.[2]

The second part of Elon's solution was electricity storage—Tesla's Powerwall. Storage unlocks the potential to use solar energy for use on cloudy days and starry nights, rather than just when the sun is blazing. To hammer this point home, the wall behind Elon once again lit up, this time showing gigantic live feeds from two power meters. The first meter displayed the amount of power flowing from the power grid into the announcement venue; the second meter showed the power flowing from a bank of batteries Tesla had installed in the building. Turning to view the screen, Elon noted with delight that the building was not using power from the grid. "Oh wow, the grid [meters] are actually zero. This entire night has been powered by batteries. Not only that, the batteries were charged by the solar panels on the roof of this building. This entire night, everything you're experiencing is stored sunlight."

Finally, Elon translated his solution into actions that individuals can take. "If you're thinking about buying a battery, what does this provide you? Well, it gives you peace of mind. If there's a cut in the utilities you're always going to have power." More importantly, he suggested that this might be an alternative to the entire power grid. "You actually could go, if you want, completely off-grid. You can take your solar panels, charge the battery packs and that's all you use." Three years later, market analysts would write:

> Under the "no boundaries" leadership of Elon Musk, the Tesla brand is trying to become the first company to truly offer an alternative to the modern day utility. You could even say Elon Musk wants Tesla to become

> the modern-day energy company and re-frame traditional energy companies as fossil fuel utilities.[3]

A few sentences later, Elon noted that Tesla's new batteries could also be used on the large power grid, combining many small batteries to create "gigawatt-scale" installations. Following up on this, a few months later, Elon spoke to the assembled CEOs of the U.S. utility industry and confidently predicted that despite his disruptive solution, two-thirds of all U.S. power would still be provided by the large power grid in 2050. But anyone attending this evening, or hearing about it through a blog or tweet, could easily be forgiven for gaining the impression that the solution to climate change was solar, batteries, and dismantling the grid entirely. One expert website wrote that the Tesla brand is trying to become the first company "to truly offer an alternative to the modern-day utility."[4]

This impression continues to spread. Millions of citizens across the world have been moved by their concerns over climate change to take immediate, visible actions to lower their carbon footprint. The solar capacity installed in the United States now exceeds 60,000 MW—half the capacity of the U.S. nuclear fleet and enough power for 11.3 million homes.[5] The respected Rocky Mountain Institute think tank predicts that power customers in New York, California, and other states will find it economical to "defect" from the grid entirely by 2025.[6] Morgan Stanley warns investors about the utility "tipping point,"[7] and news outlets from Forbes, Business Week, Grist, and CNBC have all covered the "utility death spiral" that will result from widespread grid defection.[8] At a 2015 Bloomberg conference, after demonstrating that next to no one in the room still had a telephone landline, Al Gore remarked that "we're not far off from the day when I can ask you how many of you no longer have a grid connection."[9]

Solar in the City, 2016

The idea that individual customers can self-generate enough solar electricity to leave the grid has obviously fired the modern imagination. In later chapters, we will examine several aspects of grid defection, as it is called. However, to get a sense of the purely technical need for a large-scale power system, we don't need to forecast exactly which customers will leave the grid and which will stay—we need only look at the total amount of potential local generation versus demand. If there's not enough local power to meet local needs, we know the grid will remain needed. We'll then face the question of how to make sure it stays viable for everyone who continues to rely on it.

Interestingly, national long-term forecasting models are only now starting to distinguish between distributed and large-scale power sources.[10] These forecasts are mainly used to paint big pictures of the future for policy makers, and for that purpose all photovoltaics (PV) are similar enough. Both the official U.S. Mid-Century Strategy and the international Deep Decarbonization Pathways Project (DDPP), the two most visible works to date, simply list "solar" in their reported 2050 installations.[11] Germany's DDPP study also omits any distinction, despite that fact that the growth of rooftop solar in Germany has been especially strong and disruptive.[12] Forecasts by the U.S. Energy Information Administration (EIA) and the International Energy Agency (IEA) sometimes distinguish between small- and large-scale PV generators but just as often lump them together simply as "solar."[13]

For our purposes, we can't afford to be quite so vague on this critical point. Utilities serving areas with very little distributed power will have to supply almost 100 percent of electrification-driven demand growth—a handsome scenario for big power vendors. Even with electrification, a grid with strong sun and ubiquitous net-zero buildings has a more variable and possibly negative sales future. From the standpoint of utilities' future business mission, the proportion of large- versus small-scale power seems rather important. In most of the world, looking for the amount of power that might come from *distributed* PV means looking at urban areas. The fraction of the U.S. population living in urban areas was 81 percent in 2014 and is forecast to be 87 percent by 2050.[14] By 2050, 70 percent of the world's population will be urbanized. Cities have nearly all the roof space, facades, and nearby parking lots usable for rooftop solar.[15] There is plenty of space beyond city limits to place PV, but then you are back to needing a delivery grid, maintaining the utility's traditional sales and delivery role.

Compared to most of the predictions in this book, estimating the potential for solar in mid-century cities is comparatively straightforward. A team led by Harvard researcher Lee Miller addressed this question quite elegantly by comparing the solar energy falling on cities to the total energy they used in 2016. Miller's team found that major U.S. cities used at least 10 W / m^2 of city space—New York topped out at 70 W / m^2—while global urban sunlight averages 170 W / m^2.[16] Using the global average and a current conversion efficiency of 20 percent, if every inch of a city was blanketed with PV, cities using 34 W / m^2—about half that of New York—could hypothetically make power equal to their use (the time periods of generation and use don't match; we'll get to this critical fact shortly). But obviously, much of the surface of cities is covered by pavement and trees, while other areas are shaded or unusable for solar generation.

This might seem to settle the issue of urban energy self-sufficiency, but the gap between future urban use and distributed sources isn't as large as calculations suggest. On the one hand, electrification will increase urban power demand dramatically, perhaps doubling it in places with inadequate efficiency policies. Conversely, improved solar collection efficiencies and surface options, as well as other sources of urban electricity, will help close the gap. And some cities get a lot more sun than the global average.

Arithmetically, PV generation is the product of solar radiation per square foot, the number of square feet of collection space PV panels, and system conversion efficiencies. Location-specific solar radiation won't change by 2050. So, there are really only two quantities to forecast for any given urban area: collector space and conversion efficiency.

In most cities, there are four good surfaces from which solar energy can be collected: building roofs, building facades, paved areas that can be covered, and usable unshaded urban land. We have relatively good data on current roof space and some hints as to how to extrapolate this to 2050. Google has Project Sunroof, which uses satellite imagery to create a searchable website that will estimate solar potential for rooftops across much of the United States. A MIT spinoff called Mapdwell can estimate the solar potential of "any community on Earth . . . using state-of-the-art, hyper-precise, advanced technology."[17] The U.S. National Renewable Energy Laboratory (NREL) has published careful estimates of roof space for 128 major cities and the United States as a whole, and many academic researchers have conducted similar surveys for cities around the world.[18]

The NREL results, reproduced in Table 3-1, show a wide range of present-day rooftop solar potential in U.S. cities. At the low end, current roofs in Washington, DC, and New York City can provide 16–18 percent of *current* electricity use. At the other end of the range, Sacramento, California, and Buffalo, New York, could potentially generate 71 and 68 percent, respectively, of current use from all roofs within city limits. These results are on a par with other studies of developed-world cities and countries, which find median distributed generation (DG) PV potential in the range of 30–40 percent of current use.[19]

These estimates apply only to existing rooftops using vintage solar panels from 2016 or earlier, with conversion efficiencies of 16 percent or less. To use these results for a long-term forecast, NREL's researchers point out the need for several important changes:

> Our results require several caveats. First, they are sensitive to assumptions about PV system performance, which is expected to continue improving . . . Second, we only estimate the potential from existing, suitable roof planes—not the immense potential of ground-mounted PV. Actual generation from PV in urban areas also could exceed these estimates if systems were installed on less suitable roof area, PV were mounted on canopies over open spaces such as parking lots, or PV were integrated into building facades.[20]

In other words, to extrapolate these findings to 2050, we need to examine how urban PV collection space is going to change and the size of expected increases in panel efficiency over the next three decades. (We also need to ask whether other forms of distributed urban power will become economical during this period, but the answer to that seems to be no.)

Table 3-1 Percentage of Current Power Use That Could Be Generated by Current PV Using All Suitable Rooftops, 2016

City	Installed Capacity Potential (GW)	Annual Generation Potential (GWh / Year)	Ability of PV to Meet Estimated Consumption
Mission Viejo, CA	0.4	587	88%
Concord, NH	0.2	194	72%
Sacramento, CA	1.5	2,293	71%
Buffalo, NY	1.2	1,399	68%
Columbus, GA	1.1	1,465	62%
Los Angeles, CA	9.0	13,782	60%
Tulsa, OK	2.6	3,590	59%
Tampa, FL	1.4	1,952	59%
Syracuse, NY	0.6	657	57%
Amarillo, TX	0.7	1,084	54%
Charlotte, NC	2.6	3,466	54%
Colorado Springs, CO	1.2	1,862	53%
Denver, CO	2.3	3,271	52%
Carson City, NV	0.2	386	51%
San Antonio, TX	6.2	8,663	51%
San Francisco, CA	1.8	2,684	50%
Little Rock, AR	0.8	1,099	47%
Miami, FL	1.4	1,959	46%
Birmingham, AL	0.9	1,187	46%
St. Louis, MO	1.5	1,922	45%
Cleveland, OH	1.7	1,881	44%
Toledo, OH	1.4	1,666	43%
Providence, RI	0.5	604	42%
Worcester, MA	0.5	643	42%
Atlanta, GA	1.7	2,129	41%
New Orleans, LA	2.1	2,425	39%
Hartford, CT	0.4	404	38%
Baltimore, MD	2.0	2,549	38%
Bridgeport, CT	0.4	435	38%
Detroit, MI	2.6	2,910	38%
Portland, OR	2.6	2,811	38%
Milwaukee, WI	2.1	2,597	38%

Table 3-1 (continued)

City	Installed Capacity Potential (GW)	Annual Generation Potential (GWh / Year)	Ability of PV to Meet Estimated Consumption
Boise, ID	0.5	760	38%
Des Moines, IA	0.8	1,026	36%
Cincinnati, OH	1.0	1,176	35%
Norfolk, VA	0.8	1,047	35%
Wichita, KS	1.1	1,537	35%
Newark, NJ	0.6	764	33%
Philadelphia, PA	4.3	5,289	30%
Springfield, MA	0.3	370	29%
Chicago, IL	6.9	8,297	29%
St. Paul, MN	0.8	903	27%
Pittsburgh, PA	0.9	907	27%
Minneapolis, MN	1.0	1,246	26%
Charleston, SC	0.3	407	25%
New York, NY	8.6	10,742	18%
Washington, DC	1.3	1,660	16%

Source: Gagnon et al. (2016).
GW, gigawatt; GWh, gigawatt-hours; PV, photovoltaic.

Solar in the City, 2050

In his book, *Taming the Sun,* researcher Varun Sivaram describes the 2050 urban solar future in glowing terms:

> Architects rejoiced. Today at the mid-century mark, most urban buildings are wrapped in electricity-generating solar materials that tint the windows, enliven the façade, and shrink the carbon footprint. Nearly free electricity has induced heavy industries to switch from burning fossil fuels to running off solar power. Solar PV isn't just powering glamorous urban buildings or massive industrial plants; PV materials are now light enough to be supported by flimsy shanty roofs in the slum outskirts of megacities in the developing world.[21]

Unfortunately, beyond conventional rooftop PV, we have little experience with the other collection technologies and surfaces. To complicate matters, cities in the United States and around the world have vastly different layouts, densities, and natural features, and thus very different amounts of solar resources.[22] Some cities are dense and vertical, with relatively more facades, fewer rooftops, and less open space that could be covered with PV. Many cities in the developed world (and auto-dominated cities in the developing world) have huge amounts of roadway, but we have no real idea of how much could be used for solar collection.

Perhaps the best we can do to assess long-term urban solar potential is to assemble rough but plausible estimates for two very different cities, say New York City and Phoenix, Arizona. New York City is expected to grow slowly, adding 14 percent to its population by 2050; Phoenix's projected growth (62 percent) is more than four times as large. At 3,126 people per square mile, the density of Phoenix is about one-ninth that of New York City (28,211 people per square mile).[23] New York City is a far more vertical city, with more than 6,000 buildings that have more than ten stories; Phoenix has eighty-nine such buildings.[24] With differences such as this, it should be possible to illustrate two rather different cases of urban DG potential.[25]

Table 3-2 displays the main elements of an estimate for Phoenix. Column 2 of row A shows the annual PV electricity generation that could be made now by covering all suitable city roof space with 16 percent efficiency panels—about 15,600 GWh.[26] The remainder of the 2016 potential estimates in column 2 show zero current potential for solar PV from building facades, parking, unused urban land, or roadways simply because there is a negligible amount of this sort of PV collection now. This total 2016 potential for urban PV, also shown in column 2 of row E, indicates that Phoenix *could* today generate about 60 percent of its current power use if it were to use *all* suitable roof space for PV.

To estimate urban PV generation in 2050, assume for the moment that rooftop PV becomes largely universal by 2050. The next step is to estimate how Phoenix's total roof space will change.[27] Suppose we assume that in Phoenix, both the living floor space per person and work floor

Table 3-2 PV Solar Potential for Phoenix, Arizona, in 2016 and 2050—Illustrative Numbers

		Estimated PV Usable Area (Square Miles), 2016	GWh / Year @ 16% Average Panel Efficiency, 2016	Estimated PV Usable Area (Square Miles), 2050	GWh / Year @ 28% Average Panel Efficiency, 2050
A	Suitable building roof area	Not used to derive GWh	15,624	Increased proportional to average growth of population and employment	44,936
B	Sunny facade space			10 percent of rooftop generation	4,494
C	5 percent of unshaded off-street parking			1.95	2,946
D	1 percent of city land, excluding paved surfaces			3.10	4,687
E	Total PV generation (GWh)		15,624		57,063
F	Total electric demand (GWh)		26,481		52,962
G	Total as percentage of (projected) electric use		59 percent		108 percent

space per employee don't change between now and 2050, and that Phoenix maintains its density and average building height as well.[28] Under these conditions, residential and commercial roof space will grow in the exact same proportions as forecasted population and employment: 62 and 81 percent, respectively.

The second change that will occur between now and 2050 is PV efficiency improvements. Sivaram's excellent discussion of this topic concludes with Albrecht and Rech's 2017 prediction that the best systems on the market will increase to 35 percent efficiency by 2050.[29] If we could replace every single panel in use in 2050 with the best collector then on the market, this would more than double NREL's estimates. However, because solar panels last twenty to thirty years, the average efficiency of installed systems will increase much less quickly than the efficiency of the best system sold. Aiming for reasonable guesses, suppose we say that the average collector is 28 percent efficient in 2050.

With greater roof space and better panels, potential rooftop power is more than two and a half times the 2016 potential, or almost 45 TWh per year (column 3 of row A). It seems that there is a lot of room to grow rooftop PV harvesting through the mid-century in cities such as Phoenix if rooftop PV technology continues to become cheaper, easier to install, and more efficient.

The second row of Table 3-2 (row B) shows even rougher estimates—guesses really—for the 2050 potential for PV power from building facades. Some progress has been made on PV-generating windows and other building surfaces, but at this point, the commercial market is too small to project long-term potential.[30] These surfaces also don't ordinarily receive nearly as much sun as rooftops do. Suppose we guesstimate that generation from facades could reach something in the range of one-tenth the level of generation from rooftops.[31] The result is another 4.5 TWh of PV potential in 2050—enough to put a small but visible dent in the supply–demand mix.

As it happens, the real wild cards of 2050 urban PV are the remaining collection surfaces: paved parking, paved roadways, and unused urban land. Parking lots (row C) are excellent sites for PV collection on canopy structures if they are unshaded. Off-street parking occupies vastly different-sized areas of American cities—just over 1 percent of Washington, DC, but almost 27 percent of Houston, according to one estimate.[32] According to the National Governor's Association, U.S. parking lots cover an area the size of West Virginia.[33] In Phoenix, Arizona State urban researchers Chris Hoehne and Mikhail Chester estimate that off-street parking occupies nearly thirty-nine square miles.[34] If we assume that just 5 percent of this receives enough sun to generate power, this adds another 3 TWh (column 3 of row C).

There is also undoubtedly some amount of usable urban land within the city's borders that could be used for small systems mounted on unshaded ground. This could be anything from a solar system in someone's backyard to a community system mounted over an old landfill or brownfield site. Experts estimate that about 60 percent of Phoenix is not covered by pavement. Row D shows that if 1 percent of the unpaved land was suitable and used for PV, potential generation is another 5 TWh.

We could also speculate about the use of the 200 square miles of paved roads in Phoenix for PV generation, either by paving surfaces that incorporate PV cells or by building canopies over roadways. But even without including this, the assumed potential from 2050 roofs, facades, parking lots, and unshaded land (rows A–D) adds up to more than 57 TWh—about twice 2016 use. Moreover, in-city small wind generators, waste-to-energy plants, and several other small sources might contribute something to local supplies.[35] (Urban geothermal is unquestionably on the rise, but for technical reasons, it is counted as an efficiency measure not an energy source.)

The solar potential figures for New York City tell a very different story. Its current rooftop potential is 17 percent of current power use. Roof space in New York City must be growing very slowly, as annual building square footage grows by only about 0.8 percent a year. Much of this is undoubtedly upward growth that adds no new roof space, and it may even shade other previously usable areas. There is also not a heck of a lot of surface parking or unused urban land that can be used for PV. With 14 percent more people, 29 percent more workers, and the electrification of end uses and transportation, power use is virtually guaranteed to increase much faster than PV collection space. It is unlikely that the New York City of 2050 will be able to match its current maximum local generation percentage (17 percent), much less increase it.

Local Power versus the Grid, 2050

The fact that some cities *may* generate most or all of their net power use locally by 2050—while many others won't come close—doesn't mean that they *will*. A number of key factors will have to fall into place to make local production dominant.

First, the costs of rooftop PV energy will have to remain reasonable relative to the costs of buying distant PV energy and shipping it in. Right now, small-scale PV is roughly twice the cost of large-scale PV, and the gap has recently widened rather than narrowed.[36] Moreover, local PV can both add and subtract to local distribution costs. However, technology breakthroughs and new policies could reduce this difference substantially, making local PV that much more attractive.[37] For starters, new PV installation systems and residential panels are in the works, and PV is increasingly built into new structures voluntarily or by mandate.[38]

Electricity storage is also going to play a supremely important role in influencing the growth of local PV from here on out. Small-scale local storage, such as local generation, is going to cost more than its utility-scale counterpart, but power stored in your basement provides a key benefit: partial avoidance of grid power outages. The long-term appetite for ubiquitous local solar, whether from policy mandates or simply from customer demand, is as much a function of the benefits of added resilience as it is a function of costs and carbon benefits and other factors (there is more on this in Chapter 4).[39]

In short, customers and policy makers won't be comparing just the total delivered dollar cost of PV power from local versus distant sources. In deciding the degree to which they will promote or require local production, they will consider a range of costs and benefits, including the differences in resilience to outages, local job creation, and the practical and aesthetic aspects of universal PV.

Although rooftop solar is understandably popular now, several developments could turn the tide toward *less* local generation. Net-zero and required-solar codes, now in the works in several places, will have to survive concerns that these mandates don't reduce house values or raise costs. Some communities are already rebelling against ground-mounted PV installations due to their effect on land use or the view-shed; the entire country of Italy has banned utility-scale PV on agricultural land.[40] Once solar is available cheaply from the market and local storage is cheap, the tastes of building owners who want to help the environment may shift to green roofs, helping biodiversity and adding small carbon sinks.

Last but not least, the politico-organizational dimensions of the two alternatives will also be important and will undoubtedly introduce frictions against change. Regions with an established preference for local utility control may pursue local-only options more aggressively, and vice versa. Preferences such as this are already yielding incremental measures in one direction or another in some cities—a topic that we will return to in more depth in Part II.

Of course, relying on local generation does not mean that the power grid is unneeded or can go away. Like giant net-zero buildings, net-zero cities will use the grid as an enormous battery, sending huge amounts of surplus solar power into the grid by day and importing power from the Big Grid at night.

This isn't to say that local power won't be used extensively on a much smarter, two-way grid. There's not the slightest doubt that DG and local storage will grow substantially for decades to come. In a recent survey of European utility executives, 37 percent predicted that some of their customers would leave the grid entirely by 2020.[41] Australia's energy network operators predict that rooftop solar will provide *all* needed power during sunny hours on mid-spring days in South Australia.[42]

The question raised by the future DG potential of cities is not whether the Big Grid is needed, but rather how large it needs to be. With ample economic local sources, what's the proper geotechnical scale of the power network?

4 Why We Grid

Why don't we just generate as much power as each of us needs, store our excess in batteries, and leave it at that? What exactly is it that grids provide to consumers at the end of the line?

Thomas Edison asked himself this question around 1880. Edison and his early rivals primarily sold isolated systems to wealthy individuals and businesses. J. P. Morgan had one in his mansion—now in the Morgan Library in Manhattan.[1] These systems had a generator, wiring, and a group of lights, much like systems sold today to folks who want to live "off the grid" in rural settings—though today these systems would include solar panels and batteries.

Edison soon made a critical realization. If wires could deliver power to multiple sites, the cost of supplying them with the same amount of light would be much cheaper. Two large generators able to power fifty homes cost much less than fifty small generators, even after you paid for the wires. Fifty individual generators, each supplying 200 lamps, cost him $162,500; two larger units lighting 10,000 lamps cost him only about $72,000. The additional wires and meters cost $15,000. At $87,000 the total cost of the "centralized" system was a little over half the cost of isolated power.[2]

Along with the power savings, the grid created reliability benefits to each connected customer. Before the grid, if your in-home generator broke down your house went dark. Afterward, one or even several generators could break, and the remaining connected plants would supply enough power to keep everyone lit.

The original grids made by Edison spanned a few square blocks. As the geographic scale of these system expanded, cost continued to decline, and the industry kept connecting more and more loads to larger and larger grids and generators. Economies of scale in generation and transmission, and the benefits of load diversity and reliability, continued to be so large that combining several entire utility systems into a single regional grid yielded savings.[3] This virtuous circle has continued until just recently, by which time the United States had lines about 200,000 miles of high-voltage transmission in three huge interconnected grids. These huge systems connect about 10,000 large generators to each other and to 120 million customer accounts. Two leading grid architects nicely summarize the reliability value of the large-scale grid:

> The grid's generally routine reliability is largely a consequence of the system's scale, literally its angular momentum. Titanic power flows from the Hoover Dam and its kin, an immense fleet of large-scale central power generation stations throughout the country. Small generators and loads are effortlessly swept into synchronicity by the current flowing from these huge turbines. This has proven to be a very good way to design a system, especially given the economies of scale and increased efficiency of most electricity generation technologies.[4]

Yet, today, it is again possible to treat a single structure as its own power system, otherwise known as an *off-the-grid* building. The next largest system, a microgrid, is typically the size of a college, hospital, or office campus. Next up is a grid roughly the size of a small city, then a large city, then a U.S. state or two, and so on. There is also an important second dimension to grid size: the degree of isolation from other grids. This can range from fully interconnected free-flowing webs to fully controlled connections used only in special cases.

Is it possible that the optimal size of power grids is evolving to become smaller than the continent-sized behemoths we operate today? Every grid must balance supply and demand at every moment, but with the right amount of local generation and storage, this is now almost possible at any scale. In an era of cheap solar and storage and sophisticated controls, it is not crazy to wonder what scale of power grid is best suited for a carbonless future.

Appropriately enough, world power grids are divided into balancing authority areas (BAAs). They vary greatly in size, but there are roughly 130 in the United States and one for each country in Europe. Within each BAA, a single system controller is required to balance supply and demand at all times. Within this constraint, most controllers allow power to be traded within the BAA, and many operate centralized auction markets to facilitate trading. To ensure balance, flows and trading across the lines that connect two BAAs are carefully controlled, but are rising with new technologies and markets (more on this below). As a result, it is impossible to draw a hard boundary for most power grids until you hit the physical end of the system, typically at the edge of an ocean or mountain range.

Within their BA boundaries, grids can be divided into what might be called Big and Small Grids. Big Grids, officially referred to as transmission systems, are the high-capacity, high-voltage lines seen on giant steel towers. Small Grids, or distribution systems, are the lower-voltage systems that connect the Big Grid to each individual customer. Microgrids and community grids are two somewhat special types of Small Grids, in no way exempt from the need to maintain their own constant balance.

Before distributed generation (DG) and storage came along, balancing Small Grids was easy. Nearly every Small Grid is owned and operated by some type of local utility. That utility monitored total uses on each of its distributed lines, and transmitted its total need to the Big Grid. The Big Grid provided the necessary power and adjusted whatever it needed to maintain balance. The local utility often had to make minor operating adjustments to its system, but mainly it made sure that its system was adequate and working properly. Now that small generators are popping up everywhere on Small Grids, balancing these systems is becoming more like the process conducted by Big Grid operators.

As it happens, the case for a large, free-flowing Big Grid is both strengthened and weakened by the industry's multifaceted transformation.

A few of the traditional cost-saving benefits of a huge grid become a little less clear, but other changes argue for an even larger grid. Other technological and politico-economic forces we turn to below will push the industry toward smaller networks—but if your key metric is the average cost of delivered power, Big Grids are here to stay.

This result should not be mistaken for the view that every individual customer will choose to stay connected to the grid forever. Even today, it is cheaper to supply some customers who are far away from the power lines with their own off-grid systems. As the costs of distributed power and storage continue to decline, the number of "defections" from the grid is certain to continue. The hallmark of these new defectors will be that they have chosen to disconnect, even though they are well within the footprint of the grid, not at or beyond its edges. They defect because they prefer to be independent of the grid, DG and storage have made it technologically feasible, and they prefer (and can afford to) pay the upfront and maintenance costs of the comparatively large power system they need to operate independently.

In 2014, this new type of grid defection was quite the rage. The highly respected Rocky Mountain Institute (RMI) published a seminal study that compared the projected total costs of completely independent solar-plus-storage operation to continued retail utility service through 2050 in five major U.S. cities.[5] Forecasting that retail electricity rates would go up in each city (albeit at different rates) while solar-plus-storage costs would continue to decline, the study found that defection would be cheaper for residential customers in 2022 in Hawaii, 2038 in California, and 2048 in Nevada. Two years later, a team of researchers raised eyebrows further when they projected that fully 92 percent of all seasonal households in the rural Upper Peninsula of Michigan would find it cheaper to get off the grid by 2020.[6]

Of course, these inflection points were based on many assumptions and projected costs, notably the rates that electricity customers who stay on the grid have to pay. As we'll see in Part II, the detailed structure of future electricity prices will affect the economics of defection greatly, and

RMI's projections were necessarily simplified. More importantly, from the standpoint of the role of the grid and its utility owner, the future supply choice of electricity customers will not be binary, but rather will include a wide range of partial supply scenarios with continued critical reliance on the grid for the full, continuous service. In Chapter 5, we examine how microgrids—small, potentially independent portions of the grid—represent another way in which customers will partly defect.

In the last few years, attention has shifted away from complete to partial defection, including by microgrid. In 2015, RMI itself issued a second important report that emphasized the magnitude of total utility sales losses from all forms of partial self-supply, not full independence. "While the presence of customer choice [to defect] has important implications," the report concluded, "the number of customers who would actually choose to defect [entirely] is probably small."[7]

In effect, the RMI recognized that the benefits of remaining on the grid, beyond the simple average economic costs of supply, would weigh heavily on individual customers. Defectors would either have to finance, build, and maintain their own systems or (much more probably) pay a third party to do this, adding substantially to the cost. In response to the defection studies, Berkeley professor Severin Borenstein famously remarked that "count me among the people who get no special thrill from making our own shoes, roasting our own coffee, or generating our own electricity."[8] Large-scale industries have become central suppliers of almost everything we use for a reason, Borenstein noted, and electricity is unlikely to be an exception.

But the reason to stay on or leave the grid is far from a matter of lowest expected cost. Average cost comparisons implicitly assume that both alternatives are equally (or at least sufficiently) reliable and have the same qualitative, unpriced attributes. In the case of electric power, they don't. Grids of different sizes and scales have very different vulnerabilities and benefits that aren't captured in the price of power. In this realm, bigger isn't always better, and it is for these reasons that the large-scale grid will be partly supplanted by complex smaller grids.

The Case for Big

The traditional benefits of large grids can be placed into five categories: scale effects in power production and delivery; the smoothing benefits of aggregated load; cheaper aggregation of load; the benefits of trading power; and lower costs of preventing traditional blackouts.[9] The latter is often referred to simply as higher reliability, but we will try to parse this broad concept a little more finely. In the coming era, we can add the benefits of renewable supply diversity.

Although generation is gradually shifting to small-scale renewables that work well and are often affordable, power production scale effects remain quite large for every form of power production. To put it bluntly, small *is* beautiful, but bigger is almost always cheaper, including for renewable energy. The installed cost of small wind projects is more than twice the cost of energy from a large windfarm turbine.[10] The latest U.S. government figures on solar installations show that power from rooftop photovoltaic (PV) systems costs about four cents per kilowatt-hour, while power from utility-scale plants 10,000 times as large costs an average of two cents per kilowatt-hour. Because these are all emissions-free alternatives, there are no emissions externalities that change the relative cost rankings. In fact, large and small PV plants use exactly the same PV panels, and most of the rest of the equipment is quite similar—just larger.

The same scale effects turn out to be true for storage as well. I once asked the chairman of EOS Energy Storage (a company I advise), "Which element of your storage installations is most subject to the economies of scale: the cells, inverters, or electronics?" He looked at me and said, "All of them." The data bear him out. Lazard's highly authoritative report on the cost of storage finds that lithium-ion batteries cost $268 per kilowatt-hour in utility-scale installations and about four times as much, $950 per kilowatt-hour, in home-sized systems. "As compared to in-front of the meter systems," the report notes, "behind-the-meter [small] system costs are substantially higher due to higher unit costs."[11]

The proponents of rooftop solar have argued correctly that local PV can produce some benefits for the local distribution system that cannot come from distant renewables, such as displacing some of the costs of power delivery. However, local solar can also *add* costs to the distribution system because the latter must be reengineered into a much more sophisticated, multidirectional system. These benefits and costs are complicated to calculate and highly situation specific. All of the costs have to be added up, including plants, lines, fuel, control equipment, grid operation, and oversight. Every cost and operational parameter must be projected over the forecast horizon, involving many uncertainties and judgment calls.

Due to all these differences, comparing complete alternative future clean power grids of different scales is extraordinarily difficult. One way to get around these challenges is to compare the capital and fuel costs of equivalent amounts of 100 percent carbon-free generation resources, which is relatively straightforward. Then, see whether all the rest of the costs might tip the balance one way or another.[12] This is the approach I took in a 2015 study comparing the total delivered costs of PV power from rooftop and utility-scale systems. Looking only at the costs of making the power, utility-scale PV plants were about one-half the cost of rooftop systems under exactly the same solar conditions. Today, this cost difference is roughly three cents per kilowatt-hour in reasonably sunny parts of the United States.

How likely is it that the net cost savings rooftop solar enables for the rest of the system (transmission, distribution, etc.) exceed three cents per kilowatt-hour, making rooftop cheaper overall? In general, this is quite unlikely. A local solar system can reduce the need for transmission and distribution, but these savings are significant only if it is almost entirely self-sufficient. Solar systems are also capable of providing specialized local electrical products that can help with grid management, but these typically don't add up to anywhere near three cents per kilowatt-hour. That said, the messy truth is that in some specific locations on the grid, it's highly likely that local solar can supply products and reduced costs that add up to more production cost differences. We must be careful not to make overbroad statements about system-wide total costs, even if large-scale systems have much cheaper elements.

Scale-based production cost differences therefore don't settle the question of optimal grid scale. However, in the vast majority of cases, the large-scale grid is an essential tool for keeping power costs as low as possible, and total supply costs are only the first of several scale-related benefits.[13]

Aggregation and Trading

In traditional terms, the second major benefit of large grids was load diversity. Nowadays, we call this *load aggregation,* but it's the same idea. Simply due to the law of large numbers, the aggregation of many individual loads follows a smoother pattern over time than any one of the individual loads. Along with pure scale effects, this makes the cost of service cheaper. Traditional power generators have large rotating metal armatures. As with any mechanical device, the more rapidly and more often you turn the machine up and down, the more it costs you. Smoother loads are cheaper to supply.

Even more importantly, massed loads save costs by enabling specialized forms of regeneration. If you combine a large enough group of loads, you create a level of demand below which total load never drops, even in the middle of the night. A rotating generator scaled to supply this minimum level of load, known as a base load plant, runs steadily for weeks or months. These plants used to be the cheapest way to produce power, and often they still are. For levels of demand rising gradually higher from midnight to midday or from season to season, utilities designed other types of plants (cycling or intermediate) that are cheapest to use during these periods. Peaking plants ("peakers") were the cheapest way to supply short periods of very high load.

The historic system design goal was for utilities to build the right number of each type of plant to match the particular aggregation of all their loads. Some utilities in moderate climates, with little electric heat or air conditioning, had relatively smooth loads, and could be served mostly by base load plants and a few cycling units. Other utilities with "spiky" loads had fewer baseload plants and a large fleet of peakers. Either way, the portfolio of plants was designed to enable the utility to match its supply to its customers' aggregate demand patterns more cheaply than any alternative plant portfolio.

With the important exception of PV, the main forms of renewable energy continue to use rotating-metal generators. Wind turbines, hydroelectric, nuclear, and geothermal plants all use them, as will natural gas plants with carbon capture. These generators, which all have production scale economies up to the level of at least 600–1,000 MW, also act as a lowest-cost portfolio if the right types are blended into one large system. PV's unique technical attributes notwithstanding, the massing of load is likely to continue to augment product scale economies well into the future.

However, grids do more than enable one utility to aggregate all the loads in its area and match them to a lowest-cost fleet. Suppose we change the rules so that any proper entity can go to groups of customers, aggregate their loads, and build or buy from a group of generators whose costs for serving this aggregated load are low. We can then let all these entities compete against each other to sign up customers, that is, become their power supplier. These two changes create a competitive power marketplace in lieu of the utility load aggregation monopoly. Competitive load aggregation is the core of competitive power markets.

For every power marketplace, the one grid to which customers, aggregators, and producers are connected is its sole physical trading platform. The customers and plants accessible to aggregators are limited to those physically connected to the grid. So, while consumers and producers connected to an isolated microgrid can trade with each other, those same traders have vastly larger opportunities connected to a larger grid. Also, on the larger grid, transactions can occur in very large as well as small quantities, making it worthwhile creating more sophisticated trading instruments and arrangements and searching harder for efficiencies.[14]

Of course, this phenomenon has its limits. At some point, the market gets so big that the additional trading benefits from grid expansion aren't important. Still, quite a bit of research has confirmed that the benefits of trade continue to the point where grids cover areas the size of a large portion of the United States and larger than many entire countries.[15] One remarkable illustration comes from California, which already totals one-thirteenth of all U.S. power capacity—about twice that of Sweden. Recently, several colleagues modeled the benefits of allowing better trading between California and the rest of the western United States, an area that includes thirty more BAAs within ten more states and two Canadian provinces. Their study found that this enormous market would enable cost savings of $1.5 billion a year under California's 50 percent renewables mandate. As the percentage of renewables on the system increased to 60 percent, savings doubled to $3 billion a year.[16]

Grids and Geographic Diversity

One of the reasons that power costs benefit from enormous geographic markets is inherent in the nature of wind and solar resources. Wind and solar power plants can't be controlled in the same way that we routinely adjust conventional power plants up or down. When the sun shines or the wind blows, renewable systems pump out power at whatever strength the sun or wind happens to have at that moment, regardless of whether that amount is needed. If there is not enough demand on the system to balance supply and demand, the power must literally be dumped—or *curtailed* in power industry lingo.

California named this problem the duck curve when it began projecting the daily shape of its system load over the course of a normal day and found that the demand it had to balance was going to resemble the shape of a duck's head (see Figure 4-1). One challenge created by the duck curve comes from the fact that their solar-rich system is flooded with more power than it can use during the middle of the day. It cannot turn down the base load generators it needs to run steadily through the night fast enough to match the solar power flooding in at midday.[17] There are two main approaches to resolving this challenge: storage and geographic diversification. Storage is obvious: if there's too much solar power at noon, store it for use at night. Once solar or wind energy is stored, it can be dispatched to balance load, much like any controllable power plant, until

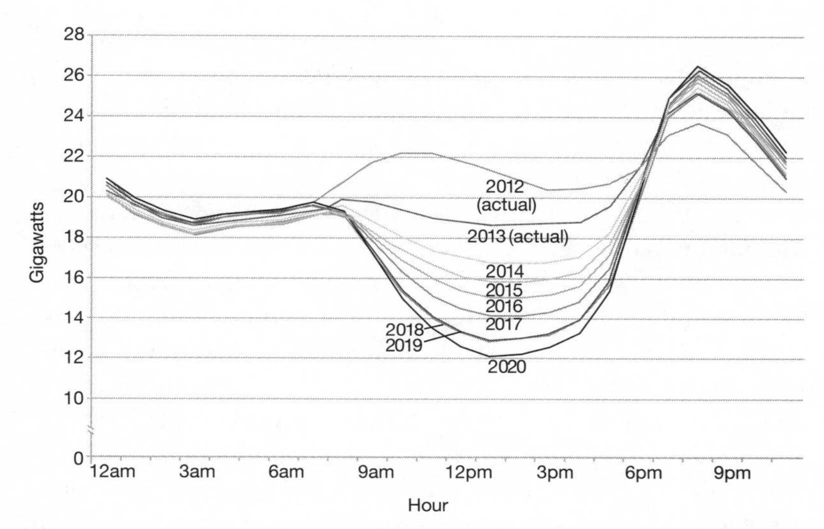

Figure 4-1 The California ISO "Duck Curve" for a Typical Spring Day, 2012–2020. This curve shows the net power that the California system operator must supply to balance supply and demand over the course of a typical day. The amount of balancing power goes down in the middle of the typical day as sunshine and therefore solar generation (which cannot be controlled) reaches its peak. Toward the hours of 6:00–9:00 p.m., when electricity load is very high but photovoltaic (PV) generation drops to zero, the system operator must increase supplies rapidly toward a peak supply around 9:00 p.m. This gives the curve of net supplied balancing power the shape of a duck. As the years progress in California, PV installations and midday PV production increases, making the duck shape more pronounced.
Adapted from California ISO (2016a).

you run out of charge. In fact, many forms of charged-up storage are *more* useful to grid operators than conventional plants because they respond to controls very quickly and can provide a variety of *ancillary services* that help support the grid.

However, using storage *alone* as a means of ensuring controllable balance is costly. Consider an entirely isolated solar-plus-storage microgrid with a peak load of 50 MW occurring (as is customary) after sunset and an average daily total use of 600 MWh.

Suppose that winter weather patterns in this location show a worst-case scenario of one day of good sun followed by eight days of clouds.

The microgrid needs to install eight times as much PV as it needs to supply a single day, so that on the one good day of sun, it gathers eight days' worth of use. Yet, all this PV capacity will not be needed under ordinary weather conditions, when the battery can be at least partly recharged each day.

Then there's the battery. To protect against a storm that lasts eight days, with no sun to recharge the battery, it must store 4,800 MWh of power. At a cost of $100 per kilowatt-hour of storage capacity, this is almost a half-billion-dollar investment that exists mainly to cover a rare, worst-case scenario. On the vast majority of days, this huge battery would just sit there, costing its owner money but providing no immediate storage value—only insurance against the next worst-case cloudy stretch.

The second approach to integrating renewables—using the grid to access renewables elsewhere—is also intuitive. Rather than relying on renewable production only in your own location, transmit power from other locations where the wind or sun happens to be producing power. This approach hinges on two crucial geophysical facts. First, there are easily identified regions of every continent where wind and solar resources are significantly stronger and more constant, so transmission lines can be built on these spots. Second, as the geographic scale covered by a grid gets larger—almost without limit—the aggregate output of wind and solar resources becomes more constant and predictable. In the words of a team of National Oceanic and Atmospheric Administration (NOAA) researchers led by Alexander MacDonald:

> Because Earth's mid-latitude weather systems cover large geographic areas, the average variability of weather decreases as size increases; if wind or solar power are not available in a small area, they are more likely to be available somewhere in a larger area. Even more importantly, access to electricity over a large region allows locations with rich wind and solar resources to supply cheap power to distant markets.[18]

Very large grids enable the trading of renewable power between regions with different natural production and demand patterns, reducing the costs of balancing loads and supplies. As this occurs, the systems need to build less backup (reserve) capacity and maintain smaller operating reserves, yet achieve the same level of reliability. In traditional and many current systems, grids provide savings in reserve capacity because they allow systems to access each other's power plants in the event that one large fossil plant suddenly has an outage. In a high-wind-and-solar era, the grid will perform exactly the same function, except that we won't worry that a million PV panels will all suddenly trip offline, we'll worry that a cloudy day will naturally bring them offline. Either way, a Big Grid enables access to renewables elsewhere in order to save the cost of storing local-only production.

Many studies have explored the effect of a larger transmission grid, harnessing the natural diversity of solar and wind production on the total costs of decarbonized supply.[19] These studies create simulation models of the power systems over successively larger regions and add up the costs of building and operating the system over the horizon of the study, typically thirty years or so. Once they are calibrated, the models are instructed to add DG in or near cities, large wind and solar plants in the best locations for their resources, and sufficient storage so that the total supply of generation plus charged-up storage at every moment is enough to match projected demand. The models are, of course, given extensive historical weather data so they can predict the probability of solar or wind production at every location in the region.

The models test out many different combinations of DG, storage, large wind, and large solar in order to find the future combination that yields the lowest total cost of electricity for different system scales. Each future configuration is checked to see whether the addition of new plants requires that new transmission lines be built to deliver distant wind and solar to the "load centers," essentially the metropolitan areas in the region. If new transmission is needed in any scenario, the cost of construction and operation is added to the tally of total costs.[20] The most sophisticated of these models also count the cost of reinforcing the distribution system when DG increases (to accommodate large local flows) and the higher cost of backup power at smaller scales.[21] Finally, the better studies also allow system operators to reduce costs and balance the system by shifting loads around, that is, demand response (DR), which inevitably contributes further cost savings.

These modeling efforts consistently find that adding large amounts of big wind and solar projects in areas where these resources are best, and building more transmission to reach them, beats the cost of adding only local power and storage without grid expansion.[22] One good 2014 study of the entire European region examined detailed scenarios in which the entire grid was constrained at its present size versus expanded optimally, with DG allowed to be built (with local backup and grid reinforcements) if that lowered total cost.[23]

Just as in our simple example above, the study found that allowing grid expansion lowered total costs, partly by avoiding the oversizing of local backup resources. Grid expansion allowed the system to install fifty million kilowatts less backup generation or storage—a saving of $50 billion at $1,000 per kilowatt.[24] The study authors explain that this occurs because the DG solar installed "is practically nil," meaning that it is not able to supply power during the time of system peak demand.[25]

The study also observes that a sort of negative feedback loop occurs as greater local DG and backup are installed. Since local DG does not yield as much surplus power for trading as large, better-sited renewables, there are fewer trading opportunities and therefore less justification for added transmission lines. "In turn," the study notes, "this [further] increases the need for building backup capacity since less transmission also reduces the scope for sharing reliability between different regions."[26] In total, the study concluded that "additional transmission greatly facilitates the integration of (variable) renewables, whereas constrained grid expansion leads to additional challenges and costs."[27]

These savings can be viewed as the modern-day equivalent of the reliability benefits of Edison's original grid. The essential idea is that stringing together more plants enables supplies to reach more customers without interruptions and less constructed capacity.

As with any projection, garbage in leads to garbage out. All these results depend on the projected level of costs of local and large-scale resources of all types and the costs of transmission—not to mention the assumptions about demand levels, energy efficiency, and many other system attributes. Were the costs of DG PV and local storage to drop substantially *relative to their large-scale cousins,* these widespread projections would turn out to be wrong. Yet, as we've just seen, there is almost no evidence that the many cost advantages of large renewables coupled to a large grid will change as the industry is transformed. This was precisely the message emerging from the study of trading across the western United States described earlier; greater transmission and trading allowed California to export its midday surplus solar power up to the cloudy Pacific Northwest while also enabling it to import inexpensive wind power produced at night in other parts of the West.[28]

It is important to note that these cost studies uniformly examine total costs for all customers combined. This is an essential metric for policy makers, but it does not mean that large-scale sources will be cheaper than self-supply for every customer. The cost of these options for each customer is a function of many local factors, including specific utility rates, the cost of local, DG installation, and local DG resource, all of which must be compared to the local cost of buying large-scale power.

Two factors in particular make the result that larger grids will remain cheaper seem plausible. First, the very same solar power systems that could be installed on rooftops or located in large utility-scale plants are much more productive in the latter. Surprising as this may seem, large PV systems are able to choose better locations, orient the PV panels accurately, install tracking devices that move the panels to follow the sun, and maintain their system for better efficiency. As a result, the output of an otherwise-identical PV panel inside a large plant is 50–100 percent greater than the output of a rooftop system using exactly the same PV panel, even though its costs are lower.[29]

The second factor is that large transmission lines, while unsightly, are inexpensive compared to all types of generators and storage, and themselves have large economies of scale. The total combined capital and operating cost of the entire U.S. transmission system amounts to about nine-tenths of a cent per kilowatt-hour, less than 10 percent of the retail cost of power.[30] As of 2013, a new 500 kV direct current (DC) line carrying about 3,500 MW of power cost $1.6 million per mile; an 800 kV DC line carries 6,600 MW (almost double) but costs only about $1.8 MM per mile, both excluding converter stations.[31] New DC transmission technologies are expanding raw scale economies into even larger lines.

The advantages of harnessing natural renewable supply diversity via transmission occur over regions so large that a new school of thought has emerged about the optimal scale of the decarbonized grid. Some experts believe that the world's future carbon-free electricity system will include "supergrids" of ultra-high-capacity DC transmission lines from renewables-rich areas around the world to urban load centers.

As far back as 2014, planners in a Midwestern U.S. grid operator modeled the benefits of a supergrid that traversed about two-thirds of the United States using nine long DC lines. This system would send wind power from the middle of the United States to the Southeast and the West at night and import solar from the South and Southeast by day. Their back-of-the-envelope calculations indicated that a system such as this would cost a cool $36 billion but would return $45 billion in total cost savings.[32] A more recent simulation by MacDonald and Clack's NOAA team simulated an even larger system of thirty-two nodes across the entire United States connected by high-voltage direct current (HVDC) lines. These researchers found that this extensive network had larger net diversity benefits than *any* smaller set of grids, even when the costs of transmission were factored in.[33] Additional studies of HVDC grids are underway in the United States by the respected Climate Institute and in Europe by several transmission agencies.[34]

The broadest vision of renewable diversity benefits comes from an organization in China called GEIDCO—the Global Energy Interconnection Development and Cooperation Organization.[35] This coalition of utilities, electrical equipment firms, universities, and nongovernmental organizations is promoting and researching the idea that improvements in HVDC cables are so large that it may be economical to harness renewable supply diversity on a global scale. Each of the earth's continents has areas that hold extraordinarily strong renewable resources, such as the Sahara Desert for solar and Patagonia for strong winds. GEIDCO's vision (Figure 4-2) is to build HVDC lines spanning the earth, so that surplus cheap wind and sun on one continent can supply power to other continents during their night or when supplies run low.

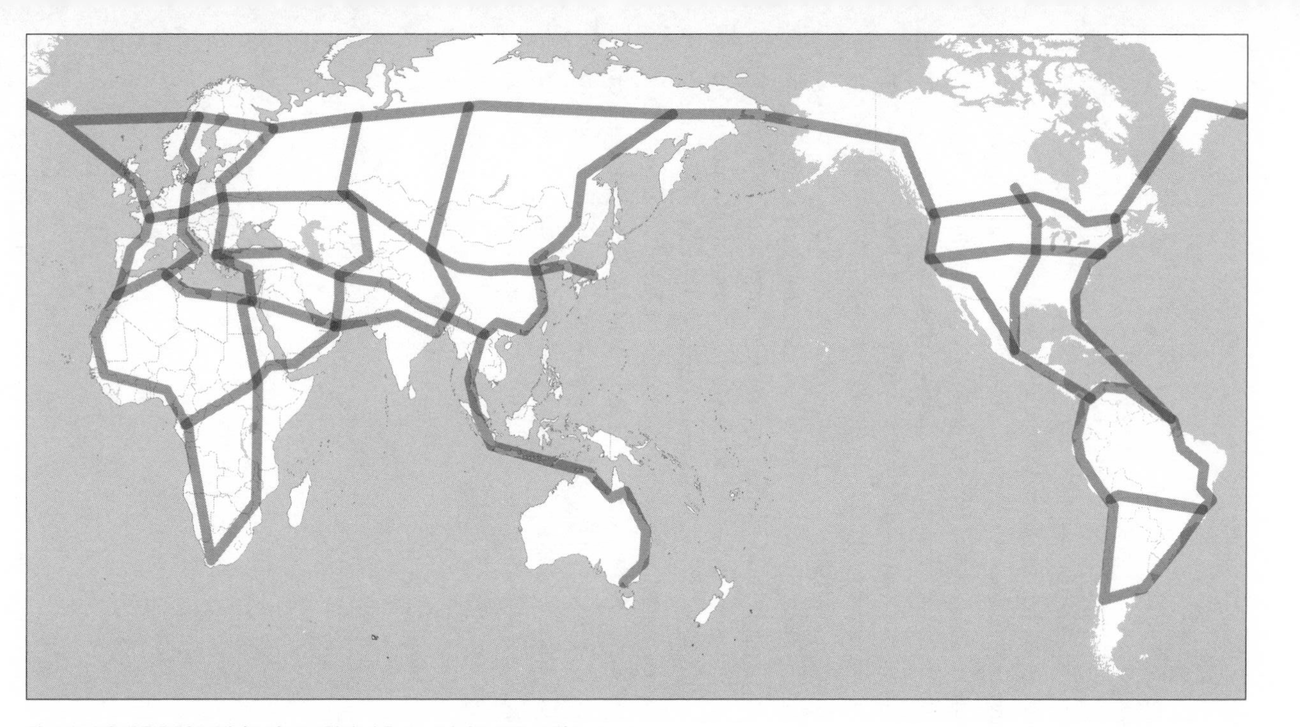

Figure 4-2 GEIDCO's Vision for a Global Energy Interconnection.
This figure shows the conceptual design of a high-voltage direct current (DC) power grid that would span the entire world, linking all the world's continents. The concept is animated by the existence of particular regions around the world where wind and solar energy are especially strong (i.e., deserts and wind belts). A grid such as the one shown would link these renewable supply basins to large urban load centers and diversify these loads and sources. Transmission lines that cross continents and oceans are becoming technologically feasible, and some long DC lines cross borders, but the economic, regulatory, and financial aspects of a system such as the one shown are far from determined.
Adapted from Global Energy Interconnection Development and Cooperation Organization (n.d.).

Not Quite Case Closed

It's a bit of a surprise that the radical changes in power system design paradigms, coupled with a complete change of the industry's generation base, don't put an end to the overall cost savings of large power grids.[36] The technical fundamentals and a huge body of research both suggest that the dollar savings from generation and storage scale effects, load and supply diversity, and large-scale trading and reserve sharing may increase rather than decline in decarbonized power grids. Moreover, these benefits appear to exceed their costs up to grid sizes even larger than the regional grids we operate today. As a result, it is almost impossible to find a projection of any future power system in which the span and capacity of the grid fails to grow by at least a little between now and 2050.

As we know, however, the social goals of power systems include more than lowest possible cost of carbon-free supplies. To fully evaluate the trade-offs between different grid architectures, we need to ask whether there are any downsides that offset the large grid's cost advantage. For starters, what we think of as reliability benefits don't automatically increase with the span of the grid. When things go wrong on a power system, bigger is not always better. And for better and for worse, there is a lot that can go wrong on a grid.

5 The Fragmented Future

In the electric power business, there are two kinds of days. On blue-sky days, the sun is shining, and the grid is delivering cheaply and reliably. On black-sky days, something has gone very wrong, and the grid is not delivering cheaply and reliably. A storm has knocked out a transmission tower or a hacker has infiltrated the system and caused blackouts, perhaps affecting millions of people. Worse still, a terrorist may have blown up parts of the transmission system that could take several months to replace.

The original grid wasn't designed for threats such as this, but we live in a world in which these dangers are real and mainly on the rise. The case for building small in the world of power grids might include reduced vulnerability to these sorts of threats and therefore greater overall reliability—even if it costs more over the course of time.

How to Damage a Grid, Part 1: Summon Poseidon

One of my all-time favorite bands—Earth, Wind, and Fire—happens to describe three of the four types of weather events that seriously damage grids. Geophysical events include earthquakes and volcanoes; wind-related events are the well-recognized hurricanes, tornados, cyclones, typhoons, windstorms, and derechos; fire or *climatological* events include heat waves, droughts, and forest fires. It would have ruined their catchy name, but Earth, Wind, and Fire should have included a fourth type of event—hydrological or floods—as the final major weather threat.[1]

Each stage of the grid has its own particular vulnerabilities to these four kinds of threats (see Figure 5-1). Traditional power plants, which make steam to turn turbine generators, can be damaged by winds, earthquakes, and floods. These include not only conventional fossil-fueled plants, but also nuclear, concentrating solar, and future coal or gas plants with carbon capture and sequestration. Although they don't use steam cycles, hydroelectric plants also belong in this group. Much of the world still remembers the March 2011 tsunami that flooded the Fukushima Daiichi nuclear reactors, triggering a meltdown, massive cleanup and protection efforts, and the permanent loss of three large plants.

While earthquakes and floods at power plants can be catastrophic—see Fukushima—they are also quite rare, and it is somewhat possible to reduce the likelihood and level of impact by "hardening."[2] The vulnerability

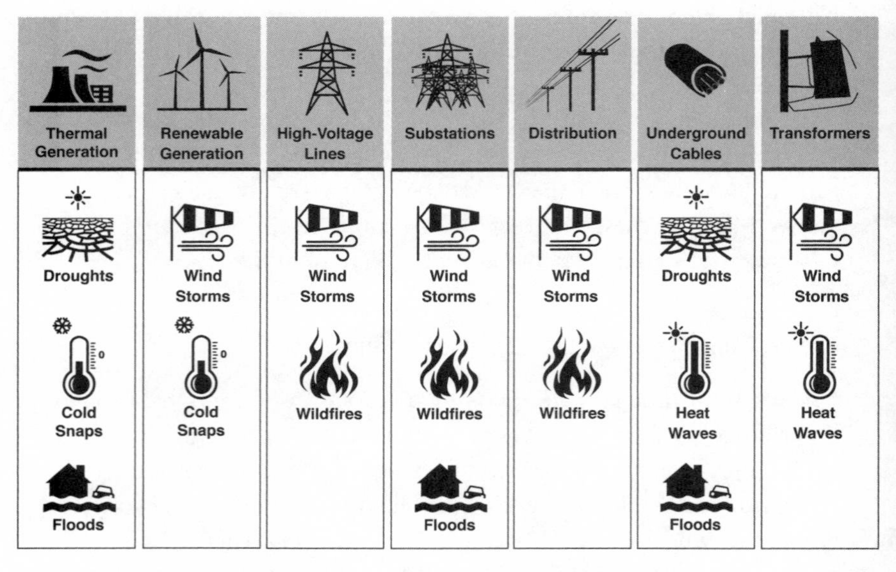

Figure 5-1 Mapping of Severe Weather Impacts onto the Stages of the Grid.
Abi-Samra (2013a).

that is more difficult to protect against, and that is shared by nearly all thermal and hydro plants, is their dependence on water.[3] Without adequate supplies of water at sufficiently low temperature, these plants must limit their output or just shut down. In a severe drought or heat wave, the problem can go on for days or weeks, and almost nothing can be done. In the 2003 French heat wave, fourteen nuclear plants had to shut down. In the U.S. heat wave of 2012, the Braidwood nuclear plant in Illinois got an environmental waiver that allowed it to operate, despite the fact that the temperature of its cooling water source had risen above 37.8 degrees Celsius (100 degrees Fahrenheit).[4] To make matters worse, the efficiency of a thermal power plant declines as temperature rises, and plant equipment becomes progressively more prone to failure.[5]

As one might expect, transmission systems are most vulnerable to wind-centered storms. Resilience expert Nicholas Abi-Samra estimates that winds are implicated in 80–90 percent of all weather-related failures.[6] The force on power lines grows with the square of wind speeds, and that's before the added weight of ice and snow, trees, and the most dangerous threat of all: flying or floating debris. When one transmission tower falls, it can start a cascade of subsequent collapses; one 1993 windstorm in Nebraska blew down 406 towers in a row.[7] In addition to contributing to line collapses, ice and snow also cause short circuits that take lines out.

The transmission system has another extraordinary vulnerability outside the four main categories—indeed, outside our planetary home. Solar flares can cause geomagnetic storms, or *space weather* as it is sometimes called. Amazingly enough, these sun storms create huge magnetic fields on earth that induce electric currents inside the earth itself. These currents can burn out the huge, highly specialized extra-high-voltage transformers in transmission grids. A single storm event could potentially knock out hundreds of these large transformers, leading to grid blackouts that could cover much of a continent. Still worse, many of these transformers cost millions of dollars or more and are customized for their location, so that replacements are almost never stockpiled and typically take months or years to make.[8] The good news—if you can call it that—is that the probability of these storms is estimated to be less than 0.02 percent for the United States.[9] In addition, new U.S. rules require utilities to study geomagnetic storm scenarios and take special action if disturbances on their systems could set off widespread blackouts.[10]

But it is the distribution portion of the grid that is by far the most vulnerable to the largest number of weather-related phenomena. Substations have high-voltage transformers and ground-mounted switchgear that is typically exposed to flooding as well as wind damage. Flooded electrical equipment can also take much longer to repair or replace than poles and wires; the latter are mostly standard pieces of equipment readily stockpiled by many utilities, while distribution switchgears must be painstakingly dried out, tested, and often replaced. More visibly, distribution lines and poles are exposed to winds, ice, snow, the aforementioned deadly debris, and wildfires.

It is often suggested that undergrounding power lines is the best solution for strengthening the weakest link in the chain: the distribution lines themselves. For some storms, such as Hurricane Irma, this has proven to be true.[11] While placing lines underground obviously removes the threat of wind damage, it increases vulnerability to flooding and heat waves, which often overheat underground cables to failure. When these problems occur, repairing underground cables takes much more time than repairing overhead poles and wires, so there is a trade-off between outage frequency and duration. Most of all, undergrounding costs five to ten times as much as overhead wires, and cables last only half as long. So, it must be used selectively for the parts of the system that benefit most in spite of the higher cost.[12]

Grid Vulnerabilities and the Climate

With the exception of earthquakes and volcanos, climate change will steadily worsen every one of these sources of power system disruption. We most often associate climate change with stronger storms, and this is fair. "Hurricanes and typhoons are projected [to increase] in precipitation rates (high confidence) and intensity (medium confidence)," according to the 2017 U.S. National Climate Assessment (NCA).[13] The increase in precipitation includes the now-familiar "atmospheric rivers"—rains such as the biblical deluge that flooded Houston in 2017 with forty inches of rain in four days.[14] In March 2015, the Atacama Desert in Chile got *twelve years'* worth of average rainfall in *one day.*[15] Though models show a lesser effect for the U.S. Atlantic coast, these sorts of storms are going to deliver progressively more rain, wind damage, and floods for the next one hundred years.[16]

The NCA also found that climate change would increase the "frequencies and magnitudes of agricultural droughts throughout the continental United States."[17] In addition to their impacts on generator water supplies, these dry conditions would lead to more frequent and intense forest and wildfires, which often threaten power lines.[18] Hydroelectric plants in the western United States would be particularly impacted; the NCA projects a *70 percent* decline in water from snow by 2100.[19]

In the urban areas that host the vast majority of power distribution hardware, steadily higher average temperatures and stronger and longer heat waves will compound the effects of more frequent wind and rain damage. In Boston, for example, the number of days with average temperatures above 32.2 degrees Celsius (90 degrees Fahrenheit) will increase from eleven in 1990 to between twenty-five and ninety by 2070, including up to thirty-three days above 37.8 degrees Celsius (100 degrees Fahrenheit).[20] These higher temperatures especially stress transformers, in part because they are designed for a climate in which cooler nighttime temperatures allow them to dissipate accumulated heat. The 2018 NCA concluded that "Climate change and extreme weather events are expected to increasingly disrupt our Nation's energy and transportation systems, threatening more frequent and longer-lasting power outages, fuel shortages, and service disruptions, with cascading impacts on other critical sectors."[21] Sadly, the trend in severe weather events is already visible in the data, not to mention the memories of millions of people who have now experienced strange and often dangerous storms. A simple look at the trends in different types of severe weather events (Figure 5-2), compiled by the UN and the global insurance industry, reveals something

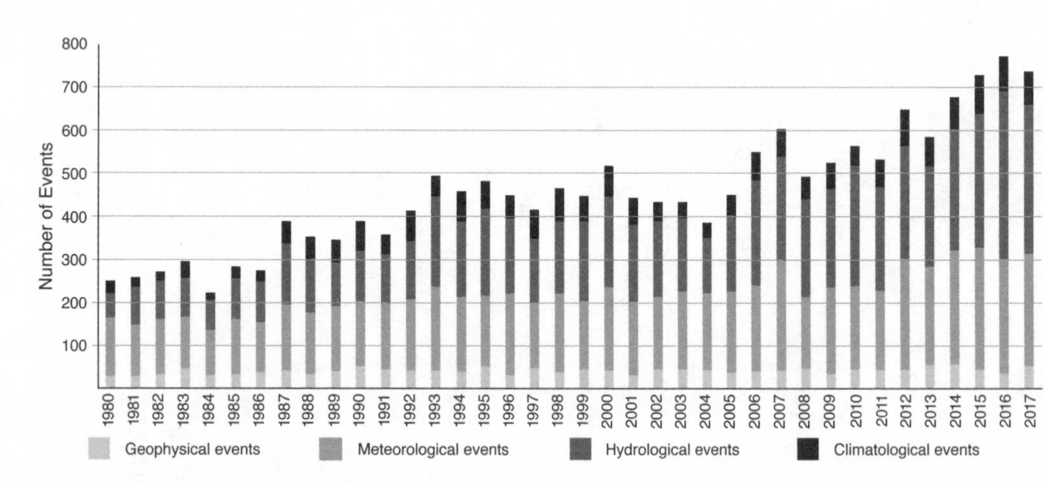

Figure 5-2 Number of Large Loss Events Worldwide, 1980–2017.
NatCatSERVICE © Munich RE.

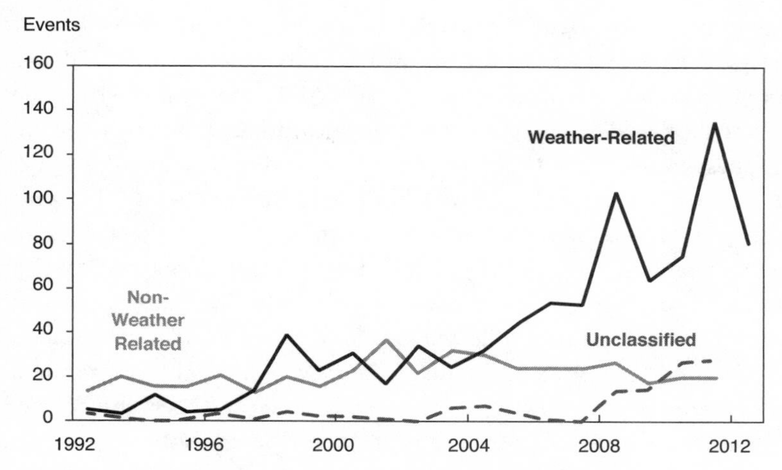

Figure 5-3 Observed Outages to the Bulk Electric System, 1992–2012.
Lacey (2014).

striking: no trend in the earthquake-type events, which are unrelated to climate change (bottom black portion of the bars), but visible upward trends in windstorms, floods, droughts, and wildfires (dotted and cross-hatched parts of the bars)—all events closely linked to global warming. The five most damaging storms in U.S. history have all occurred since 2005, with three during 2017 alone, making this the most expensive year in history for weather damage ($316 billion).[22]

The same trends clearly emerge from reports of electric industry outages over the past thirty years. Figure 5-3 shows that large outages unrelated to weather have held rather steady at about twenty to twenty-five per year, while weather-related outages have already trended steadily upward to much higher levels by 2012.[23]

Grid Coping Skills

The traditional cornerstone of electric reliability has been *preventing* outages through design and engineering. Once an outage occurred, speed of restoration was the next most important metric. To this day, the most common utility reliability metrics are the average length of time the average customer loses all power, which goes by the acronym SAIDI, with the average duration of outages known as CAIDI, and the average number of annual episodes of power loss per customer using the acronym SAIFI.[24]

These measures were intentionally designed to benchmark utilities' system performance outside the effects of severe weather, so they ordinarily exclude outages from these events. In other words, these measures try to determine how skillfully utilities design, maintain, and operate their system to provide good service on blue-sky days. Excluding storm outages from the routinely collected metrics of utility system performance made sense because storm-related outages were (and still are) comparatively rare and, as we'll see, extremely difficult to predict and prepare for. Nowadays, the metrics sometimes include severe weather events, but more commonly regulators conduct customized inquiries into utilities' performance during and after such events. Many regulators require utilities to report these statistics, and reward or punish utilities that perform poorly.[25]

In recent years, the language has shifted to the more complete concept of system *resilience*. The resilience of grids is not merely prevention via engineering, but also the "ability to survive perturbations, repair themselves, and adapt to new situations."[26] Summarizing a National Academy of Sciences panel, industry expert Sue Tierney called a resilient grid

> one with the following characteristics: It is one where the grid planners, operators and regulators assume that they cannot foresee and avoid every type of event that could take out the system in a very big way; where they therefore plan for how they will ride through big-impact events with as much of the system still intact as possible; where they can mobilize the resources to restore the system safely and quickly, especially to support the provision of critical services; and where the industry players learn lessons from prior disruptions and plan for how to better handle the next hit on the grid.[27]

It is fitting that the idea of energy system resilience was popularized by Amory and Hunter Lovins in *Brittle Power*, a study originally commissioned by the Pentagon.[28] While the book has received much less popular attention than Amory's works on distributed power, it correctly foreshadowed the future security threats to a highly complex, centralized, and (later) intelligent grid. Lovins and Lovins noted that the power grid was designed for scenarios where "everything happens according to blueprints. If such a place has ever existed, the world emerging in the coming decades is certainly not it."[29]

In the intervening decades, the loose concept of grid resilience has been expanded into four tactical dimensions. *Hardening* is the ability to absorb shocks and continue operating as designed; *ride-through* is the ability to manage a crisis as it unfolds and maintain some basic level of electrical functionality without totally collapsing; *rapid recovery* is the ability to get systems and services back as quickly as possible; and *adaptability* is the ability to learn and incorporate lessons from past events to improve resilience in the future.[30]

With respect to hardening, the federal government and most states have adopted the National Electrical Safety Code, which requires that utilities use poles, wires, transmission towers, insulators, and other system elements designed to withstand high winds, ice loads, and other threats. These standards, which are updated every five years, are indexed to a map of U.S. wind speeds and winter temperatures, and are intended to protect against about 95–97 percent of the worst-case scenario as calculated at the time of the update.[31]

Many utilities that have now coped with prolonged outages from severe storms are starting to go beyond these requirements and install metal or concrete utility poles, elevate equipment above expected flood levels, trim tree branches near lines more aggressively, and configure their networks to have greater redundancy and backup service options. Several public service commissions in coastal states are overseeing utility plans to spend billions of dollars on system hardening, such as New Jersey utility PSEG's multi-billion-dollar Energy Strong program.[32]

While hardening is inherently *before* the event, ride-through is the ability to adapt operations and continue to provide service *during* a weather event. This too is the result of system design and construction in advance, but it also requires real-time execution skills. Smart transmission and distribution control systems add sensors and computing that allow grids to "sectionalize" the grid into "islands" that continue to operate or reroute power to "self-heal" (more on this in a moment). Ride-through also includes items such as storm readiness planning and drills, prepositioning workers able to do real-time response on key pieces of equipment, and anything else that enables services to continue.

The third element of resilience is rapid recovery. Although there have been several notable failures, it has long been the posture of utilities to mobilize all available resources as quickly as possible to restore service after an outage. Utilities have increased programs to stockpile and share spare transformers as a form of physical insurance.[33] Each of the three major groups of U.S. utilities have created "mutual aid" groups that temporarily dispatch skilled repair persons and trucks to any member of their group that has been hit by a severe storm.[34]

Following Hurricane Irene, these groups sent a total of 70,000 workers from all over North America to fourteen states affected by the storm, and they mounted similar efforts for Superstorm Sandy and Hurricane Harvey.[35] Planning and practice drills, stockpiling and pre-positioning fuel and spare parts, calling in extra equipment and repair persons in advance or quickly after the storm, and creating backup communications systems are all part of rapid response.

Technologically, smart meters give utilities something they've never had: real-time information on exactly which customers have no service (and therefore exactly which lines are down). Florida Power and Light (FPL), for example, credits its smart meter system with allowing it to restore customers four times more quickly after Hurricane Irma (in 2017) than it could after Wilma (in 2005).[36] Many utilities already merge smart meter data, software that dispatches their repair crews, and customer communication platforms so that customers can see the progress of nearby utility work crews and get updated restoration time estimates.

FPL was one of the first utilities in the country to try storm hardening. After Hurricane Wilma hit Florida in 2005, cutting power to more than three million FPL customers for several weeks, the Florida Public Service Commission passed a rule requiring the utility to do storm hardening under a regularly updated plan.[37] During the hurricane, 11,000 power poles fell or were snapped, 241 substations were severely impacted, and 100 transmission towers were damaged.[38]

FPL began hardening its system in 2006, and by 2017 it had spent $3 billion and made significant progress. It strengthened 572 main power lines to hospitals and public safety offices, upgraded 223 substations and 2,000 traffic signals, equipped 120 gas stations and 250 grocery stores with backup power, and inspected 1.2 million power poles to ensure they could withstand winds of 150 mph, replacing the duds.[39]

Then, in September 2017, Hurricane Irma hit FPL's service area. Although FPL did not lose a single one of its new concrete poles to wind and lost 10,000 fewer wooden poles than it did during Wilma, more customers lost power than during the prior hurricane, even though Irma was a less intense storm. Nearly 90 percent of FPL customers—about 4.4 million people—lost power. Nonetheless, hardening appeared to pay large dividends in restoration times, cutting average outage times by 50 percent and reducing damage to poles and substations.[40]

The reasons why FPL's hardening efforts did not reduce total outages from Irma is still a matter of debate. Most experts agree, however, that we'll never anticipate all the ways in which combinations of harmful acts, weather elements, and other threat vectors come together to affect a system—and where precisely the worst impacts will center. Moreover, hardening will become progressively more costly and less successful as climate change intensifies. "If . . . weather events become more severe," a National Academy of Sciences committee concluded, "basic assumptions about the cost-effectiveness of design tradeoffs underlying the electric power infrastructure would need to be revisited."[41] Dr. Abi-Samra says it more bluntly: "What has become clear is that completely hardening the power system to extreme weather . . . is close to impossible."[42] Fortunately, hardening isn't our only line of defense.

The Microgrid Revolution

Electric grids can come in any size, from a single building to a single continent or more. The only essential feature of a freestanding grid is that it lives within a continuous, highly precise supply–demand balance. Severe weather events and many other things cause outages, creating a break at one point in the grid that quickly and unpredictably undoes the balance. On the downstream side of the break, demand and supply don't balance because there is no longer any supply—it was cut off by the break. Since demand downstream of the break must match a now-zero supply, demand is reduced to zero—everyone is turned off automatically, and there's a local blackout.

The remaining grid (upstream of the break) continues to function after a sudden loss of a downstream part by quickly reducing generation to a lower level of demand that matches the loads of what remains connected after the broken section is lost. If the adjustment is small compared to the amount by which generators can be quickly ramped up or down, it is easy for system operators to rebalance—in most cases, it happens automatically. However, if one or more very large plants or high-voltage lines are lost, the large and sudden supply–demand gap may overwhelm the ability to rebalance. These sorts of *cascading failures* are responsible for the huge North American blackouts in 1965 and 2003, each affecting well over three times as many customers as Superstorm Sandy.[43]

Until recently, there were very few generators of electricity anywhere on the grid other than at the far end of the transmission system. Downstream breaks—whether caused by an isolated squirrel shorting out a line or a Category 5 hurricane—uniformly caused blackouts, while the vastly larger remaining upstream grid easily rebalanced, and its customers felt no effects.

With the rapid spread of distributed generation and storage, this is no longer true. Thus was born the idea of a microgrid—a portion of the downstream grid that can be disconnected from the rest of the grid and immediately *island,* that is, automatically take over balancing its own small-scale generation and storage sources against the loads connected to this isolated section.[44] A grid designed in this fashion is called *segmentable* or *agile.* It is also sometimes called a *fractal* grid, named after fractal crystal structures that replicate themselves.[45]

As noted just above, there's no inherent limit on the size of a portion of the grid that can balance itself. We usually use the term microgrid to refer to a network roughly the size of a small village in Africa, or a college campus or military base, but the sections that function independently could be as large as a large city or small state. In fact, as noted in Chapter 4 but otherwise unbeknownst to almost everyone, all Big Grids are already divided up into million-plus customer sections known as balancing authority areas (BAAs) that can function independently if they need to prevent cascading failures. When Hurricane Gustav hit Louisiana in 2004, BAA operators were able to disconnect a large undamaged portion of the grid from the damaged part, creating a giant ad hoc microgrid.[46]

Semi-independent sections of the grid can come in many different sizes, including multiple sections embedded within larger areas. Experts have devised a series of names that match the size of the portion of the grid capable of functioning on its own. Community, utility, or milligrids are large portions of a distribution system that can serve thousands of customers over entire communities within cities. Microgrids are typically thought of as serving something the size of a college campus, a shopping mall, or a hospital complex, while microgrids that serve single buildings or even a part of a building are sometimes called nanogrids.[47] Initially, utilities will probably own most of the community-scale microgrids; they already own the wires in these areas, and replacing them with private, non-utility owners may be tough. This is a question of epic importance to the future utility business model—a topic we return to at length in Part II. With or without community microgrids, there are innumerable locations on most urban grids where private developers may try to insert microgrids they own between the utility wires and a customer group.

The resilience benefits of a microgrid are pretty clear. So long as the physical and electrical integrity of your microgrid and your fuel supply remain undamaged, you have power service. If the microgrid is limited to one customer, it can only improve the resilience of that customer—hence the term *private microgrid.* If multiple customers in a community, campus, or city form a microgrid, this *community microgrid* shares resilience benefits among all its "members." In addition to preventing local blackouts, microgrids can help restore larger parts of the grid that experienced a blackout.[48]

For cost reasons, many microgrids aren't sized to supply as much power as they receive under normal conditions from the Big Grid. So, "nonessential" equipment is automatically turned off when the microgrid takes over. Nonetheless, the difference between some power and no power can be the difference between life and death in a hospital, military base, or community center. During Superstorm Sandy, New York University became a vivid illustration of microgrid-based resilience. The university's campus had a microgrid that operated straight through the storm, creating the only area in lower Manhattan where the lights stayed on. At its hospital, located elsewhere and without power, a gigantic flotilla of ambulances was required to move its patients to safety.[49]

Princeton University also has a campus microgrid that operated successfully (after a very brief outage) for two days during and after Sandy. During this period, "the University served as a 'place of refuge,' with police, firefighters, paramedics, and other emergency-services workers from the area using Princeton as a staging ground and charging station for phones and equipment," according to Princeton's press office.[50] It is worth noting, however, that Princeton's system is designed to supply only about half the university's maximum load and to trade power back and forth constantly with the surrounding grid. "We don't look at the utilities . . . as the opposition," says Tom Nyquist, a Princeton official. "In a lot of ways we are shoulder to shoulder."[51]

As we learned in Chapter 3, the Big Grid enables generation (supply) economics of sale and geographic diversification that are likely to continue to be substantial for some time. And while the Big Grid has its own major vulnerabilities, it also has substantial resilience-enhancing qualities, such as the ability to rebuild quickly using mutual aid and pooled equipment. To get the best of both worlds, most microgrids want to be a seamless part of the entire grid during blue-sky days, be able to operate independently (island) when events cut off outside supplies, and then rejoin the grid when the latter is back to its cheap, functional self. Cutting-edge systems research suggests that a sectionalized grid will not only give each section added resilience, but also make the entire grid less vulnerable to cascading failures.[52]

While the basic idea of segmenting the grid into areas that can island isn't entirely new, the ability to do it community by community on the local distribution grid is downright revolutionary. "Rather than building the grid as a monolith or hierarchy that operates as a unit," four cooperative grid experts write, "it is viewed as a collection of independent or semi-independent systems operating in a coordinated way. This concept of segmentation is the fundamental change in the concept of how the grid was built from the beginning."[53]

Depending on their form, microgrids may enhance or disrupt the shared aspect of the service that the grid has provided from its inception. A community or city-scale microgrid may provide improved, shared resilience for everyone in its area, enhancing the original purpose of the grid: cheaper shared reliability. However, smaller private microgrids within larger communities may increase the gap between the haves and the have-nots. On one side of the street, customers served by a microgrid may enjoy excellent reliability, while those on the other side of the street, served by the normal distribution system, endure an outage. Microgrids will also affect the revenues the surrounding utility can collect from customers on and off the microgrid, widening differences in what utility customers pay. These differences have their own pros and cons.

As another clear positive, microgrids raise the possibility that the industry can start selling, as a product, different levels of power quality and reliability (PQR) to customers who value them differently. Businesses that need power that keeps a very pure waveform without even small interruptions can pay for higher power quality delivered via a microgrid, while customers who don't need a perfect waveform or super-high reliability need not pay for it. In effect, the basic utility grid will deliver the baseline level of PQR, and customers can pay for higher levels if they want it, much as in health insurance markets.[54]

On the negative side, maintaining fairness in apportioning the costs of power service has now gotten far more difficult. Today, regulators mainly just set and enforce a PQR baseline that their utilities must meet or exceed. With a segmented grid, they must figure out how much to charge the microgrid, and in what form, for its continued interactions with its surrounding grid. They will have to mediate when and where microgrids can be placed into service, as the Illinois Commerce Commission recently began and Hawaii is now considering.[55] As explained in a moment, there are also tough questions about the degree to which microgrids can operate solely for the benefit of their owners as opposed to helping the system of which they are a part.

The new technologies that enable grid segmentation and microgrids also enable a closely related idea generally called grid *self-healing.* A self-healing grid can analyze itself in the wake of the loss of an element (due to severe weather or otherwise) and determine whether it can rapidly rearrange and reconnect various parts of itself near the problem so as to restore service automatically. Figure 5-4 illustrates how a self-healing grid might automatically repair a break in a local distribution line.

The value of grid agility was driven home to me through a chance encounter with an amusement park in Branson, Missouri—one of America's entertainment capitals. Silver Dollar City bills itself as an 1880s theme park, "a special place in the heartland of America where your family can gather together in fun, where traditions are created, memories are shared, and where time stops for you to celebrate each other." It is also a place that uses a heck of a lot of power, with up to 24,000 daily visitors taking more than forty rides. Over the holidays, it puts up 6.5 million lights.

Silver Dollar City is served by the White River Valley Electric Cooperative. As the park grew, the co-op realized that a power outage on a busy night could cause a safety and evacuation nightmare. Thousands of families with young children could be stranded high atop amusement park

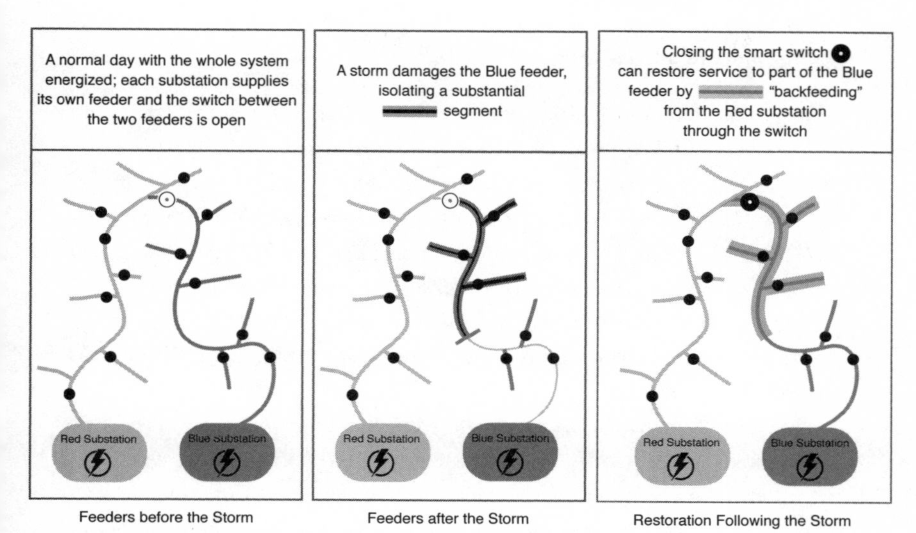

Figure 5-4 A Simplified Illustration of Self-Healing Grids.
Reformatted from Craig Miller, Maurice Martin, David Pinney, and George Walker, *Achieving a Resilient and Agile Grid*, figures 2 and 3, © 2014 National Rural Electric Cooperative Association.

rides, perhaps for hours, while thousands more would have to be evacuated in near darkness.

The co-op concluded that it needed a self-healing grid to serve the park. It installed a simple version in the form of a ring of supply wires around the park with switches much like those in Figure 5-4. White River engineer John Bruns told me about one night when he sat in their control room and saw that a break had occurred on a very busy night in the park. Perhaps 6,000 visitors were inside, many on rides or in remote structures. The self-healing system had not yet been used, leaving him on the edge of his seat. The system thought for a few seconds about what to do and then automatically rerouted the park's power to keep the lights on. No one at Silver Dollar City that night ever knew what almost happened.[56]

The implications of an agile grid go far beyond its technological capabilities—a topic we return to in a moment. Meanwhile, modern grids face a second, even more potent threat.

How to Damage a Grid, Part 2: Hire a Hacker

> *Move 1 began on the morning of November 15, 2017. Operators and IT staff observed anomalous behaviors and some failures of energy management and other systems. Adversaries launched coordinated physical attacks at predetermined sites using vehicles to deliver explosive packages to damage and disable generation and transmission facilities. The attacks caused local and large-scale outages across the U.S. bulk power system, but the attacks were not impactful enough to cause cascading outages or interconnection-wide effects. News and social media reports regarding the physical and cyberattacks increased dramatically.*
>
> *In Move 2, it became clear to participants that the attacks were large scale and coordinated, with facilities affected in multiple regions across North America. The U.S. cyber authorities led an industry-wide situational awareness conference call to provide a summary of shared information and to report on impacts to the grid. Other critical infrastructure sectors (e.g., natural gas and communications) were also impacted, further complicating crisis response and recovery efforts. Participants continued to share information to identify vectors of the cyberattack. Industry also coordinated with local law enforcement and emergency services to respond to the physical attacks.*
>
> *Move 3 then began on November 16. News videos and cyber-authority broadcast calls supplied information summarizing the night's events and set the stage for new scenario injects. In some large metropolitan areas, system operators implemented emergency procedures such as rotating load shedding and voltage reductions to maintain reliability. The cyber authorities conducted conference calls to review the situation, provide support to participants, and issue bulletins and portal postings as new information became available. By midday, repair crews made progress in switching and isolating facilities in an effort to restore or bypass damaged and destroyed equipment, activating mutual assistance arrangements.*

This multi-day attack on the American grid never happened. It is rather the scenario given at a 2017 training exercise for North American electric control room operators in McLean, Virginia. The exercise, known as

GridEx IV, was conducted to find weak spots and recommend improvements in the industry's response to cyberattacks. To improve the realism of the exercise, participants began each day viewing a simulated newscast, shown in Figure 5-5, intended to convey the public's understanding of the event.

The electric system is rightly considered the single most essential infrastructure for national security, health, and safety. It has also rapidly become one of the most frequent targets of cyberattacks by foreign adversaries and private hackers. Like other digitally enabled industries, utilities have many digitally connected technologies that operate the grid (OT) as well as the usual business IT systems. This makes them doubly attractive as attack surfaces. For two years running, utility executives have listed cyber threats as their number one concern.[57] As far back as 2014, 70 percent of all utilities in the world reported at least one security breach.[58] In contrast, successful physical attacks on the grid have been surprisingly rare in the developed world, despite the fact that the electric system has hundreds of thousands of miles of exposed facilities.[59]

Originally the province of non-state activists, hackers, and criminal networks, cyberattacks are increasingly initiated by well-trained state organizations. "State-sponsored Russian hackers appear far more

Figure 5-5 The Five O'Clock News?
NERC (2018), figure 4.1.

interested this year in demonstrating that they can disrupt the American electric utility grid than the midterm elections," *New York Times* reporter David Sanger wrote in mid-2018.[60] PwC's annual cybersecurity survey reveals that state-sponsored attacks have more than tripled in the past three years.[61]

National security agencies and the industry have responded to these attacks with a huge array of standards, procedures, training exercises, and mutual assistance institutions. In the United States, a federally regulated organization—the North American Electric Reliability Corporation (NERC)—issues and enforces cybersecurity standards, known as Critical Infrastructure Protection (CIP) rules, that apply to all large generators and the transmission system. Due to the layered structure of U.S. electricity laws, these standards are not mandatory for distribution system operators (i.e., local utilities) or microgrids—a critical gap we return to in a moment. NERC also runs a round-the-clock national monitoring center for grid cyber intrusions called the Electricity Information Sharing and Analysis Center (E-ISAC). When the E-ISAC detects or is alerted to a possible hack, it works to diagnose the problem and quickly warn all other utilities that might be hit.

But this is only the beginning of the story. Writing in 2015, three leading computer security experts argued that electric utilities are subject to too *many* different layers of cybersecurity requirements, not too few:

> First, if we are to achieve a defensible grid, we must put each layer of complexity to the test. To do this, we'll have to gauge whether the layers in question (physical, technological, human, and regulatory) deliver business and system risk-reduction value commensurate with the level of effort expended to maintain it, while we grapple with the ripple effects of the complexities it induces.
>
> A quick look at just the regulatory layer, for example, finds utilities [in 2015] operating on a playing field strewn with standards, policies, guidelines, compliance audits, and the regulatory and oversight organizations that enable them. A not nearly exhaustive list for just the electricity subsector would include the National Institute of Standards and Technology's

> (NIST) Cybersecurity Framework (CSF) and NISTIR 7628, the Department of Energy's (DOE) electricity subsector cybersecurity capability maturity model (ES-C2M2) and risk management process (RMP), innumerable standards promulgated by industry technical groups, the torrent of security alerts and bulletins from the Electricity Sector Information Sharing and Analysis Center (ES-ISAC) and ICS-CERT, and of course, the mandatory security requirements for reliability of the high-voltage bulk electric system: the NERC CIPS.[62]

Opinions certainly differ as to the degree to which all of these protective measures are adequate and / or working well. On the one hand, utility cyber experts pride themselves on the fact that outside the notorious two Russian hacks of the Ukraine power system, there have been no publicly reported electric grid cyberattacks that succeeded in causing significant outages.[63] At the same time, they recognize that they will never be able to stay one step ahead of hackers, especially well-resourced units from adversarial states. From here on out, cyberattacks on the grid are simply a routine fact of life.

What part of the vast American grid should a bad actor target? There's no doubt that immobilizing a large power plant hurts the grid, but there is typically quite a bit of spare generating capacity on most Big Grids, and most large plant owners have invested in cyber defense and response plans. There is also redundancy in much of the transmission system—the exception being transmission substations, which are a subject of great concern. It is the distribution part of the system—seemingly a low-value target—that has been the focus of much cyber warfare.[64] Each distribution network, serving roughly 50,000–5,000,000 people, is fed by a handful of distribution substations. These substations are the only transfer point between the bulk power system and the distribution grid, and they typically do not have much redundant capacity. They are the ideal choke point for the traditional, non-agile grid—knock one out and everyone whose power flows through that facility goes dark. This is precisely why the Russian cyberattack on the Ukraine targeted substations, not power plants, transmission lines, or transmission substations (although their second attack, in 2016, did in fact target the latter).[65]

Utility grid equipment tends to last for decades, and much of it was installed before digital control, the Internet, and wireless digital communication were widespread. This older equipment is controlled manually or via analog circuits (e.g., hard-wired or radio remote control) and is therefore inherently unhackable. Only half-jokingly, three security writers recently reminded us that the *Battlestar Galactica* TV series begins with the entire Twelve Colonies fleet being attacked by an enemy cyber strike. Every other ship in the fleet is rendered inoperable, but the Galactica survives to fight because its analog control systems have not yet been digitized.[66] These and other experts advise utilities to maintain their ability to control key equipment on their grids manually or with backup analog controls—seemingly primitive measures they call the "gold standard of cyberprotection."[67]

Ironically, the ongoing digitization of the electric power system, culminating in what is loosely referred to as the smart grid, only makes matters worse.[68] The essence of the smart grid is that network sensing and control enables the flexible and efficient joint optimization of hundreds of thousands of distributed points of generation and use. To do this, you need sensors, power sources, and control devices scattered over an enormous geographic area, all of which can communicate with the rest of the system.

Every piece of smart, networked equipment is a possible point of entry and disruption for a dedicated hacker. The entire supply chain that manufactures and installs this equipment, from the computer chips inside each device to the local maintenance contractor's smartphone, is a potential target for malware or intrusion. This supply chain includes a vast number of software products, from standard Windows operating systems to highly customized apps. Smart grids also need huge amounts of fast communication bandwidth over multiple systems; each of these offers its own attack surfaces.[69] "Ask any power company whether moving to a smart grid will improve reliability by itself and they'll tell you no," says cybersecurity guru Erfan Ibrahim. "Taking something as simple as a wire and then adding . . . a very complex system to control electricity is *not* going to make that wire a more reliable electricity carrier."[70]

In the context of these developments, an agile grid provides a unique form of cyber resilience. As with a physical destruction due to a severe storm, if a cyberattack interferes with outside supplies, disconnecting and operating as an island may be a good temporary fix. Smaller grid sections also create safety in numbers; there are so many of them that a hacker could not individually target them all, and each individual microgrid is a much smaller disruption prize than an entire transmission or distribution system. Accordingly, every full cybersecurity strategy references islanding as an important layer of defense, and U.S. military bases and other essential users, such as water systems, are increasingly putting themselves on cyber-secured microgrids.[71]

While grid agility provides unquestionable cyber as well as physical resilience, this does not mean that sectionalized grids are themselves invulnerable—indeed, quite the contrary.[72] Microgrids can function on their own only because they have extensive sensing control and communication networks that work without long latency or interruption. As Accenture consultants recently noted, "Microgrids (and smart grids) present a broad attack surface for cyber-criminals. Assets range from industrial network components such as programmable logic controllers and remote terminal units, all the way to mass market components such as smart meters and in-home displays," and "As the number of system components of the microgrid increases, so too does the reliance on the distributed, active control of the network increasing the potential impact of an attack."[73] The devices and systems that microgrids use are subject to the same supply-chain security concerns as the large utility grid, but (as noted) this portion of the system is not yet subject to enforceable cybersecurity standards.

In many cases, microgrids use commercial off-the-shelf hardware and software and unsecure Internet connections. It is easy to envision cyberattackers penetrating hundreds or thousands of microgrids by injecting malware into a piece of commercial equipment present in all these systems, such as a commercial router. Cisco and the FBI have already found that someone has infected more than half a million household and commercial routers with malware waiting to be triggered to deny service.[74] Smart (automated) metering systems, which are necessarily part of any microgrid as well as a larger smart grid, have also been successfully hacked many times.[75]

As sections of the grid become owned by local communities and governments, large real-estate developers, and others, these stakeholders will have to invest in cyber defense as a cost of doing business. If cyber threats become more common and impactful at the distribution level, states will face greater pressure to conform to a nationwide rules framework.[76] Much of grid cybersecurity involves ongoing monitoring, quick response to attacks, and constant updating of equipment and training. This is a major challenge, even for large utilities that can afford to train and retain specialized resources; small grid owners will find this extremely challenging (if not impossible) without assistance networks designed to help them. There may no longer be economies of scale in power generation, but there are clear and necessary economies of scope and scale in cyber defense.

New Architectural Paradigms

The ability to create an agile, highly segmented grid raises some existential questions for the power industry. Right now, each of the roughly 4,000 owner-operators of local distribution systems in the United States use communication and operating protocols that follow strict engineering codes and federal rules. These rules are designed to make each large regional grid operate as seamlessly as possible, maintain reliability, and allow each distribution system to serve its own customers as cheaply as it can—but not by actions that raise cost or reduce reliability for other systems. In other countries, it is largely the same story.[77]

There is also a well-specified control hierarchy in every country that has a grid. In the United States, it starts with a single national reliability organization and extends down to eleven regional reliability coordinators that span the continent; it then goes to the 100+ balancing authorities, and so on, down to the far end of each distribution system. Every Big Grid operates according to a single set of rules and a layered authority structure, both designed with the stated goal of jointly maximizing reliability and minimizing costs.

The operating and control paradigm for a future, deeply segmentable grid may be much more complex and remains far from settled. First and foremost, there are several very different ideas of how authority to make decisions should be set for an agile grid. Some experts propose a modification of the current hierarchy; while connected to the rest of the grid, each segment will be run by a more flexible and layered version of today's top-down monitoring and control.[78] Other experts believe that each segment of the grid—whether a microgrid or a whole distribution system—should routinely act as its own controller, signaling its willingness to transact with the rest of the grid. There are many variations on these control suggestions, each posing its own set of technical and institutional challenges.[79]

The control architecture of an agile grid is closely related to the question of the goals and trade-offs inherent in its design and operation. The design goal of today's grid is to deliver all power demanded, operate safely, and enable delivery at lowest reasonable cost via any of several alternative scales of grids. These goals lead utilities to expand or manage system capacity anywhere it is constrained so that one customer's use does not displace or jeopardize that of another. Similarly, huge efforts are expended to keep outages from spreading and to prevent "leaning on the system"—abusing use of the system in a way that raises overall prices. Communications and control systems that enable all of this are a given.

In contrast, Maurice Martin, NREL grid architect, notes that the decisions of a microgrid to cooperate and support nearby peers "may not be binary." Referring to each islandable section of the grid as a control area, he writes:

> The algorithms making the decisions may want to factor in reliability and economic parameters. Also, just how local decisions are made when control areas collaborate is an open question. Do all the collaborating control

> systems operate in lockstep? Or does a single control system still have the option to make local decisions (beyond the decision on whether or not to collaborate)? Much work is still needed to determine the optimal algorithm (or algorithms) that will govern the behavior of control areas.[80]

Miller et al. note that current microgrids are mainly built by non-utilities and have different goals:

> Current microgrids generally are focused on the needs of the developer—often military, institutional, or large industrial—rather than on the grids' needs, leading to a technical emphasis on generation rather than control and integration. The agile grid will require a shift in the development of microgrids from a developer and generation focus to one in which the broader grid's interest is an equal factor, recognizing that a robust, shared agile grid is in everyone's interest.[81]

We return to this important idea at the end of this chapter.

The third major agile grid challenge is the delineation of each grid's partitions that can separate and run autonomously. In a fully agile grid, the boundaries of each segment are not fixed; the grid analyzes the pattern of outages and then decides how to divide itself up. A fully agile grid might reconfigure itself any of several ways after a hurricane destroys a portion of the system. The sensors and software that analyze and control each part of this grid must be flexible enough to reconfigure themselves on the fly—an exceedingly tall order for a grid that must operate on a microsecond basis with highly secure communications links. Grid control software and communications systems are now extensively configured to follow well-established boundaries: Changing them and recalibrating control systems to balance a new configuration is a huge technical challenge. At least for the near future, this will be addressed simply by limiting the allowable reconfigurations to boundaries that the communications and control systems know how to handle.

Even in a world where artificial intelligence (AI)-assisted computation is exploding, the control architecture challenges of the coming agile grids should not be underestimated. Following ninety-two pages of dense analysis, a distinguished group of experts from the National Academy of Sciences concluded that

> it is fairly straightforward to recognize that different grid architectures present different mathematical and computational challenges for the existing methods and practices. These new architectures include multiscale systems that range temporally between the relatively fast transient stability–level dynamics and slower optimization objectives. They consist, as well, of nonlinear dynamical systems, where today's practice is to utilize linear approximations, and large-scale complexity, where it is difficult to completely model or fully understand all of the nuances that could occur, if only infrequently, during off-normal system conditions but that must be robustly resisted in order to maintain reliable operations at all times. In all these new architectures the tendency has become to embed sensing / computing / control at a component level. As a result, models of interconnected systems become critical to support communications and information exchange between different industry layers. These major challenges then become a combination of (1) sufficiently accurate models relevant for computing and decision making at different layers of such complex, interconnected grids, (2) sufficiently accurate models for capturing the interdependencies / dynamic interactions, and (3) control theories that can accommodate adaptive and robust distributed, coordinated control. Ultimately, advanced mathematics will be needed to design the computational methods to support various time scales of decision making, whether it be fast automated controls or planning design tools.[82]

In other words, this is a difficult problem, and we don't nearly have it figured out. Moreover, these technical issues also pose important questions for policy makers. Should each segment of the grid (microgrid, small city, large city, and others) be allowed to design itself for different levels of reliability? How about different levels of assistance to adjacent parts of the grid that might need its help to stop a blackout? What if offering self-healing to an adjacent segment means reducing service within your microgrid? Questions like these strike at the heart of what it means to be a public utility—a topic we consider at length in the coming chapters.

The Fragmented Future

The grid today is monolithic—nearly everyone is connected to it and nearly everyone relies on it to receive all of their power at about the same level of reliability. Whether we like it or not, the future grid will be closer to Swiss cheese. The Big Grid will be the cheapest source of carbon-free power, justifying its existence in a world racing to moderate climate change. Local grids will have customers here and there who have dropped off completely, but partial suppliers (prosumers, as we'll call them in Part II) will be extremely common. In each locality, the grid will have a patchwork of microgrids, minigrids, and nanogrids, nearly all electrically a part of the overall local grid, though perhaps separately owned or controlled.

Like the *Battlestar Galactica* analog, there is a good chance that the combined effects of global climate change, cyber threats, and improved small-scale grids will *eventually* lead to the obsolescence and shutdown of major portions of the Big Grid. If the effects of climate change on the power sector are anything like current scenarios, there are many places where thermal power plants won't be viable, even if they emit no carbon. Eventually, some catastrophe is going to knock out a part of the Big Grid that will take months or more to repair. Before repairs can occur, experts will run their AI-assisted computer models (or vice versa), and policy makers will be told that it simply isn't cost-effective to put the entire grid back together again the way it was. While I can easily foresee this happening, I also suspect that there will be many areas where the numbers show the opposite, and most of the Big Grid will be around for a very long time to come.

But it really doesn't matter whether these predictions eventually come true. For the next three decades or so—the crucial period in which we must tame the forces of climate change—we are definitely going to need the Big Grid. During this period, we will have to improve its resilience and protect against cyberattacks as best we can. It will take decades—and many billions of dollars—to analyze each part of our 4,000 distribution systems, determine the boundaries capable of segmenting into workable sub-grids, and install the hardware and software necessary to make it all work. "By beginning now," a thoughtful white paper by California regulators concludes that "*within a generation's time* we may have an electric grid that is more flexible and agile, and would accommodate new plug-and-play technologies mindful of maximizing security and resilience."[83]

At the same time, public and private microgrids will slowly diffuse across every system, and the rest of the grid will steadily become more segmentable and self-healing. As agility increases, so will the pressure to resolve the existential questions surrounding the grid's purpose for being. Do we still want an electric utility grid that offers roughly similar service at similar levels of resilience to everyone? How will we achieve this in a world where cities can defect for good, while others wall off parts of their grid whenever the going gets tough? Who will own what, and who will be in charge?

We know that millions of customers will be able to make some of their own power, and that the modern power industry will offer many new applications and management options. But we need to know much more than the fact that large and small grids are going to have to co-exist. Starting at the top, we'll need a reliable and economically viable Big Grid able to supply huge amounts of 100 percent clean power to our hungry cities. Do we have affordable technologies able to do this by 2050 or sooner? Can we design them into real-world systems connected by transmission lines? Will the trillions of dollars of new plant needed earn enough revenue to make investment viable? The answer has to be yes—but how can we be sure of it?

PART II

The Grid and Its Challenges

6 Decarbonizing the Big Grid

The picture is changing steadily, but right now, most major power systems in the world rely on coal-, oil-, or gas-fired power plants for much of their energy. As of 2016, the United States and Europe obtained 65 and 43 percent of their power from fossil fuels, respectively. China still gets almost 70 percent of its power from coal.[1] In the developed world and China, these fractions *should* be declining, but tragically they have barely budged in the last few years. The fraction of U.S. generation from gas, coal, and oil declined by only 1.7 percent between 2014 and 2016, from 67 to 65 percent, while Europe's fraction didn't decline at all (at about 24 percent) and China's fell by only one half of one percent to 72.5 percent.[2] Clearly, we need much sharper declines to zero-carbon emissions as quickly as possible.

Since the energy these plants provide will continue to be needed, there is an obvious need to replace or decarbonize all this capacity, as well as create room for growth.[3] For decades now, scientists and engineers have been at the drawing board, designing hypothetical generators and systems with no carbon emissions. But showing that at least one set of carbon-free generators (with lines, storage, etc.) can be made large enough to supply all demand—if only someone would build it—is only the first step toward a real solution.

The next step is to examine these technically feasible systems using more detailed feasibility criteria, especially cost. For any one country or region, there will be several future configurations that seem likely to work.

Each of these systems will have its own estimated dollar costs, environmental costs, job impacts, and other attributes. Each will also have its own set of risks—known and unknown—that hampers performance or triggers higher costs. In the run-up to moments of policy choice and system change, someone has to evaluate the trade-offs and recommend a forward path. If the first step is engineering feasibility, the second step is true system planning—a topic we return to at length in Chapter 7.

This takes us to the third step, which is retiring or retrofitting current fossil-fueled plants on a schedule that matches the plan. The fourth step is determining who will build, own, and operate the new capacity, and how they will get paid enough to finance their activities. This is sometimes confused with the simpler metrics of average total or levelized cost of electricity, but the fourth step is all about the quintessentially tactical questions of who exactly will pay what to whom, who gains and who loses, and whether this outcome is financially and politically viable. We'll get to steps 3 and 4 in the chapters ahead. While step 1 discussions are great directional guides, all four steps of activity are essential for sustained action.

The Old Design Paradigm

The primary challenge in the first technical feasibility step is how to design a system of power resources that has the correct size and technical adjustability to match the immediate demand for power in real time. In any balancing area (BA), total power demand will remain mostly (though not entirely) outside the system's control, so you mainly have to build supply that is adequately sized and can be adjusted properly to "follow the load." Under some circumstances, system operators can also control (i.e., reduce) demand to reach a balance; we'll get to this in a moment.

In the developed world, the annual time pattern of demand for power in a BA has a shape similar to the jagged gray line in Figure 6-1. This picture, which shows the actual 2018 hour-by-hour load for the mid-Atlantic portion of the United States known as PJM, is what the PJM system operators had to match perfectly on a continuous basis with power from generators of all types within their control.

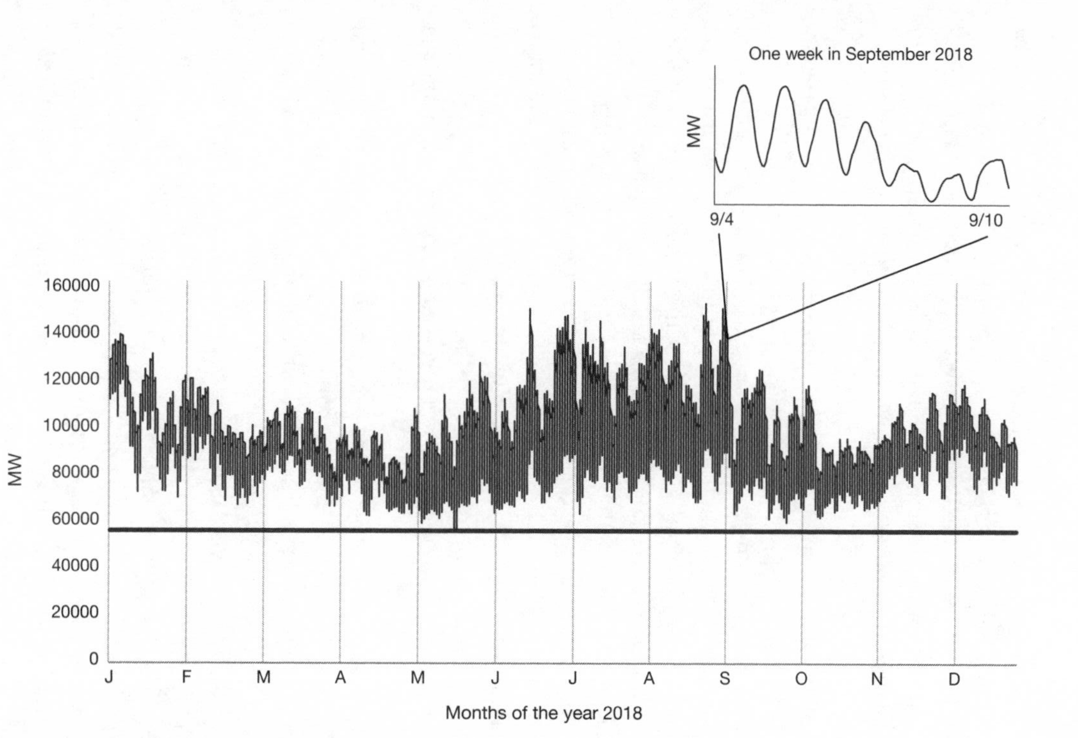

Figure 6-1 Electricity Load for 2018 in the PJM Region.
Data source: PJM (2019).

This annual load pattern has several features common to all modern economies. First, load levels never drop below 55,496 GW (the thick black line in Figure 6-1). So, there is effectively a fixed part of total demand (the area below the line) that is needed around the clock throughout the year. Because this portion of load never varies, it could come from *baseload* plants that stay on continuously, though it doesn't have to. The baseload portion of demand was 486 TWh or about 60 percent of all PJM electrical energy for the year. For perspective, this is well in excess of 10 percent of all annual power used in the United States—a downright huge block of 24 / 7 electrical energy.

The next portion of demand rises and falls with the seasons. In hot climates with much air conditioning and mild winters, electricity use rises gradually toward the hottest part of the summer and then falls gradually into the milder fall before rising less sharply to a lower winter peak. Using a rule of thumb that peak loads are about the top 10 percent of each day's use, the seasonal portion of energy, labeled above the baseload part of Figure 6-1, is about 243 TWh (30 percent of the total) and peak usage is 81 TWh. In cold climates with a lot of electric heat but little summer cooling, such as the Scandinavian countries, the size of the summer and winter peaks are reversed, but the general pattern holds.

The final dimension of balancing occurs each day as demand rises and falls (see inset to Figure 6-1). Whereas the changes in demand usually increase gradually across the seasons, within a single day, demand often changes very quickly as families awaken, businesses open, and temperatures rise. System operators need to know they have enough *fast-ramping resources* or power sources that respond fast enough to increase their output in lockstep with the rapid rise in load. During the average day in PJM, the daily peak period (highest twelve hours) usage rises about 22,000 MW—the average daily amount operators must ramp the system up and then down. Since this is the same general period during which solar generation also rises and falls, operators perform a delicate dance between the natural uncontrolled profile of the PV generation curve and the adjustments of fast-ramping resources to fill in the gaps.

There is a final dimension of balancing a grid—transient stability—that is more easily illustrated with Figure 6-2. In this figure, the analog to the power system is a series of balls of different sizes suspended from a ceiling via a series of elastic threads. The balls symbolize power plants, with the size and weight of each ball proportional to the size of each single power source, while the elastics symbolize the transmission lines that connect the plants. Under normal conditions, the system is stable, and the balls hang at rest. When one plant suffers a sudden outage or something else disturbs the system, it is like knocking one of the balls into motion or cutting off its string entirely, leaving the rest of the system to react as the ball drops to the floor. The changes in tensions on the elastics from the rearrangement of weight start the other balls moving in their own ways. If all the balls return to rest, the system is stable. If the first ball's disturbance causes wild swinging that prompts more elastics to break,

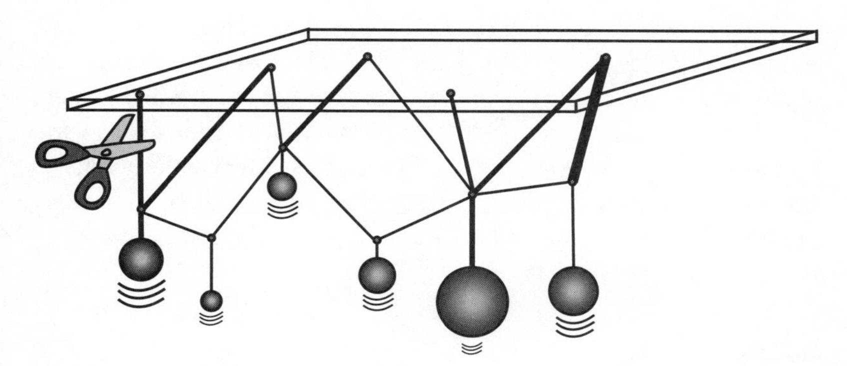

Figure 6-2 A Simplified Illustration of Electric Grid Stability.
The concept of electrical stability on an alternating current power grid is illustrated by a collection of weighted balls of different sizes interconnected by elastic bands, as shown here. The balls symbolize power plants, with the mass of the ball analogous to the capacity of the plant; elastics are the transmission lines, with their diameter analogous to transmission line capacity. The system meets the required level of stability if any two balls can suddenly be cut off (analogous to two power plants failing at the same time) and the remaining balls rearrange their heights and the distribution of their weights on the elastic lines without any of the elastics breaking—that is, no transmission lines overload, even if two plants are lost.
Adapted from Miller (2018).

the system has become unstable and suffers the mechanical analog of a blackout.

Traditional grids addressed these four dimensions of load balancing using a design paradigm that was reliable and cheap—provided no one measured the environmental costs of fossil fuels. The large fixed part of demand was supplied by nonadjustable baseload power plants, which produce large amounts of energy at the lowest measured cost. Seasonal loads and some of the daily changes in demand were balanced by system operators controlling *cycling* power plants—mainly oil, gas, and some coal plants, but also hydroelectric and other types. These are the workhorses of system balancing, ramping up and down to meet most balancing needs today. Finally, rapid load changes during a single day are often balanced by a third type of super-adjustable power plant, *peakers,* which are fired up only as needed, usually on especially hot or cold days.

For more than a century, this layering of three types of power plants—one cheap and nonadjustable and two adjustable—was the heart of power system design. The approach yielded systems that reliably balanced just about every kind of demand pattern you could throw at them, almost always at a cost (excluding environmental) lower than the realistic alternatives. Of course, you have to build sufficient transmission to allow the system to work—a topic we return to in a moment. More to the point, the very thing that made this design paradigm so robust and cheap—the use of fossil-fueled generators to power both cheap baseload and adjustable plants—is now its fatal flaw.

The New Paradigm

The new design paradigm starts with the premise that solar and wind are going to be major sources of energy because they are already the cheapest power in the world and come from abundant, widely distributed natural sources. Hydroelectric energy, still the largest source of renewable electricity, will also be critical. In most places, the question is how to build a system around these generation sources that has the right size for larger, more variable demand and is both sufficiently adjustable and stable.

The difficulties with using the old design paradigm on this kind of system are immediately apparent. Wind and solar plants aren't controllable, except to turn them off when nature has already turned them on, nor do they put out large steady blocks of energy. This is illustrated in the hourly graphs of daily wind and solar generation in California shown in Figures 6-3 and 6-4. Indeed, since winds never blow with perfect steadiness and solar power doesn't produce anything during the night, creating large blocks of firm energy from these resources is sort of like mixing oil and water.

This challenge is even larger than it seems because the momentary output from wind and solar plants turns out to be quite unpredictable. The output from a single solar panel can vary wildly from minute to minute, during what appears to be a sunny day, due to passing clouds or other factors. Figure 6-5 shows the momentary output of one solar installation measured every minute rather than averaged over an hour. As the figure shows, passing clouds cause load drops as large as two-thirds of maximum output from one minute to the next. It turns out that the momentary output of wind power plants also varies just as much with Aeolian gusts. These intra-hour fluctuations are just as damaging to grids as any other form of imbalance. So, system operators must compensate for these generation spikes with resources large and fast enough to create generation profiles that follow a smooth path.

Then, there are seasonal differences. As we saw in the PJM load curve, 30 percent of all its energy (243 TWh) rose and fell with the seasons. This is the equivalent of seventy-seven typical gas-fired power plants running 60 percent of the time during the year. *Something* other than carbon-emitting fossil plants has to be ready to provide much more controllable power in the summer and winter than in the spring and fall—wind and solar alone can't do this.

To illustrate this challenge, Figure 6-6 details the retail customer load and hydro, coal, and solar generation from NorthWestern Energy, a utility

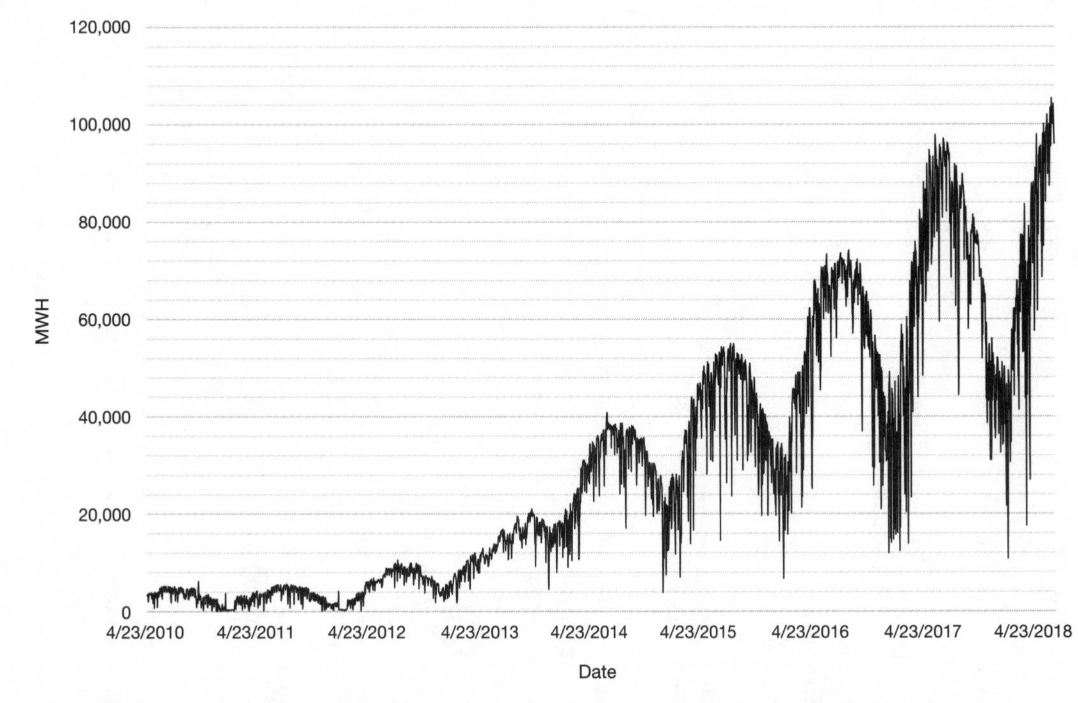

Figure 6-3 Daily Total Grid Solar Photovoltaic (PV) Generation in the CAISO Region, April 2010–April 2018.

Adapted from California ISO data as reported by Joskow (2019).

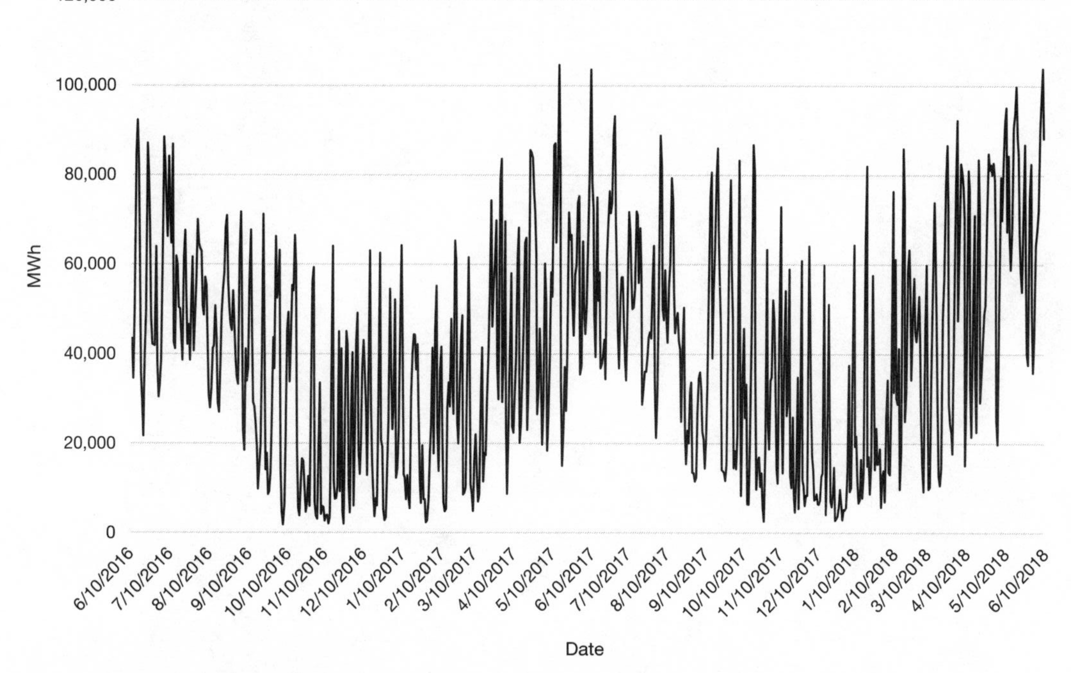

Figure 6-4 Daily Total Wind Generation in the CAISO Region, June 2016–June 2018.

Adapted from California ISO data as reported by Joskow (2019).

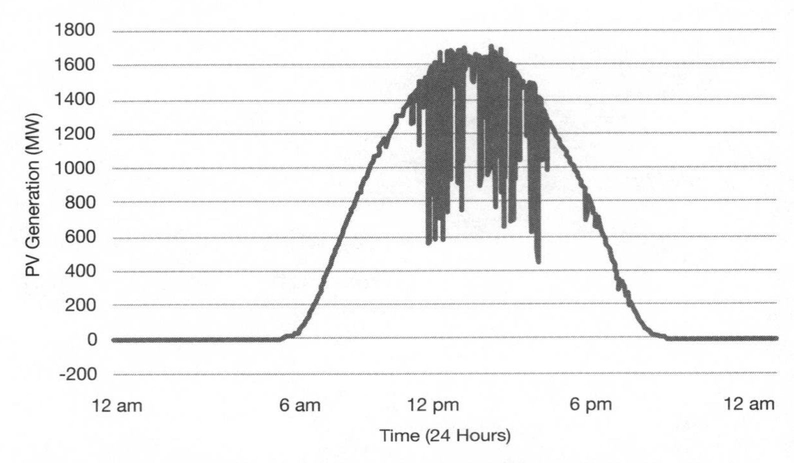

Figure 6-5 Minute-by-Minute PV Generation during a Day with Passing Clouds.
Federal Energy Regulatory Commission (2018).

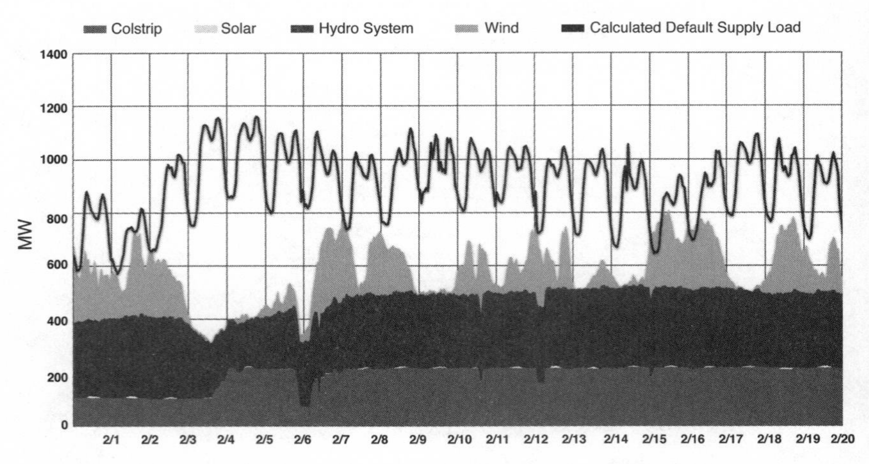

Figure 6-6 Hourly Generation and Load during February 2019 from NorthWestern Energy.
William W. Thompson, NorthWestern Energy (2018).

in Montana, during the month of February 2019. Retail customer demand, illustrated by the wavy black top line, follows the usual daily pattern of increasing each morning, with a little dip in the middle of the day before increasing again in the evening and then decreasing overnight. That's the retail load NorthWestern has to balance.

NorthWestern's baseload resources include their 222 MW share of the Colstrip coal plant and a fleet of ten hydro plants. The output of these plants, shown as the bottom two bands of gray, is pretty steady. The plants are also used to help balance load and can be dispatched to increase or decrease generation. NorthWestern's system doesn't have much solar installed (yet), but it currently has about 378 MW of wind generation (with another 159.5 MW targeted to be online by the end of 2019) whose actual generation is shown in the lightest shade of gray above the lower two gray bands. When February winds are strong, wind can generate as much energy as either type of baseload resource, but as Figure 6-6 illustrates, wind generation is highly variable and has no relationship whatsoever to load levels. NorthWestern balances the gap between the top of the green generation line and the black load line with owned and contracted generation—a mix of fossil and renewable power.[4]

As we learned in Chapter 2, climate change and climate policy itself will affect the level and shape of electricity demand for many decades to come. This phenomenon is illustrated starkly in work performed by Boston University professors Cutler Cleveland and Michael Walsh, who modeled electricity use in 2050 in Boston after about 75 percent of all buildings were converted from gas or oil heat to electric heat pumps. The left panel in Figure 6-7 ("Baseline") shows projected 2050 peak hourly natural gas (black line) and electricity (light gray) use in Boston. In this panel, there is *no* special climate policy that causes Bostonians to shift from carbon-emitting heat sources to 100 percent carbon-free electric heat. The graph, which runs from January to January, shows that gas for heating will spike up greatly during the projected cold Boston winter of 2050, while peak electricity use rises only a little during the relatively mild summer due to air-conditioning loads.

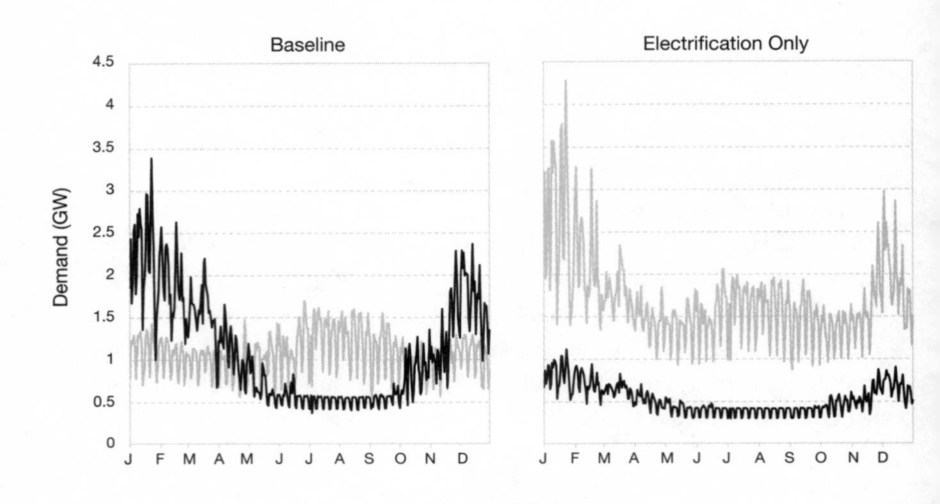

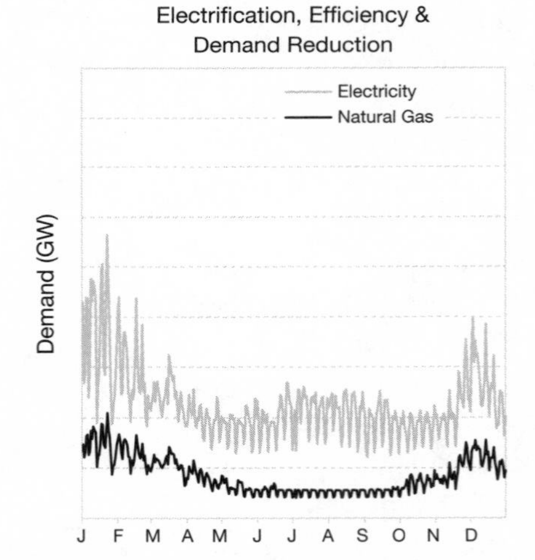

Figure 6-7 Projected Electricity and Natural Gas Use for Heating and Cooling in the City of Boston under Alternative Climate Policies.

Cleveland et al. (2019a), figure 39.

The middle panel of the figure shows what happens to electricity demand on the local grid when most housing is converted to electric heating and energy-efficiency improvements are *not* made. Peak electricity use *triples* or even *quadruples,* creating much higher and more volatile seasonal and daily peaks to manage throughout the winter. For the rest of the year, electricity use grows a little but presents no special balancing challenges.

The essential role of energy efficiency (EE) is highlighted by the third (bottom) panel of the figure, which shows peak demands after homes and offices are heavily insulated and other cost-effective efficiency measures are adopted. Both electricity and gas use drop dramatically throughout the year, demonstrating the tremendous value of efficiency in reducing emissions and bill savings. But a new steep winter electricity peak is still there. Now, it is "only" two to two-and-a-half times as large as baseline winter peaks—still an enormous change for system operators. Contrarily, EE has lowered peak usage during all other seasons, including the former summer peak, which now barely pokes its head up between June and October. In fact, projected total annual 2050 electricity use in the city with zero-carbon emissions and strong EE policies is up only 12 percent from the current total use of about 6,600 GWh, despite a projected 17 percent increase in population and 80 percent growth in economic output from 2015.[5] This reflects the balancing of the forces that promote electricity growth (greater population, housing, economic activity, and climate policies) with the huge potential for efficiency improvements that reduce use, as we saw in Chapter 2.

Disturbingly, climate change will itself make the challenges of electric system balancing even tougher. Higher CO_2 concentrations will make summers in temperate climates much hotter, increasing air-conditioning demand while reducing winter and water heating demand by much smaller amounts. Three researchers led by University of California professor Max Auffhammer calculated that relatively moderate climate change would increase electricity peak loads 3.5 percent by 2100 and the most extreme peak hours by 7.5 percent, while increasing total electricity use by only 2.8 percent.[6] Another team found summer-month consumption rising 15 percent, indicating higher summer peaks.[7]

There are three takeaways from all these results. First, we once again see why EE is called the *first fuel:* good energy policy applies it first and as much as is cost-effective.[8] The second takeaway is that climate-change policies that promote electrification, including electric vehicles (EVs) as well as heat, will make both daily and seasonal electricity demands larger and more volatile. Future Big and (especially) Small Grids have to be designed for much larger seasonal and intra-day load changes, foreshadowing our upcoming discussion of grid optimization in Part III.

The third takeaway is that we should not make the load balancing task harder than it needs to be by adding highly volatile loads to the grid if there's a better carbon-free energy source that does the job. To consider this, we need to plan climate policies, fuel sources, and energy infrastructure holistically, not one system at a time. For example, there are more than 300 large-scale solar district heating systems in Europe today, converting sunlight directly to heat without involving electrification, some with seasonal storage.[9] Where cost-effective, this might make system balancing a lot easier and cheaper.

The new design paradigm can't be described very simply because it involves a large basket of design options that must be custom blended for any actual real-world system. If it can be summarized in one sentence, it is something like this: *use the best possible mixture of generation and electricity storage options that are physically and financially viable in your area, connected by all necessary transmission, to create the lowest-cost system that's reliable, balance-able, and stable.* It's a mouthful, and involves several unsettled challenges, but the signs are that it's ultimately going to work.[10]

The new approach doesn't try to map each new carbon-free resource into each type of old-paradigm power plant, though sometimes this happens to work nicely. It seeks to combine carbon-free resources that can't be adjusted with those that can to get to a reliable, cost-effective solution. However, it is helpful to think about how combinations of new technologies will yield the same types of energy services as in original systems, even if they are not coming from the same types of plants. We'll continue to need blocks of energy that are huge, steady, and as cheap as possible; seasonal energy should be almost as large as these steady blocks but a lot larger in summer and winter, as well as adjustable and as cheap as possible; weekly and daily energy comes in relatively smaller amounts but must be highly adjustable; and stabilizing services ideally come from the carbon-free resources already installed for providing energy or balancing.

The Clean Power Toolkit

An enormous number of different carbon-free generation and storage technologies are already in commercial use or under development. Apart from PV and wind plants, clean generation sources include nuclear, geothermal, concentrating solar, hydroelectric, biofuels, renewable hydrogen, renewable gas, power plants, fossil power plants, and fuel cells with carbon capture. Many experts no longer use these three traditional categories, but for our purposes they keep the discussion most manageable.[11]

The list of storage technologies is even longer, including dozens of different types of battery chemistries. In total, there are far too many options to describe in detail, but it is possible to see how the new paradigm is taking shape, as well as what remains to be done to fill the gaps and lower costs, by reviewing a handful of the main building blocks.

Table 6-1 shows several of the major options likely to be combined in future carbon-free grids. Although the table classifies options by their main contribution to the old categories of peaking, cycling, and baseload energy, it is important to stress that most options contribute in multiple categories and will migrate further as their technologies improve. The remainder of this section contains a brief discussion of each of these options.

Table 6-1 Some Major Carbon-Free Power Resources and the Main Types of Energy Services Provided

Peaking energy (weekly, daily, and intra-day balancing energy)	• Flexible demand (demand response) • Pumped storage hydroelectricity • Electrochemical batteries • Heat storage
Baseload energy blocks	• Large hydroelectric plants • Decarbonized fossil-fuel plants (carbon capture, utilization, and storage) • Nuclear power plants • Combine cycling resources below
Large adjustable blocks of seasonal energy	• Reservoir hydroelectric plants • Overbuilt wind and solar diversified by transmission • Overbuilt wind and solar stored as hydrogen • Decarbonized fossil-fuel plants

PEAK LOAD RESOURCES

Demand Response The first design imperative is to match the momentary and daily output of variable renewable energy sources with momentary demand blips and the more predictable up- and down-ramp of daily loads. One way to do this is by flipping the old design paradigm, in which power demand was outside the system operator's control but supply was not. In the new paradigm, some of the demand in each BA can be shifted around by system operators, that is, reduced for a few hours when renewable output is low and allowed to rise a few hours earlier or later when the sun or wind is stronger. The term for this sort of action is flexible demand or *demand response* (DR).

System operators don't balance the grid by casually pushing buttons that reduce the amount of energy being used by assorted customers. Instead, they operate a market that will pay the same price for one unit of intended, measured reduction of use as it will for pay for one added unit of supply. The offer to purchase load drops alongside added supply is renewed every hour; each hour's auction pays that hour's market-clearing price to everyone who offered to drop load at that price or lower. Needless to say, no one offers to turn off their entire power service; they offer to reduce it a little bit. Each DR seller must offer to supply at least 100 KW of load reduction (roughly twenty homes' worth), but the administrative costs of doing this only make it economical if you offer much more.[12]

DR has been in use in the power industry for decades, but improved technology and more widespread use of customer control technologies is taking DR to never-before-seen levels. Hundreds of thousands of households have purchased "smart" thermostats that communicate directly with utilities or private energy managers. As an example of their ability to implement DR, these thermostats can predict that midday prices will be high on a coming day, cool your house down by an extra amount before prices go high, and then reduce cooling during the high-price period, saving money and energy in total. Commercial buildings have even more sophisticated control systems that remotely monitor temperature and light levels and can be programmed to cycle lights and heating, ventilation, and air-conditioning equipment on and off strategically during high-price hours without allowing temperatures, air quality, or lighting levels outside a comfortable range. DR programs are also starting to incorporate EV chargers, that is, enabling chargers to defer or slow down charging a little during high-price hours, or to use price signals to induce customers to shift charging altogether to low-load hours.

The amount of DR any system operator can count on to balance load obviously depends on the level of system load before DR begins—you can't very well drop a load you don't have. This creates some uncertainty around the use of DR during different times when balancing is needed, but operators are getting better and better at using DR as technology gives them better visibility and measurability of the loads they can reduce. Meanwhile, the size of this resource is getting every bit as large as other forms of balancing. A Rocky Mountain Institute study estimated that by 2050, the state of Texas could reduce its peak load by 24 percent with flexible demand.[13] There is already a maximum adjustable load of about

160 GW of DR in the United States—more than twice the 64 GW of total installed solar PV capacity.[14] Andy Lubershane, research director at Energy Impact Partners, estimates that total DR will grow to a maximum reduction level of 237 GW by 2030—nearly one third the peak demand of the entire United States.[15]

Importantly, the nature of DR is such that it can't be used for balancing needs lasting more than a few hours. It is also the most complex, distributed, and context sensitive of all balancing technologies. Nonetheless, its tremendous size and low costs make it all but certain to become a large, widely used short-term balancing resource.

Pumped Storage Pumped storage hydro (PSH) takes electricity from the grid at chosen times and uses it to pump water from a geographically lower to an adjacent higher reservoir—the exact equivalent of charging a huge battery. When power is needed, the water from the upper pool is released down a spillway through a turbine back into the lower pool, equivalent to a huge battery's discharge. The lower reservoir can be integrated into the natural water system. However, more commonly in the United States, the two reservoirs are an isolated "closed loop" system.

PSH plants are usually smaller than reservoir hydro, charging and discharging over six- to twenty-hour periods. As of 2012, the estimated 150 GW of PSH plants in the world stored about 1.5 TWh; for the United States, forty-three current PSH plants have an aggregate capacity the size of twenty-two large coal-fired plants and store a total of about 0.25 TWh.[16] While this isn't nearly enough energy to handle seasonal storage, they have very high levels of adjustability and are ideal for balancing wind and solar output during cloudy or calm days, and also for daily and intra-day ramps and fluctuations.[17]

After a long period of little to no growth, there are now suddenly forty-eight new PSH projects in the U.S. development pipeline, including a $2.5 billion, 1,200 MW project in the State of Washington and a $1.3 billion, 1,300 MW project in Southern California.[18] Outside the United States, China has expanded PSH capacity at a torrid 16 percent average annual growth rate since 2005 and plans to continue, all to help balance its equally fast additions of wind and solar power.[19] Europe added about 1,200 MW of PSH in 2017.[20] In short, PSH is a carbon-free peak storage resource whose reliability and cost are 100 percent proven; its limitations are geographic suitability and the pace of grid decarbonization.[21]

Storage in Chemicals and Heat The third major daily and intra-day balancing resource is electricity storage. Storage technologies are usually grouped into four categories: electrical, mechanical, electrochemical (i.e., batteries), and thermal. Electrical storage is mainly supercapacitors, while mechanical storage forms include compressed air, flywheels, and some new technologies. These technologies will contribute, especially in the area of stability, but absent breakthroughs, they will probably not—with the possible exception of compressed air—play a large role in energy balancing.[22]

Electrochemical storage includes several types of lithium-ion batteries as well as less common zinc-air and sodium sulfur.[23] Tesla's Powerwall is the most widely known example of this type of battery, though there are many highly qualified manufacturers and installers of these batteries in sizes as large as several hundred megawatts all over the world. Tesla's largest installation, for example, is 100 MW at the Hornsdale Wind Farm in Australia, and the Pacific Gas and Electric Company has announced its intention to contract with a 300 MW short-term storage plant in its area.[24]

In the popular imagination, batteries such as this are the silver bullets that unlock an all-renewable grid. What could be more natural than pairing wind and solar energy with a battery to store, smooth out, and deliver power on demand? Isn't this how millions of off-grid village power systems will work in the parts of emerging nations not served by unified grids? Tesla's CEO, Elon Musk, encouraged this sort of thinking when he revealed the Powerwall in 2015, saying that this battery alone would beget "a fundamental transformation of how the world works, about how energy is delivered across the Earth."[25]

Perhaps someday it will, but for now, essentially all the foreseeable forms of electrochemical or mechanical storage will be limited to providing intra-hour and daily balancing, perhaps someday extending to several days or a week.[26] These technologies simply cannot hold the dozens of terawatt-hours of energy needed to balance seasonal loads, much less the scores needed for baseload service. The latest Powerwall 2, for example, holds 13.5 kWh and is designed to discharge completely over about three hours. Nearly every large-scale battery storage installation on the Big Grid now provides about four hours' worth of energy, with some technologies stretching their discharge cycles to eight. With storage amounts and discharge cycles so limited, you would need millions of batteries like this chained together to create seasonal or baseload storage. A back-of-the-envelope calculation by a Stanford research team estimated that it would take 6,500 years' worth of output from the Tesla Gigafactory to supply enough lithium-ion batteries for an all-renewable U.S. grid.[27]

Beyond the sheer logistical unworkability, the economics are forbidding. Electrochemical battery technologies now cost about $200–600 per kilowatt-hour of capacity, which is economical for many applications but not for seasonal balancing. Daily and hourly storage is charged and discharged roughly once a day, sometimes more; each cycle is a chance to earn money in a power market or otherwise provide value. In contrast, seasonal storage by its nature should charge and discharge only a few times a year across the seasons. With so few annual cycles, the cost would have to be closer to something like $5–10 per kilowatt-hour of storage capacity to make electrochemical storage economical. While preparing this chapter, I took an informal poll among the storage experts I know, asking them, "In what year do you think electrochemical batteries will achieve a cost of $5–10 per kilowatt-hour?" From every expert, the answer was the same: "Never."

But enough myth busting. Electrochemical (and thermal storage as well) are already *extremely* good technologies for weekly to momentary balancing, and they are getting better and cheaper every year. Utility-grade lithium-ion batteries are expected to cost less than $100 per kilowatt-hour—perhaps even $75 per kilowatt-hour—by 2040.[28] Even at current cost levels, this sort of storage is growing at a predicted rate of 1,700 MW (20 percent) a year at this point, about half in the United

States, and roughly evenly split between electrochemical and thermal technologies.[29] There are reportedly ten more battery Gigafactories being planned around the world, and the United States storage order book reportedly stands at 33,000 MW.[30] For short-term uses, electrochemical storage is going nowhere but up.

Though lithium-ion dominates current sales, there are several other technologies widely believed to have an even better long-term potential to provide days-to-weeks balancing.[31] Lithium-ion chemistry is good for giving the sort of rapid bursts of power electric autos need, but it is not inherently suited to ultra-low-cost bulk storage of power over long periods of time. One proprietary electrochemical battery is under development by a startup called Form Energy. Form claims that its technology will allow storage for hundreds of hours at costs less than $10 per kilowatt-hour.[32] Zinc-air technologies are also showing promise.[33] Another technological alternative is known as a *flow battery.* Flow batteries pump the power-storing electrolyte past the battery's anode and cathode; the more fluid you pump, the more energy you store. The capacity of these batteries, and the duration of their discharge, is therefore mainly a function of the size of the fluid tanks that are part of the battery complex. These batteries are also attractive because their electrolytes are fully recyclable, do not degrade over time like lithium-ion cells, and do not have lithium's explosive fire hazard. The service lives of these batteries are expected to be twenty years or longer, at least twice the life of lithium batteries, making them much like other Big Grid utility assets. According to market researchers Frost & Sullivan, developments in flow batteries are creating "a silent revolution" in which flow batteries will eventually overthrow lithium-based cells.[34] This seems possible, as a report by the International Renewable Energy Agency (IRENA) predicts that the cost of flow batteries will decline to as little as $108 per kilowatt-hour by 2030, below predicted utility-scale lithium-ion costs.[35] Regardless of whether this technological *coup-d'etat* ever occurs, the fact that flow batteries create more pressure on other chemistries to reduce costs and boost performance can only be good news.

The use of heat rather than chemicals to store electricity for short periods is now also experiencing a surge in R&D interest. Putting aside small-scale thermal technologies to focus on the Big Grid, companies are now mainly using and experimenting with molten salt attached to concentrating solar power plants, which use mirrors to create high-temperature solar heat. Perhaps surprisingly, 3,700 MW of thermal storage is already in use today, mostly in Chile, South Africa, Spain, and Israel, and China is making a big push in the area with 500 MW under construction.[36] The largest U.S. installation, Solar Reserve, holds 1,100 thermal megawatts of heat, substantially larger than the giant vanadium flow batteries announced, and can discharge for ten to twelve hours—again, longer than almost any current electrochemical technologies.[37] In addition, the costs of thermal storage are reportedly comparable to pumped hydro, and stored heat loses only one degree per day, making it potentially viable for seasonal use.

These factors have spurred a revival of interest by a variety of technologists. A spin-off from Google X, Malta, has received an investment from a Bill Gates–backed venture capital fund. Siemens Gamesa already advertises thermal storage units that it says can be as large as 12,000 MWh—a tenfold increase over electrochemical storage. The company's website claims that its thermal storage costs

> will be significantly lower than classic energy storage solutions. Even at the pilot stage, a commercial project at GWh scale would be highly competitive compared to other available storage technologies. Economies of scale will bring substantial reductions in capital expenditure, while increasing the storage rating.[38]

Many other thermal storage companies using salt and other materials are also working on the problem, including a German proposal to convert a coal-fired plant into a large thermal battery.[39]

While heat technologies look promising, many are still considered not yet ready for commercial operation. But even if we limit ourselves to demand flexibility, lithium, zinc-air, sodium sulfur, and some flow batteries commercially available at current costs, there is an emerging consensus that carbon-free peak period balancing is entirely possible. "Thanks to Lithium Ion batteries," says Ted Wiley, CEO of Form Energy, "storing energy for less than a day is a problem solved."[40] Coming from the head of a flow battery firm competing with lithium, this is high praise indeed. Even more convincing evidence comes from lithium's old-paradigm competitor: peaking power generators. At current and projected prices for storage, solar, and wind, power system planners are finding it cheaper to use carbon-free peaking resources than to build and run an old-style peaker. Ravi Manghani, energy storage director at Wood Mackenzie Power and Renewables, recently forecast that if battery costs continue to decline at 10–12 percent a year for the next eight years, eight out of ten new peaking power generators would be replaced by short-term storage.[41] The boldest prediction of all comes from James Robo, CEO of NextEra, the world's largest wind and solar firm, with $40 billion of planned investment through 2020. "Post-2020, there may never be another peaker built in the United States," Robo said way back in 2010. "Very likely you'll be just building energy storage instead."[42]

BASELOAD ENERGY BLOCKS

The middle row of Table 6-1 displays resources that can provide huge baseload energy blocks. As we saw in the case of NorthWestern Energy, hydroelectric plants can provide power in amounts steady and large enough to be considered baseload. In the developed world, the issue with large hydro plants is primarily that the acceptable sites have all been used. So, the potential for expanding this source lies mainly at existing dams where turbines can be added and upgraded.[43] The United States added the equivalent of about two large coal plants' worth of hydro between 2006 and 2016, while Europe added slightly more. Almost all of the additions were in the form of small upgrades to existing dams.[44]

The second baseload energy option is decarbonized fossil-fuel plants. By now, nearly everyone following energy issues knows that it is possible to remove carbon dioxide from coal or gas plant emissions and use it either to make an inert product such as cement or store it somewhere deep below the planet surface. The official name for this, of course, is carbon capture, utilization, and storage (CCUS).

The three elements of CCUS—capture, utilization, and storage—each has its own development and / or cost issues. Capture of CO_2 before or after combustion is technologically proven, but so far it is much too expensive to install profitably on power plants. Even so, there are already eighteen post-combustion carbon-capture projects in operation and twenty-five more under development.[45] Several factors are acting to reduce the capture costs barrier, though not quickly enough. First, there are cost-reducing technology breakthroughs that seem to be making steady progress.[46] One company, NET Power, is testing a 25 MW zero-emissions fossil plant in Texas that, according to its promoters, will eventually produce power at the same price as traditional natural gas plants.[47] Another, Inventys, just won a Global Cleantech 100 award for a new capture technology that uses new materials known as functionalized-silica or metal-organic frameworks, which purportedly can achieve much higher capture efficiencies and lower costs.[48] CCUS economics will also improve if a carbon fee of any type is adopted, and through utilization of captured CO_2 to make revenue-generating products. This, too, is the topic of much current research.[49]

Storage of CO_2 in underground and undersea formations has proven itself in a number of sites around the world, including depleted oil wells, salt caverns, and underground aquifers.[50] There appears to be no shortage of sites that scientists think are geologically suitable for storage—the U.S. Geological Survey has estimated U.S. storage capacity at "thousands of gigatons."[51] So far, storage seems to have occurred with little leakage or catastrophic failure, though one facility in Algeria was shut down prematurely due to fears that it would begin to leak.[52] In addition to leaks, induced seismic activity has been observed at this and other injection wells.[53] As of 2017, there were thirteen large-scale CO_2 storage projects operating—five in the United States, four in Canada, two in Norway, and one each in Australia, Germany, and China. They are about evenly split between storage in depleted oil and gas reservoirs and saline aquifers.[54]

This is not to say that either all the scientific or regulatory issues are anywhere near resolved. Creating carbon storage facilities capable of storing millions to billions of tons of CO_2 will require massive levels of investment in transport and storage facilities that must be guaranteed to be safe over a century or more. No government has the capacity to make this sort of investment, and private developers require a regulatory framework and project economics that unlock private capital.

Several major steps in the direction have occurred recently. In 2018, the United States adopted a tax credit of $50 per metric ton for CCUS projects that began construction by 2024. Several countries are addressing the grant shared infrastructure costs by exploring whether they can start CCUS hubs that combine multiple nearby capture sources with shared storage facilities.[55] Perhaps most significantly, California recently issued formal rules under which a commercial CO_2 storage facility can receive a "Permanence Certification" that entitles the facility to tradeable low carbon fuel standard credits that have traded for more than $122 per ton in the last year.[56] These rules set requirements for monitoring, insurance, financial reserves, and other project features, putting in place a framework that enables private developers to consider investment.[57] If cost reductions and regulatory certainty continue to develop, adjustable gas plants with CCUS will become important balancing resources, especially in the developed world.[58]

Nuclear fission power plants also produce baseload power, and (apart from plant construction) they are carbon free. Currently available plants can't be adjusted up and down to balance loads, so they just run steadily between refueling, producing huge blocks of carbon-free power. This is an enormously valuable contribution to a decarbonized system where these plants exist. The United States has fifty-nine operating nuclear plants that avoid more than half a gigaton of CO_2 emissions each year—far more than any other renewable energy source to date.[59]

For better or for worse, utilities and governments in Western countries are probably going to build no more than a handful of large nuclear plants in the foreseeable future. The last several new nuclear plants built in the United States and Europe have had momentous cost overruns. The only new nuclear plant under construction in the United States, the Vogtle plant, is now estimated to cost $25 billion and is still four years away from completion.[60] Europe is adding two new nuclear plants this year, each about ten years behind schedule and two-and-a-half times their original budgets.[61] Without government cost guarantees, most utilities I know simply cannot afford such costs and risks.[62]

Novel new forms of nuclear power, such as thorium reactors, show some promise but are likely to happen only well into the future. However, small modular nuclear reactors (SMRs) may become commercially available in the coming decade, creating what could be a genuinely new option for baseload power. One U.S. company, NuScale, recently announced that it was "on track" to receive design approvals from the U.S. Nuclear Regulatory Commission by 2020, and a consortium of Utah municipal utilities announced its intention to install twelve new SMR units in their area "during the 2020s."[63] The commercial viability of these plants won't be known until several plants are successfully constructed and have operating histories that yield safe, reliable power at costs that match other baseload options.

Recall from the last section, the new paradigm doesn't chain itself to the old classification of power plant types. As a result, one final way we will undoubtedly get more baseload energy is by combining the right mix of resources in the next category.

CYCLING AND SEASONAL ENERGY

The third row of Table 6-1 shows resources that can provide energy that is adjustable over days and months to match seasonal energy needs, or for shorter periods of days or weeks when the sun and wind aren't strong.

The first of these resources is hydroelectric plants with water storage reservoirs. Where they already exist, or can be added with low economic and environmental costs, they provide the largest and cheapest seasonal storage available today. One single huge U.S. hydro plant, the Grand Coulee Dam, generated about 21 TWh per year, or about one tenth of PJM's seasonal energy block, at a price in the range of $35 per megawatt-hour.[64] The massive reservoirs of Hydro Quebec hold 176 TWh, and all of Europe's reservoir hydro plants total 180 TWh.[65]

The second seasonal / cycling option in the table is to combine wind and solar generation over very large geographic areas, along with some energy storage, to create a more shaped seasonal energy block naturally. As we saw in Chapter 4, "Why We Grid," one of the roles of transmission is to harness the natural diversity of wind and solar resources. The idea is conceptually sound, but to get large blocks of energy with approximately the right shape, you need a very large area over which to collect wind and solar, much more capacity than you would build only to serve local load, huge amounts of transmission, and substantial storage. Using thirty-six years' worth of weather data, a team led by Matthew Shaner of Stanford calculated that you could balance the entire demand of the United States throughout the year if you built surplus wind and solar plants across the continent, had no limits on the transmission connecting these plants, and had enough power storage to hold somewhere between twelve hours and thirty-two days' worth of total U.S. demand.[66] (For reference, storage large enough to hold twelve hours of U.S. power demand is 150 years of production from Tesla's Gigafactory and twenty times existing U.S. short-term PSH.[67])

As you can imagine, these assumptions represent gigantic leaps of faith. First, the wind and solar plants have to be dispersed across the whole country, and in most of Shaner's scenarios, you have to build two or three times as much wind and solar capacity as the level of peak demand. Much of this capacity would only be used to supply infrequently used peaking energy, and paying for large amounts of largely fallow capacity would be a huge challenge. In addition, you have to be able to build a massive transmission system crisscrossing the country. As we'll see in Chapter 7, this would require a gigantic shift in the weak grid expansion processes in place, particularly in the United States. These facets and their associated costs make this approach unlikely as an option unto itself, but elements of the approach will undoubtedly be part of any low-carbon future. In other words, expanded transmission is very much part of the toolkit, but it is a bit of a special case, which is reserved for Chapter 7.[68]

One variation on this idea that is likely to occur is the use of wind and solar capacity to generate hydrogen gas. The power from spare or purpose-built renewable generators could be used to split water molecules into hydrogen and oxygen (electrolysis) or methane into hydrogen and CO_2, which must then use CCUS to decarbonize. The resulting hydrogen can be stored (and, if necessary, transported) either as a pure gas, a pure liquid, or in other chemical forms such as methane or ammonia. Storage can occur in tanks or in large salt caverns, much like today's gas storage fields, for periods of months, which is ideal for seasonal storage. When electricity is needed, stored hydrogen can be converted on demand back into carbonless power exactly when needed in a generating plant or fuel cell. All these variations on an electricity–hydrogen–electricity cycle are often shorthanded as P2G2P, or *power to gas to power.*

Hydrogen presents an interesting case because every element of the P2G2P technology chain has long since been proven to work, though mainly at relatively small scales. Even more so than CCUS, the key issue with this option is that costs must drop significantly as the size of installations increases by orders of magnitude and a highly capital-intensive new infrastructure capable of widespread production and delivery is established.

There are four main stages of the P2G2P approach: production, transport, storage, and reconversion to electricity. Depending on the cost of the renewable energy used for water splitting, U.S. production via electrolysis costs about five times as much as natural gas per unit of energy produced. This is the Achilles heel of P2G2P, and entrepreneurs are working hard to bring this cost down substantially. One company called HyTech, chronicled by clean energy blogger David Roberts in 2018, claims to have invented an electrolyzer that "can produce hydrogen at three to four times the rate of electrolyzers with similar footprints, using about a third of the electrical current," Roberts wrote. "That represents a stepwise drop in costs."[69] Meanwhile, transport and storage of hydrogen are already being done cheaply and at large scale, so they add only a little to cost. Further innovation and the use of new storage and transport materials are likely to reduce costs even further.

This brings us to the final stage: reconversion to electricity. There are no commercially operating hydrogen power stations as of yet, but the German company Kraftwerk Forschung claims that current 200 MW turbine plants can burn hydrogen-rich gases, and General Electric claims that its turbines can also run on hydrogen-rich fuels.[70] There is a bit of a chicken-and-egg problem here because hydrogen is now much too expensive to make a large power plant that burns hydrogen worth building. However, hydrogen has been burned successfully in boilers for decades, and there appear to be no technical barriers to a hydrogen power plant when fuel costs make it economical.[71]

From this discussion, it is evident that the issue with P2G2P is scaling up and reducing costs. After a recent careful study, two German researchers concluded that P2G2P was "not profitable under current market conditions; even under the most optimistic assumptions . . . PTG plants need to run for many hours . . . [using power with] very low prices that do not currently exist in Europe."[72] However, a more recent study by two German researchers in the journal *Nature Energy* examined the projected costs of a full P2G2P cycle in Germany and Texas if cost reductions continue their current trends. The authors concluded that "renewable hydrogen is already cost competitive in niche applications and is projected to become competitive with industrial-scale supply within a decade *if recent market trends continue and current policy support mechanisms are maintained.*"[73] We should not forget that establishing a whole new infrastructure almost from scratch involves high entry barriers, as infrastructure by its nature requires high levels of throughput to be economical.

One possible saving grace comes from the fact that hydrogen is a very versatile fuel that can also be used for clean transportation in fuel-cell vehicles and as a heat source for many industrial processes. This makes hydrogen infrastructure unlike most other electricity system investments, which are built and paid for almost exclusively by and for the power grid. The economics of producing hydrogen at very large scale, with a widespread supply and delivery infrastructure, will be supported by revenues from sectors and applications beyond electricity. This makes apples-to-apples comparisons of the cost of hydrogen and other seasonal storage options more difficult and highlights the importance of integrated multi-sector energy planning, a point we return to momentarily.

The multiple use cases for the fuel and the many forms in which it can be stored and converted lead hydrogen proponents to argue that its economics will steadily improve to the point of full commercial feasibility in the 2030s. There is certainly increased commercial investment in hydrogen technologies, especially in Australia and Japan, and a lot of bullish sentiment in the hydrogen community. Reviewing recent developments in renewable hydrogen, the editors of *Nature Energy* concluded that they "lend weight to the idea that hydrogen could play an important role in energy system decarbonization"[74]—or, as Pierre-Etienne Franc from the Hydrogen Council puts it, "The years 2020 to 2030 will be for hydrogen what the 1990s were for solar and wind."[75]

From Lab Bench to Toolkit

As evident from this discussion, there are already many carbon-free generation and balancing options that are already economical, reliable, and steadily replacing fossil generation. However, there are also some critical gaps in baseload energy and seasonal storage, where many tools are too new or expensive for use today.

This highlights a critical need for continued research and development, especially for these technologies, to make more of them widely available and much cheaper. A clean power toolkit usable everywhere, at costs close to current levels, is not quite here right now. We have prototypes and demonstration pilots for many tools, but you can't use new technologies to make real Big Grid systems until they are very large, tested for reliability and durability, and reasonably priced.

The policies that accelerate technical progress and cost reductions in clean power technologies are by now well understood and politically uncontroversial. Public funding for basic energy science and R&D is irreplaceable, and contributes to greater overall competitiveness, improved STEM education, greater economic competitiveness, and better job opportunities, in addition to decarbonization. Carefully monitored public co-funding and risk guarantees for early stage projects with unproven or as-yet-uneconomical technologies are also essential. Governments also play a critical role convening stakeholders, forming consortia and fostering collaboration, helping to set or revise standards, protecting intellectual property, providing technical assistance and access to shared facilities such as testing labs, and many other roles that sustain new technologies until the market can take over.[76]

The critical role of R&D in climate policy has been recognized for decades, but it is gradually now being taken as seriously as it should. Some countries, such as China and Germany, have been strong investors in energy R&D for many years. In the wake of the Great Recession, the United States established the Advanced Research Projects Agency—Energy, modeled after a counterpart U.S. Defense Department agency, whose mission "advances high-potential, high-impact energy technologies that are too early for private-sector investment . . . [by] . . . developing entirely new ways to generate, store, and use energy."[77]

Shortly after this, Microsoft founder and multi-billionaire Bill Gates started getting interested in climate change and energy technologies. Gates came to the conclusion that the level of energy R&D was well below what was needed to address climate change. Along with a pledge to invest $2 billion of his own money in energy R&D, he helped form the Breakthrough Energy Coalition, a group "committed to building new technologies that change the way we live, eat, work, travel and make things so we can stop the devastating impacts of climate change," along with an interlocking, multi-billion-dollar private investment fund.[78] Gates explained at the time that "investing in radical, 'wild-eyed' energy tech companies is the only way that the world is going to get the solution to climate change at an affordable . . . cost."[79]

But Gates didn't stop there. Understanding the critical role for public sector R&D, he embarked on a personal mission to convince Western

Figure 6-8 Mission Innovation, Launch Event, 2015.
The leaders of fifteen countries join Bill Gates onstage at the Paris Climate Summit to launch Mission Innovation, a shared pledge by each member country to double its energy R&D budget by 2023.
Mission Innovation (n.d.).

governmental leaders that they should each *double* their public funding for energy research. In a testament to his tenacity, and to the breadth of political leaders' recognition of the importance of R&D, he succeeded. On November 30, 2015, at the edge of the Paris Climate Summit, Gates stood on stage with President Obama and fifteen other national leaders and launched Mission Innovation—a multinational effort united by a pledge by every member country to double energy R&D by 2023 (see Figure 6-8).[80]

From Toolkit to Reality

As essential R&D continues, how can the rest of us be confident that an affordable, decarbonized Big Grid will be feasible soon enough to keep moving toward this goal with current technologies?

First off, several countries that are blessed with good hydro resources already have almost carbon-free power systems, such as Norway (98 percent fossil free as of 2016) and Brazil (80 percent); France, which invested heavily in nuclear power, is also at 80 percent.[81] In 2016, the national system of Portugal ran for 107 hours straight on wind, solar, and hydro power, even though its annual share of renewable energy was only about 48 percent.[82] Germany set a national record in 2015 when it ran for an hour on 78 percent renewable power, and Danish wind turbines produced more wind energy than all of Denmark used for 317 hours in 2016.[83]

This progress is prompting many state and national leaders to pledge to create 100 percent carbon-free power systems by 2050 or earlier. As of this writing eight countries, four U.S. states, and thirty-two large cities have made formal commitments to become carbon neutral by 2050 or sooner.[84]

Beyond the real world, a bevy of researchers have performed computer simulations of future national electric systems that have no fossil-fueled power plants, including many for the United States.[85] These simulations predict wind, hydro, and solar generation for every hour during a future year and determine the amount of storage, transmission, and other resources needed to match this production to simulated hourly loads. The simulation models are programmed to add enough storage, transmission, and other balancing resources, as well as "build" more solar, wind, and transmission, until every hour is balanced. The ability to simulate a carbon-free, balanced system for one or more years is offered as proof that countries like the United States can power themselves without fossil fuels.

These simulations have garnered a huge amount of controversy in scientific and energy policy circles.[86] Criticism has been especially strong for simulations that include only renewable forms of electricity and storage, categorically excluding fossil with CCUS and nuclear power, such as those by Stanford professor Mark Jacobson and his colleagues. These simulations have been criticized because they achieve balance economically only if huge amounts of seasonal storage can be built much more cheaply than today, many new transmission lines can be built, and other adjustable carbon-free generation such as biofuels becomes widespread. Critics argue that because these assumptions are far from certain, it is misleading to assure the public that 100 percent renewable systems will be feasible and affordable.[87]

One widely recognized feature of all these simulations is that achieving complete decarbonization is especially difficult and expensive compared to, say, 80 percent. Nearly all simulations show that getting the last 20 percent of fossil-fueled electricity off the system will be most difficult and costly, as this is the energy that is needed for seasonal peaks and requires huge amounts of storage, overbuilding, and / or transmission. Matthew Shaner's all-renewable simulation, which we met earlier in this chapter, required thirty-two days' worth of energy storage—6,500 times the annual output of the Gigafactory; without his assumption that new transmission could easily be added, this amount would undoubtedly jump.[88]

As the discussion in this chapter has hopefully made clear, technical change and cost reductions *are* required for feasible no-carbon grids. However, simulations that demonstrate feasibility *if* technical change occurs are not, in themselves, misleading. Like any other forecasting exercise, they are pictures of the future under one set of assumed circumstances. One can quarrel with the particular assumptions each researcher makes—I am sure I would not agree with every one—but as long as the assumptions are made clear, the simulations are what they are. Furthermore, the fact that one can achieve a balanced system without nuclear power or fossil with CCUS doesn't mean that policy makers will exclude these technologies in the race to decarbonize under real-world constraints.

If anything, both these simulations and their criticisms are valuable precisely because they crystallize the most important changes needed to make any of these carbon-free systems a future reality. Some of these changes, such as readily available long-term storage, will require continued investment and the full suite of R&D policies to pull them forward. Others require non-R&D changes in markets and planning institutions. None of these simulations are designed to reflect the business structures, financing, planning, or regulation of future Big Grid systems, nor do they close down existing fossil-fuel plants. Indeed, some of the criticisms of the simulations are simply that they don't incorporate realistic constraints on how quickly energy systems can be retooled, whether for financial, logistical, or political reasons.[89] Nevertheless, in view of the rate of technical progress and the changes we see in real-world systems, these simulations and the rest of the evidence clearly point to one conclusion: technologically speaking, it is more a question of *when* rather than *if.* But remember, we're still only at stage 1.

7 Not in My Backyard-State-Region

Transmission is the third rail of electricity system expansion. It differs greatly from all other Big Grid resources for two major reasons. First, it is fully rate regulated like traditional utilities and subject to very extensive oversight beyond rates. Second, transmission must be planned and built as an integrated regional system. No matter how large or exclusive your government-sanctioned service area, there is no way you can simply decide to add a large new line to your system on your own. Almost every balancing area (BA) has several transmission owners whose systems are strongly interconnected, so that every major addition affects flows and facilities across the region. To add to the system, the transmission owners and the BA operator must all agree well in advance, after studies and simulations, that an addition will not harm system function anywhere. If weaknesses are found, fixes must be made in the parts of the system that are harmed, perhaps in another owner's territory.

The cost of most transmission lines is modest compared to large generators, and there are relatively few risks of technology failure. Provided that superstorms don't knock them down, modern transmission lines are long-lasting and highly efficient, and thus a relatively cheap way to integrate generators, storage, and distributed energy resources (DERs). Each customer's payment for its allocated share of the entire transmission system requires only about 10 percent, 9–30 percent, and 5–12 percent of average power bills in the United States, Europe, and Australia, respectively, and there is evidence that even these costs can be reduced significantly.[1]

Accordingly, getting the grid we need for full decarbonization is less an issue of cost than a challenge to get the right system planned, financed, and built. In most countries, and especially in the United States, the planning and enablement processes for Big Grid expansions are fractured and weak. In the United States, this includes all three elements of the larger process: planning, siting approvals, and the mechanics of financing large lines.

More importantly, transmission planning grew up as an engineering exercise with little formal link to energy and climate policy. Transmission planning is mainly done within the power industry and shies away from attempting to reflect—much less to help set—changes in future energy policies. It views itself as a passive reflection of whatever electricity needs emanate from the larger economy and body politic. Especially in the United States, this decoupling between a proactive climate agenda and power grid planning must give way to a more integrated energy planning framework.

The three direct impediments to grid expansion—planning, siting, and paying—are all symptoms of the politics of the power industry in its current state. An unusually large set of stakeholders tends to line up against most large new transmission lines, each for its own reasons. Together, these groups usually have enough political power to slow down or scale back many proposed new lines. The coalition isn't monolithic, and many groups support specific projects that *do* get built, but the aggregate effect is to limit process improvements and the scale of expansion ambitions greatly. Indeed, the title of this chapter comes from a statement by an environmentalist that she didn't want to live "under aluminum sky" created by many new overhead power lines.[2]

Despite these hurdles, a fair amount of transmission gets built in the advanced economies each year. Not counting governmental or rural electric projects, U.S. utilities are spending about $20 billion a year on transmission; all U.S. utilities were building 2,800 circuit miles of new lines as of 2017—almost enough to reach from coast to coast.[3] Europe added only 164 miles of new alternating current (AC) lines in 2017. In the United States, many of these lines are short reinforcements that bolster reliability or replace aging plants, rather than being aimed at emissions reductions.

Meanwhile, as we saw in Chapter 6, transmission is a key part of one of the main seasonal / baseload balancing options in the clean power toolkit. Nearly all the relevant studies suggest that substantial new transmission is needed alongside balancing resources and DERs for lowest-cost decarbonization.[4] For example, a team headed by Alexander MacDonald found that 21,000 miles of new high-capacity lines would yield the cheapest way to decarbonize the U.S. power system.[5] (Interestingly, this is about the length of new lines added in the United States between 2012 and 2017.[6]) European studies also find the need for continued grid buildout to meet EU carbon targets; one study concludes that meeting climate targets would cost €43 billion *per year* more by 2040 if no additional transmission was built.[7]

This research points to the need for more transmission, but they are almost all research studies, not sufficiently detailed or implementable plans. As a result, we can't even seriously assess the gap between current expansion plans and an affordable, reliable zero-carbon system. Especially in the United States, we need a clearer picture of what we need to do to arrive at a decarbonized Big Grid as soon as possible and a legitimate, actionable process to get there.

Planning the No-Carbon Future

Because grid facilities last many decades, and new lines influence power flows on other lines over a wide region, Big Grid plans must compare several possible region-wide expansion plans over at least the next twenty years. The process starts with forecasts of practically every aspect of the system: future demand patterns, DER installations, the cost and attributes of balancing resources, the cost of fossil fuels, other large plants or lines to be added, changes to power market rules, and many other inputs. These assumptions are used to construct multiple scenarios that involve a proposed grid addition, along with new generators and other resources that planners guess will together enable ongoing balance at reasonable costs. These scenarios are usually checked by sophisticated simulation models that test how the Big Grid will behave in each scenario under a huge range of weather, demand, and outage conditions. In addition to verifying reliable operation, planners do cost–benefit comparisons between different scenarios with and without the proposed new line(s).

The challenge of these planning exercises is that they must merge the highly technical work of simulating power flows, which only grid engineers and other specialized experts really understand, with future energy and climate policy assumptions and predicted costs of many types of power resources. The first stage of the process is developing scenarios, involving scores of assumptions twenty or more years out. Ideally, this involves consultation with regulators, policy makers, subject matter experts, and a wide range of stakeholders. The second stage is the highly technical simulation modeling by the engineers, along with benefit–cost calculations. The third stage is discussing all the results and choosing what new facilities to include in the official final plan.

Because the grid is the one system integrating everything that makes use of power, transmission planning is also generation planning, climate policy, and DER market policies and planning all rolled into one. Grid expansion scenarios reflect assumptions regarding policies enacted over the coming decades by all levels of government, including policies into which transmission systems and electric regulators may have little or no input. Properly done, they are multilevel, multi-governmental plans for the decarbonization and structural evolution of the Big Grid.[8]

Planners and policy makers are starting to recognize this and create more wide-ranging, transparent processes that better integrate technical studies with future policy scenarios. Rather than think of transmission planning as its own exercise, best practices treat it as one part of an integrated energy and climate plan. But irrespective of whether it is recognized as such, highly technical grid studies must continue. These exercises cannot help become an arena in which essentially every major debate over the structure, function, or environmental impact of the industry is reflected in the assumptions and scenarios grid planners use.[9]

The controversies usually begin with the degree to which scenarios should assume policies and outcomes that maximize DERs and energy efficiency and therefore show the smallest need for Big Grid expansion. Stakeholders who favor the "soft path" push planners to make sure that DER costs are not overestimated or their potential to reduce the need for new lines underestimated. As an example, the Sierra Club's long-standing policy is that "the siting of large, energy-related facilities should not proceed unless a definitive need for them has been demonstrated, through open public disclosure and certification of need, which cannot be met through conservation and smaller-scale alternatives."[10] Yet, DER forecasts and their ability to substitute for transmission are both evolving rapidly, and factoring these into already-complex grid simulations is not easy.[11]

The debate over future policies and technologies continues with every other assumption, including whether to assume that nuclear plants continue to operate (or new ones are built), whether carbon capture and sequestration becomes available for fossil plants, the amount of offshore wind that can be developed at various cost levels, whether carbon limits or taxes are enacted, and of course the forecasted costs and availability of power storage. Within the planning process, each of these issues has constituencies that want planners to make assumptions favorable to their preferred technologies or policies.

Estimates of benefits and costs that accompany the simulations are also sources of controversy that often ultimately yield weaker plans. A 2015 Brattle report for the U.S. transmission industry listed twenty-six benefits from transmission additions, including system reliability, more competitive power markets, and lower renewable power prices.[12] Beyond reliability, many of these benefits are not measured by planners because stakeholders can't agree on how to measure or set a value for them without first agreeing on even more contentious economic and policy assumptions.

While arguments continue over the definition of benefits, transmission imposes costs on the land and communities it traverses. These costs are also difficult to quantify, especially when they impact important natural areas. These costs are usually considered after the planning stage, when permits are considered, but they inevitably weigh on the planning process as well. In this era, the primary reason we need transmission plans is to prevent the unthinkably large environmental costs of climate change, but it is hard for citizens and policy makers to grasp and then trade off the climate benefits of transmission against other environmental costs.[13]

Finally, the diffused economic impacts of long transmission lines naturally lead to local opposition. In addition to their unpopularity with NIMBYs, lines to distant resources create less economic activity and fewer jobs than local resources of any type, whether nearby offshore wind plants, local DERs, or anything else that might be built in place of a new line. It's no different in Europe, where international projects backed by EU planners face opposition from local citizens who ask "what is the real benefit of that line to our community?"[14]

In the United States, transmission planning is done by regional groups of system operators and transmission owners. The process has gotten significantly better in the last decade, as federal regulators issued a series of detailed orders requiring more openness and other improvements.[15] Nonetheless, these exercises remain lengthy, technical, and cumbersome, and they are also limited to single regions. No official body even attempts to do a multiregional much less nationwide or continental plan.[16] Most importantly, U.S. regional plans are entirely "non-binding," which means no utility or anyone else is required to build the lines in the plan, nor does inclusion in the plan earn you any permits or approvals.

In Europe, the power grids in the EU's twenty-eight-member bloc were built with few connections to neighboring countries, creating systems too small for lowest-cost supplies or the integration of variable renewables. Well before climate policy was recognized as critical, one of the EU's goals was to increase transmission interconnection across the bloc and create a single huge pan-European power market.

The incorporation of climate goals as a centerpiece of EU policy has prompted the Union to combine its prior goal of market integration with climate policy. This has created the best large-scale example we have of a strongly integrated energy system planning and policy framework. The biannual process, known as the Ten Year Network Development Plan (TYNDP), follows the canonical three planning stages described above: scenario development, comparison of plan options for cost, and selection of plan elements. However, all of the scenarios chosen for evaluation explicitly reflect achievement of the EU's goal of 80–95 percent reduction of carbon emissions by 2050 (recently updated to carbon neutral).[17] The three long-term scenarios—Distributed Generation, Sustainable Transition, and Global Climate Action—have very similar levels of electricity use in 2040, differing mainly in the amount of distributed solar that is installed and the amount of power-to-gas and biomethane used.[18]

As in all good processes, the TYNDP's developers obtained broad stakeholder and expert input to inform forecasts of macroeconomic trends, transport technology shifts such as EV adoption, and building and industry power use. They also made the necessary assumptions regarding the availability and cost of carbon-free generation and storage—their own low-carbon toolkit. Significantly, the entire effort was integrated with the EU's parallel planning effort for the natural gas system, allowing planners to analyze more fully options such as hydrogen and biogas and their link to the power system.[19] Also of significance, EU policy makers were deeply engaged with planners and EU electric regulators, reportedly meeting with them every four to six weeks throughout the two-year process.[20]

As Oxford energy researcher Alexander Scheibe explains, this result of the process yields an EU-wide plan in which "all projects . . . are effectively contributing to the achievement of EU energy policy goals."[21] Scheibe goes on to note another extremely important aspect of the TYNDP process. The EU backs up the plan—which is officially non-binding—with the co-funding transmission lines or storage facilities that are "projects of common interest." This funding, amounting to as much as 50 percent of a project's cost, is accompanied by a streamlined process that reportedly reduces permitting time from about ten years to roughly three-and-a-half.[22]

The TYNDP process has not produced any miracles of grid construction—at least not yet—but it does herald a new era in comprehensive policy-driven power grid planning. It is a major step forward. Moreover, the EU clearly intends to deepen its commitment to planning to meet climate goals. Its road map to the next biannual plan, due in 2020, is now holding workshops to "identify . . . how social values and corresponding political objectives can be mapped in a model, and ultimately how the most ambitious TYNDP scenario can be moved closer to compliance with the Paris Accord."[23]

Not in My Backyard

Transmission siting approval processes are legendary for the number of different permits required, government agencies involved, length of time required, and low overall success rate. In the United States, so many distinct federal, state, and local approvals are needed that long, detailed instruction manuals have been prepared for would-be line developers. Table 7-1, from the state of Utah's official guide, lists federal approvals that are required for most significant new lines, omitting dozens of state and local requirements. The second line of the table shows that any line involving federal action or lands needs an environmental impact statement, which by itself can take many years to complete. Beyond this, the table shows possible approvals required from eight more federal agencies under nineteen distinct federal provisions, including the Endangered Species Act and the Clean Water Act. One pro-transmission trade group created a version of the popular table game *Chutes and Ladders* modeled after the transmission permitting process.

After federal approvals, there are numerous state, county, and city approvals. Many states have energy facility siting boards or councils, which must grant a state permit after their own extensive inquiry. Oregon's Energy Facilities Siting Council requires a seven-stage application and review process that can include judicial-style hearings, all culminating in an order that can be appealed to the Oregon Supreme Court.[24] If there is no statewide siting agency, a new line must obtain separate permits from

Table 7-1 Federal Permitting and Regulatory Requirements

Agency	Division	Regulatory Framework	Required Study / Permit / Consultation
Advisory Council on Historic Preservation	*Not applicable*	National Historic Preservation Act 1966	Section 106 consultation
Council on Environmental Quality	*Not applicable*	National Environmental Policy Act (NEPA)	Environmental Impact Statement—major federal actions
			Environmental Assessment—minor federal actions
			Categorical exclusion
Federal Communications Commission	*Not applicable*	Communications Act 1934; 47 CFR 15.1	Consultation to avoid line-of-sight obstruction
U.S. Department of Agriculture	U.S. Forest Service	Federal Land Policy and Management Act 1976	Special-use permit
			Temporary-use grant
	Natural Resources Conservation Service	Farmland Protection Policy Act	*Not applicable*
	Rural Utilities Service	Rural Electrification Act 1936	Borrower's request for funding
U.S. Department of Defense	U.S. Army Corps of Engineers	Clean Water Act	Section 404—discharge to and fill in waters of the U.S.
		Rivers and Harbors Act	Section 10—work affecting the course, location, or condition of a navigable water
	Military Facilities	10 U.S.C. 2668	Right-of-way authorization

U.S. Department of Energy	Federal Energy Regulatory Commission	Sections 201, 205, 206, and 216(a) of the Federal Power Act	Rate filing
	Western Area Power Administration	Established 1977 as part of Department of Energy; ARRA Section 402	Requires NEPA compliance
U.S. Department of the Interior	Bureau of Indian Affairs	CFR 25 Part 169 (Rights-of-Way Over Indian Lands)	Requires NEPA compliance
	Bureau of Land Management	Federal Land Policy and Management Act 1976	Right-of-way grant
	Bureau of Reclamation	43 CFR 429.3	Right-of-way grant
	Fish and Wildlife Service	Endangered Species Act	Consultation with Fish and Wildlife Service (Incidental Take Permit if applicable)
		Bald and Golden Eagle Protection Act	Coordination with Fish and Wildlife Service (Incidental Take Permit if applicable)
		Migratory Bird Treaty Act	Coordination with Fish and Wildlife Service (Incidental Take Permit if applicable)
		National Wildlife Refuges	Right-of-way authorization for crossing and special use permit
	National Park Service	15 U.S.C. Director's Order 53 (Special Park Use Permit) Section 10.2.1 for linear rights-of-way	Right-of-way grant

Source: Utah Office of Energy Development (2013) and California Public Utilities Commission (2009).

a dozen or more state agencies analogous to the federal agencies in Table 7-1. Then, of course, there are county and city approvals: zoning and land-use approvals, engineering and building permits, excavation and road permits, and others. These are needed for every county and city a transmission line crosses.

Though permitting processes are often fatal to new line development in many countries, the United States stands out as having the most layered and lengthy approval process among Western nations. The official permit approval timelines for new lines are five and seven years within Utah and California, respectively, but the real timetables are always longer.[25] Development and permitting periods of ten years or longer for lines that cross multiple states are not unusual, as in the thirteen years needed to develop the TransWest Express line.[26] The EU has recently mandated that each member create an expedited siting process with a maximum timeline.[27] A review of EU transmission expansion keyed to decarbonization found that "lengthy permitting procedures" were one main reason why major projects started in 2012 were an average of three years behind schedule.[28]

The final hurdle new transmission must clear is its rate approval. Transmission is a wholesale service charged only to those who use, make, or buy power on the Big Grid—namely, distribution utilities and deregulated retailers / energy service companies. They pay a single transmission tariff, usually per megawatt-hour purchased, which includes the amortized costs of all transmission system plants on the system to which they happen to be connected. However, every system is connected to every other system, and a buyer who purchases from a source one or two systems away must also pay a charge to traverse these other grids or a regional charge that covers the entire geographic path.

When a new facility comes online, the owner of that facility petitions regulators to add that system into the pool of lines that serves as the basis of the calculated rate. If the cost is large enough, the regulator updates the system's calculated rate. If the facility is all within one owner's system, the only tariff with any significant adjustment is that system's rate. This is comparatively easy because lines that stay within one system tend to produce local reliability benefits or are used to integrate power resources that are nearby, yielding local development benefits. In most cases, the battles over whether this line could have been avoided entirely or another route could have been chosen have all been settled by the time the line enters service and thus needs to have its rates established.

But none of this is easy if the line is *interregional,* that is, long enough to cross many states and transmission owners. In this instance, grid regulators must decide how much to adjust the transmission rate of each system that is connected to the new line because every system traversed may gain some benefits from the line and every system should presumably pay something for its use of the line. This *interregional cost allocation* sets off another round of prolonged negotiation or litigation between regions, transmission owners, and others. As U.S. plans are prepared by regional agencies, transmission rates are set within these regions. So, any line crossing these regional borders has two formal rate-making arenas with which to contend.

As noted earlier, the EU has found a partial solution to this by creating a fund to help pay for projects of common interest (PCIs)—their term for transmission that spans multiple EU countries or provides region-wide benefits.[29] The Union has also established several other vehicles to provide funding for PCIs.[30] Nonetheless, cross-border lines continue to be "hindered by conflicting national interests and the administrative and regulatory complexity of multinational projects," according to a 2017 report for the EU parliament.[31]

All of these factors combine to yield transmission plans and implementation processes that are understandably weakest when it comes to large, long transmission additions. Yet, these are exactly the type of additions that connect the world's cheapest carbon-free supplies to distant, power-hungry cities. This brings us, quite naturally, to the subject of "supergrids."

Searching for Supergrids

The need for bold climate action is causing planners and policy makers to begin considering the idea of adding a continent-scale, largely new network of huge power lines on top of the existing grid as an alternative to line-by-line expansion. Power geeks call network additions like this an *overlay,* but in the popular press, they're usually referred to as *supergrids.* A supergrid proposal from the Chinese that spans the entire planet has already been shown in Figure 4-2. Another conceptual proposal imagined by the National Renewable Energy Lab within the United States is shown in Figure 7-1. The latter shows nineteen large-scale delivery hubs (white circles) connected by twenty high-capacity transmission lines. The hubs (in black) are envisioned as being located mainly near major U.S. cities or areas with concentrated renewable generation, such as the wind belt in the center of the United States. The hubs on the Atlantic and Pacific coasts play two roles as both urban delivery hubs and collection points for offshore wind production. The network in the figure is a series of loops, allowing power to be routed along multiple pathways, depending on production

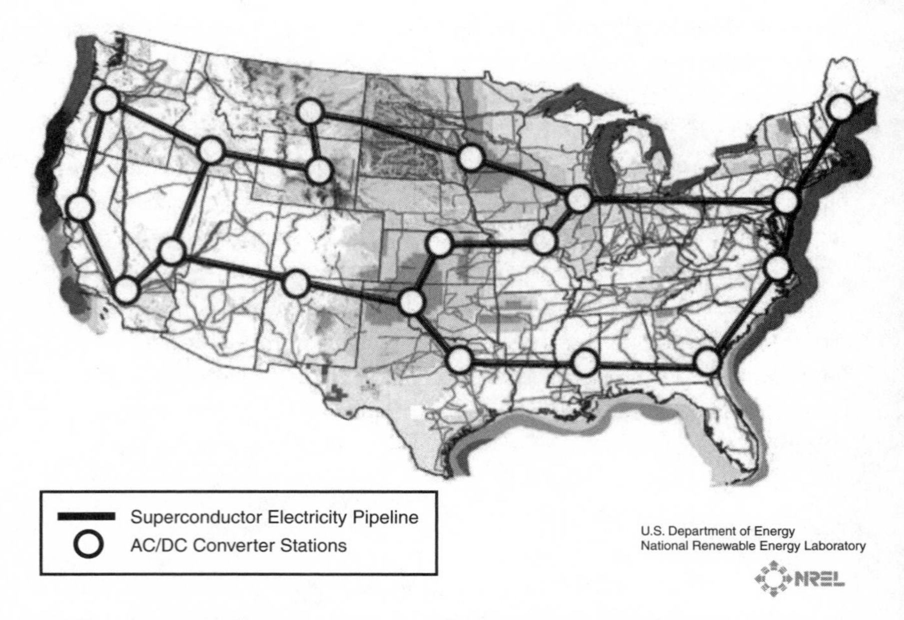

Figure 7-1 One Proposed Nationwide Transmission Supergrid for the United States.
Reformatted from U.S. Department of Energy National Renewable Energy Laboratory (2012).

and demand patterns. This entire network would sit "on top of" the entire current grid, much like superb highways sit atop the arterial road system—in effect a high-capacity interstate highway system for electricity.

Supergrids are usually proposed to use direct current (DC) rather than AC lines because the former are cheaper when sending huge amounts of power long distances, and this is the case for the grid concepts in both supergrid figures.[32] It costs almost $1,500 per mile to install a 345,000 V AC line large enough to carry 1 MW of power; the cost of equal capacity on a 600,000 V DC line (excluding converter station) is only $537.[33] The distinguishing feature of a supergrid is that it is a network of big DC lines, not just a point-to-point delivery system, so that it provides all of the benefits of reliability and diversification described in "Why We Grid." Somewhat ironically, DC line technology is not well suited to interconnecting as a meshed (i.e., strongly interconnected) network, though engineers are making steady progress toward this goal.[34] On the flip side, DC lines can also route power around a grid better than AC technologies. So, a DC supergrid seems ideal for sending large amounts of renewable power between production basins and cities.

Meanwhile, quite a number of point-to-point DC lines are being built, many of which deliver renewables from rich resource areas to the rest of their grid. Europe is building sixty-two DC lines, many of them undersea cables between countries, to help integrate renewables and hydro storage.[35] Several U.S. utilities on the West and East coasts operate older DC lines, a newer one traverses the San Francisco Bay, and one was just added between Newfoundland and Nova Scotia.[36] The leader by far is China's main electric utility, State Grid, which now uses twenty-one of these giant lines, some extending the length of the United States, and it is reportedly planning to approve three more in 2019 alone.[37] These lines are not interconnected as a grid, but China plans to do this as the system expands and DC technology matures.

The best areas of the world produce wind and solar at one to four cents per kWh less than half the costs of production areas with middling wind and sun. Eight percent of the solar energy falling on the Sahara Desert equals the momentary electricity demand of the entire planet[38]—a fact noted by Liu Zhenya, the former head of China State Grid and the world's foremost supergrid visionary. If long-distance, high-capacity transmission adds only a few pennies per kWh , large DC lines could potentially deliver wind and solar power anywhere cheaper than locally produced renewables—and, of course, there are no roof-space or other local constraints. Even with maximized distributed resources, climate policy makers ask themselves if it's more important to decarbonize power with large cheap supplies, even if it means overcoming all the potent objections to a supergrid.

By their nature, supergrids cross dozens of states or countries, affect thousands if not millions of landowners, and cost hundreds of billions of dollars ($540 billion for the MacDonald proposal). With all the delays individual interregional lines encounter in their planning, approvals, and cost allocation, how could increasing these problems exponentially in one giant multi-line expansion possibly be a good idea?

If there's an answer, it stems from Dwight Eisenhower's famous quote: "If a problem cannot be solved, enlarge it."[39] A North American supergrid would be a project so large and bold that it might come to life if championed seriously by an American president, perhaps as a part of the much discussed "Green New Deal." Indeed, President Obama's 2008 campaign platform included references to a supergrid, though his administration did not seriously adopt the cause once in office.

Outside the United States, several nationwide or transnational supergrids have been proposed and a few, such as the Chinese, have built precursor sections.[40] The most heavily studied proposal in the United States is the North American Supergrid (NAS)—a proposal advanced by the Climate Institute and several other nongovernmental organizations, not, interestingly enough, by utilities or renewable energy companies. The proposal is a fifty-two-node network spanning the entire country and costing $540 billion, excluding the costs of acquiring right-of-ways.[41] Its proponents say their idea will be "a resilient backbone to the existing system . . . [and] would provide the flexibility and reliability to enable expanded use of electricity across the economy."[42] The grid would be financed by "private, user-based fees that require no public funding," though government loan guarantees and other federal support is mentioned as a possibility.[43]

Eisenhower's quote notwithstanding, this and any other overlay proposal face utterly daunting implementation challenges. There is no legal mechanism or institutional history for binding nationwide energy or transmission planning, and getting the actual plans made and through the inevitable scores of legal challenges could require decades. Even NAS proponents admit that their idea "comes with inherent feasibility challenges, and its operation might result in environmental consequences that range from minimally adverse to highly beneficial."[44] Millions of government dollars would be required for planning, engineering, and permitting before the private sector would be willing to step in and put its capital at risk. And settling legislative issues such as who would have final say over line routes and how decisions could be appealed without enormous delays will require political compromises that have so far eluded every administration that has tried.[45]

The enormous cost of a supergrid, along with the potential for planning or implementation errors, is more than enough to prompt many power experts to recommend continuing to add large DC lines, but one or two at a time. This could evolve into a supergrid in stages and would provide greater access to cheap renewables in the meantime. China is proceeding exactly this way, and GEIDCO's conceptual proposal for a global supergrid is envisioned as occurring in multiple explicit stages.

All in all, supergrids are giant, all-or-nothing bets in which (1) the amount of transmission they provide is roughly the amount needed for lowest-cost decarbonization several decades from now; and (2) the larger-than-life implementation challenges are manageable, if not easier, as a national undertaking. National and transnational leaders will have to assess that these are both likely to be true, and then have the courage, skills, and luck to place the bet. Supergrids also shouldn't be analyzed or planned independently of an integrated national climate strategy involving all economic sectors. An integrated plan is necessary to lay the foundation for the size, location, and timetable for electricity demand growth by sector well into the future.

The Future of Grid Expansion

If decarbonization is to proceed as quickly as it must, integrated transmission and climate planning is a critical accompaniment to a large downstream DER sector, more balancing plants, and the new business models we'll explore in the next several chapters. Without better integrated planning, we can't even guess at the amount of transmission we need and where and how it should be built. Europe, Australia, and other countries are starting to get a good handle on these questions, while the United States lags well behind.

Many suggestions for U.S. transmission planning improvements have been made over the years. In the 1960s, U.S. federal agencies actually prepared a nonbinding long-term national grid plan. As regional planning groups have become institutionalized in the United States, most attention is now focused on improving these processes and adding a planning layer that looks beyond the region. I and others have proposed regional compacts and collaboratives that broker economic compromises between the states in a region at the planning stage, so that permitting begins with a high-level signoff from the main affected states.[46] Proposals to improve siting also focus on creating single permitting agencies within states and giving eminent domain authority to transmission lines that get federal approvals.[47]

The political reality in the United States is that the best hope for any of these improvements will occur in the context of strong presidential leadership on broad-scale climate action. A transmission plan and its resulting grid are not ends in themselves; they are critical parts of any complete, achievable climate plan. As the European process shows, a sound integrated plan leading to actionable projects guided by climate and other social goals is within the realm of possibility. If U.S. leaders can create a way to mediate economic interests firmly and fairly so as to produce reasonable investment returns, regulated private capital, public capital, and regulation will do the rest.

8 The Big Grid Bucks Stop Here

In addition to advancing the necessary technologies, designing integrated systems, and installing the grid we need, someone has to pay for all these new investments. Every resource on the Big Grid has owners who won't build unless they believe, or are assured, they can recoup their investment. Financing the new assets in a decarbonized Big Grid is every bit as large and important a challenge as designing them to work as a system or getting permission to build.

From the exceedingly important perspective of ownership, the large power plants and other resources that populate the Big Grid come in three flavors. Investor-owned regulated utilities (IOUs), found in the United States and some countries outside Europe, distribute power over their Small Grids and also own many of their own generators, large storage facilities, and transmission lines. Others are owned by governmental authorities such as the U.S. Tennessee Valley Authority or state utilities in Europe and Asia.[1] The third category is independent power producers (IPPs), resources owned by independent firms that specialize in building and operating generators and other large facilities. IPPs operate in bulk power markets that have been price deregulated or "liberalized."

Owning deregulated power assets has become very big business. IPP firms you've probably never heard of own and operate billions of dollars' worth of power facilities all over the world, some fossil and some renewable. Two that grew out of U.S. utilities are in Platts' top 25 energy companies in the world: NextEra (assets of $99 billion) and Exelon

($116 billion).[2] The European giant Engie operates in no fewer than forty countries around the world.[3] Together, companies in the IOU, public, and IPP sectors own almost 20,000 power plants of all types in the United States and even more in Europe.[4] In the United States, utilities and publics own 1,804 operating *fossil-fuel* plants, and IPP companies own another 631.[5] In Europe, where IPP ownership is the rule rather than the exception, the comparable numbers are 8,121 and 2,839, respectively.[6]

Until retrofitting fossil plants with carbon capture systems becomes commercially viable, decarbonizing means shutting them down and replacing them, creating a very different kind of power system. As we've just seen, there is much work to be done finding feasible and affordable carbon-free power and transmission to replace them. In the second stage of Big Grid decarbonization, each area must evaluate its indigenous and external resource options and its political and organizational preferences and create a customized plan. The third stage of any actionable plan is fiscal: who and how will the new resources in the plan be financed, and how much must be paid to retire essentially all the current fossil plants?

A Power Plant's Early Retirement Package

Financially speaking, every power plant is an asset on some entities' balance sheet. It has associated debt to pay off and explicit or implicit equity holders (including governments) that depend on plant revenues for earnings and/or to fund essential services. For IOUs and publics, which have the obligation to provide adequate supplies, early shutdown does a double whammy on their finances, zeroing out a formerly valuable asset and requiring a new investment or contract for alternative supplies. When the alternatives are more capital intensive or expensive than current sources, as is the case for some baseload- and seasonal-energy options, the financial challenge only grows.

The fiscal mechanics of retiring a power plant early depend quite a lot on which of the three ownership categories the plant is in. There is a wide range of government actions that can force a fossil plant to close early. The Rocky Mountain Institute (RMI)'s guidebook to coal transitions lists ten types of policies, from direct government orders to close with no compensation, to policies that make plants highly uneconomical, to outright government takeovers.[7]

If the plant is owned by regulated IOUs, the financial implications of early closure principally involve five groups. Two of the groups, utility stockholders and bondholders, are represented by utility management, while customers are represented by their regulator and sometimes a separate consumer advocate. In addition, the workers at the affected plant and their unions have understandable concerns with a shutdown, and so do the leaders of nearby communities, which depend on tax payments and economic activity created by the plant.[8]

To avoid delays and litigation, these five groups need to agree on a plan for retiring a local fossil plant, which includes financing the replacement power, continuing to pay off all bonds, providing acceptable profits to equity holders, creating a transition plan for workers, keeping rates reasonable, and ideally offering some sort of tax or development compensation to the community. When these groups agree on a plan that addresses all these needs, a path to an equitable transition has been found.

Several financial tools have been developed that enable regulated utilities to reduce or even eliminate the impact of early closure on earnings. Regulators can approve changes to utility rates that accelerate plant depreciation, create a smaller "regulatory asset" that continues to act like a profit-generating plant after the real plant closes, and / or issue special "securitized" bonds that pay off the pre-closure stream of projected earnings from the plants.[9] These measures each add incremental amounts to customer bills to compensate investors, but often the total bill impact can be moderated by combining these payments with new investments in renewable power cheaper than power from the plants being retired. The financial measures can also include money for worker retraining and economic development.[10]

In human terms, the most difficult aspect of an early shutdown is its impact on displaced workers and communities. In many rural counties that host large plants these facilities are the largest local employer and taxpayer and the core of the local business ecosystem. The economic impacts of closure extend far beyond the plant's workers, rippling through their families, local businesses, and essential community services such as health care. In the United States, many communities hosting coal plants are already declining from reduced economic activity (including coal mining) and are struggling with the opioid epidemic and other social challenges.

These realities are prompting climate policy makers to search hard for strategies that reduce, if not eliminate, the local economic impact of shutdowns. Following extensive worldwide studies of early closures, an international group of researchers known as the Coal Transitions Project

Table 8-1 Taxonomy of Community Resilience Strategies in the Wake of Coal-Plant Closures

Strategy Name	Brief Description
"Related diversification"	Developing closely related industries that do not involve coal
"Smart specialization"	Developing industries that leverage local competitive advantages (e.g., rail infrastructure, local specialties)
Strengthening of local entrepreneurial networks	Technical and financial assistance to start new companies
Improvement of local infrastructure	Add infrastructure designed specifically to attract new business (e.g., office space or broadband)
Improvement of "soft attractiveness factors"	Add amenities for families, such as schools and parks
Location of public-sector activities in the region	Locate military bases, government offices, schools, or hospitals in the area
Location of nationally relevant innovation or energy transition projects in the region	Locate carbon capture, utilization, and storage plants, industrial pilots, and other large new energy projects in the area

Source: Based on Sartor (2018).

identified seven kinds of "economic resilience and transition" strategies, shown in Table 8-1. Other groups have assembled case studies of how communities have handled the closure of coal plants and mines.[11]

There are many examples of the first two strategies in Table 8-1—related diversification and smart specialization—under discussion or even underway. Converting coal generators to burn natural gas is occurring in the United States, Europe, and China. Needless to say, this is not full decarbonization, but it does keep the plant going for a while.[12] Other economic development strategies include repurposing plant sites for other industrial power sector uses, since these sites often have well-developed infrastructure advantages such as deepwater ports and strong transmission connections. Former AES executive Robert Hemphill has proposed using closed California power plants for solar desalinization, while authorities in Massachusetts are converting the former Brayton Point coal plant into an offshore wind support center.[13]

Beyond repurposing sites, transition strategies become somewhat more diffuse and usually more difficult. Many national governments have established programs to retrain and aid workers displaced by the coal transition, such as the U.S. Appalachian Regional Commission (ARC).[14] In additional to publishing assistance material and developing local strategies, ARC funds retraining programs, new businesses in affected areas, and infrastructure that aids resilience and transition. In France, the government has proposed "transition contracts" to provide aid to each of the regions that host the four remaining coal-fired plants in that country.[15] One especially attractive retraining program, featured in Al Gore's 24 Hours of Reality in 2018, retrains unemployed oil and gas workers in Alberta, Canada, to become technicians on wind power plants.[16]

For a variety of reasons, early closures of IOU-owned plants are a little easier to effectuate than they are for the other two types of owners. The utility has many other assets, an ongoing business, and revenue stream, and the replacement power investment can become a new source of business and profit. As just noted, IOUs can combine cheaper new sources of renewable power with measures that pay off investors and end up with total ratepayer bills quite similar to those under continued fossil operation, easing the concerns of ratepayer representatives and investors while also enabling a just transition.[17]

Xcel Energy's 2018 plan to close Commanche 1 and 2, two coal plants in Pueblo, Colorado, provides a good example of an IOU plan with these ingredients. The utility proposed to close the 660 MW plants about a decade ahead of schedule and replace them with 1,100 MW of new wind capacity, 750 MW of PV, 380 MW of adjustable gas capacity, and some battery storage—a classic combination of clean grid tools from Chapter 6.[18] Because the costs of wind and solar in Colorado have dropped below the cost of power from these plants, the utility was able to create a plan that repaid investors through accelerated depreciation without too large an impact on rates.

Xcel's plan was fortunate in its ability to ameliorate the economic impact on Pueblo, which was set to lose eighty high-paying jobs at the Commanche stations.[19] It was able to site two of the three new solar plants that are part of its plan within Pueblo County. In addition, Pueblo has a manufacturer of steel wind plant towers that was expected to benefit from Xcel's new wind purchases. Pueblo's economic development director, Chris Markuson, calculated that the net overall impact on Pueblo County's tax revenue would be a *positive* $1.4 million—a remarkable win-win outcome for a community losing two large generators. Markuson expressed hope that the new plants could "add up to making Pueblo prime for renewable development."[20] Still, Xcel's plan was opposed by the IBEW and local customer groups, the latter concerned with the possibility that the plan would cause rate increases.[21]

Many challenges of early closure apply even more strongly to plants owned by the publics, including rural electric cooperatives. These utilities are owned by their communities, so it is more difficult for them to impart the economic blows that early closures can bring. There are no third-party equity holders that can help absorb losses, and no federal or state tax liability, and thus no tax-related financial strategies that cushion rate impacts. Even so, the declining costs of solar and wind, along with widespread concern over the local and global effects of coal generation, are prompting consideration of early closures. A case in point comes from

Florida's Orlando Utilities, which is studying the early closure of its Stanton units one and two.[22]

The most difficult type of plant to close early is an IPP. The owners of these generators are unregulated, for-profit companies driven by the difference between their cost of making power and the price they earn from their long-term contracts and spot market sales. If operating one of their fossil plants becomes unprofitable on a sustained basis, they are likely to move to close it as soon as possible. However, many conventional IPPs are either coal plants that were purchased used at low prices and therefore remain profitable to operate or are gas-fired plants that are highly profitable. Moreover, IPPs have neither the obligation nor the built-in opportunity to replace a closed plant with a new revenue-generating plant; they must compete for each power sales opportunity one at a time.

This situation leaves policy makers with few carrots to incentivize early IPP shutdown. In the RMI's guidebook to early plant closures, most of the tools applying to liberalized markets involve ordering closure or changing the economics of plant operation to the point where low profitability prompts the owner to take action. Since IPP companies do not operate under the same compact as IOUs, publics, and cooperatives, there is also sometimes less perceived obligation to assist with labor force transition and community economic impacts.

Recent actual and forecasted plant closures follow a pattern consistent with these sectoral differences. In the United States, 63 percent of all coal plant retirements between 2007 and 2016 were regulated units, and at least nine regulated utilities have announced further plant closures.[23] Xcel CEO Ben Folke was speaking for most of the industry, and certainly the regulated part, when he recently said, "I tell you, it's not a matter of if we're going to retire our coal fleet in this nation, it's just a matter of when."[24] Coal IPPs in the United States are closing at a slower pace, with some reportedly *extending* their planned retirement duties. In Europe, where most plants are either IPPs or state owned, government mandates are the main tool for plant closures. Eleven countries, accounting for 21 percent of Europe's coal capacity, have announced national phaseouts by 2030 or sooner, and five more countries are considering the measure.[25]

Quite recently, the plant transition conversation has changed to include early closures of natural gas plants as well. Michael Bloomberg used his June 2019 commencement address at the Massachusetts Institute of Technology to announce that his foundation was changing the name of its funding program from Beyond Coal to Beyond Carbon—an unmistakable signal that gas plants are now "on the table."[26] As we saw in Chapter 6, system planners and modelers are working hard to plan fully decarbonized systems that don't rely on gas-fired watts.

Power Markets and Plant Financing in a Carbon-Free Future

Finding investors willing to finance the new combinations of plants, lines, and storage that will make up carbon-free grids would seem to be the world's largest non-problem. Provided the demand is there, who wouldn't want to invest in carbon-free assets that provide an absolutely essential service at an affordable price?

At the conceptual level, this is true—there is no shortage of investment capital that is interested in making clean power investments. The catch is that the risks and rewards have to be sufficient to prompt investors to write the check. When it comes to real-world financing, regulation and public policies shape the risk / reward profile, and public capital and risk guarantees must fill the gaps.

In the modern industry, power assets are financed in one of three ways. Regulated and public utilities do "balance sheet financing" by issuing stock (in the case of IOUs) and bonds. Because regulators oversee utility investments, there is something of an implied—though never firm—guarantee that they will keep utility revenues large enough to pay investors a fair return.[27] If a carbon-free grid asset makes it into a regulator-approved utility investment plan, there is rarely any real difficulty raising money from utilities' huge investor base.[28] In fact, utilities and other electric retailers that operate under public service mandates aren't just encouraged to invest—they are required to do so. Most of these "load-serving entities" must demonstrate to regulators that they own or have contracted with generators equal in size to their expected peak loads. Whether by ownership or contract, generators that help meet these "resource adequacy" requirements rarely have trouble obtaining financing.[29]

In deregulated (restructured, liberalized) markets, financiers are much more particular. In these markets, each power asset is supposed to compete in the market to justify its existence, so financing occurs one project at a time. To obtain equity or debt for a clean grid asset, you must convince investors of their return, which is not guaranteed by anyone. To forecast return, you must estimate both project costs and revenues over a multi-decade asset lifetime. Estimating construction and operation costs for a familiar asset like a wind or solar plant isn't hard, but what about the revenues?

In liberalized markets, power resources earn revenues by selling three different kinds of products using three different kinds of sales mechanisms. The traditional products are electrical energy, electrical capacity, and a much smaller set of specialized electric services purchased only by the one grid operator in each region. As we'll see in a moment, it is important to expand and redefine the services in the latter category, nowadays referred to as ancillary services. But for the foreseeable future, nearly all revenues will come from energy and capacity sales.

These products can be sold in short-term or *spot* markets—the first sales mechanism. Every part of the Big Grid that has been deregulated (liberalized, restructured) has created centralized spot markets where power plants can trade their megawatt-hours. For valid technical reasons, all megawatt-hours allowed into this market are electrically equivalent. Energy can be sold by the hour, day, month, or sometimes longer. Independent market operators, overseen by national regulators, match the bids of buyers and sellers and set the market price for each hour, day, and month.

If spot market energy prices are high enough, and believed to remain high enough for the next five to ten years, then plant builders sometimes decide to take the risk of borrowing money and building a plant, battery, transmission line, or other grid asset and simply selling the output one hour at a time. In the industry, these are known as *merchant* plants or lines.[30] Hourly electricity prices always move around, but as the number of available power plants on a system dwindles, energy prices in these centralized markets trend higher. One school of experts led by Harvard professor Bill Hogan believes that the key to inducing new plant investment is allowing spot prices to rise to very high levels during high-demand hours—an approach loosely labeled *scarcity pricing.*[31]

The political issue with scarcity pricing is that regulators have to be prepared to allow prices to spike up frequently to levels around $10 per kWh or more, forty times the current retail price, to signal would-be plant builders that there's enough revenue to make construction profitable. Meanwhile, regulators must reassure themselves that the very high prices customers are paying during price-spike events are truly going to cause new capacity and aren't just giving existing plant owners a windfall.[32] So far, regulators almost everywhere have been unwilling to allow anything close to uncontrolled price spikes, instead allowing prices to rise only to caps closer to $1–3 per kWh.[33]

The primary exception to the no-scarcity-pricing approach in the United States has been the Texas spot power marketplace, where there is a cap at an unusually high $9 per kWh. There has been an ongoing debate in the state over whether new power-plant construction has been adequate, with almost all new capacity since 2015 coming from wind and solar plants anchored by outside contracts, but the state has been steadfast in supporting scarcity pricing, with ongoing refinements.[34] Even without full-on scarcity pricing, several merchant plants have been built in Germany and Spain, and some EU power industry investors expect greater merchant investment over time.[35]

These exceptions notwithstanding, the majority view among stakeholders is that relying on energy spot market revenues and scarcity pricing alone is not a rapid or sure enough path to decarbonization. In part, this is because the risks of selling all output into short-term spot markets extend well beyond the variability of price in these markets. Spot market risks include the fact that your bid may not be selected at all, you may not be able to find any buyer for your power, or you may find one but be unable to sell to them because there is no transmission available to transport your power. Spot market rules and other policies that affect prices or sales opportunities are almost certain to change over the years as well. "In practice," one expert group from U.S. national labs concluded, "the uncertainty and volatility of scarcity price events makes investing in new generation difficult and risky."[36]

Ironically, the difficulties of financing new carbon-free resources from spot energy prices are further weakened by those very resources offering their power in current spot energy auctions. Because wind and solar plants have zero fuel costs, deregulated wind and solar sellers are able to bid their power into energy spot markets at prices near zero. This naturally drives the market price of energy down, creating even less revenues for new would-be sellers. To keep our discussion a little easier, the details of this phenomenon are explained in Appendix B. For now, the key takeaway is that spot market energy prices are going to fall to ever-lower levels as the amount of wind and solar steadily increases, unless these markets are fundamentally redesigned.

Even before considering the need to build a new carbon-free system, the fact that price-capped energy market revenues often don't support new plant constructions has prompted regulators and policy makers to worry that there may not be enough electricity plants of *any* kind to keep the lights on. The idea of power deregulation was that the demands of individual customers, aggregated by large retailers, would have to push prices high enough for suppliers to want to build supply. This is how normal markets work, but for a variety of reasons, including price caps, electricity spot markets aren't producing prices high enough to trigger reassuring levels of construction. Following a recent penetrating review of the state of power markets, Professor Paul Joskow—dean of all power economists and one of the main architects of electric deregulation—concluded that tweaked spot markets alone would not supplant other climate policies. "We might as well face this sooner than later," he says.[37] So, in the majority of restructured Big Grids, two other mechanisms that provide more assured resource adequacy are being used.

Pros, Cons, and Trade-Offs in Long-Term Markets

The second sales mechanism—and the fastest and surest way to trigger an investment in a restructured market—is a long-term contract. For generating plants, contracts are often called *power purchase agreements* (PPAs). PPAs usually buy most of both the energy and capacity from a plant with payment tied to various performance requirements. An "off-take" PPA paying for the majority of the output of a renewable plant at a contractually specified price virtually guarantees that the plant will be built.[38] The same goes for storage resources of all types, transmission lines outside regulation, and many other long-lived power assets. For this reason, the vast majority of IPP power is sold under PPAs rather than one hour at a time on the spot markets. In many restructured markets, long-term contract trading accounts for 90–95 percent of all revenues.[39]

One of the most successful U.S. clean-energy policies to date, renewable portfolio standards (RPS), works by requiring electricity retailers to sign long-term contracts for green power that quickly enable new plant financings.[40] Another increasingly large fraction of renewables are purchased directly by corporate and other buyers voluntarily, also under long-term PPAs that beget new plants. Of the 11,500 MW of large-scale solar and wind added in the United States in 2018, more than half (6,500 MW) was purchased by large corporate buyers using contracts.[41] Contracts are especially important for accelerating the uptake of technologies when they are relatively new and untested, or the demand for their services isn't easily projected. In the first generation of carbon-free and near-carbon-free grids, this will frequently be the case.[42]

Nonetheless, the revenue-assuring feature of contracts that makes them so good at enabling investment also represents their weakest link. When prices or market conditions shift—and especially when technological change or oversupply causes prices to drop over time—contracts are often seen as locking buyers into costs that are too high or technologies that are inferior. Indeed, some coal-fired IPPs are able to remain in operation because they have long-term sales contracts that continue to pay them rates they could not otherwise get in the market. Even a corporation buying solar energy at today's prices has to resign itself to the fact that a few years from now, prices will have dropped, and they will then be paying more than whoever is signing a fresh contract. And because contracts often contain unique features negotiated between buyer and seller, these bilateral documents are difficult to compare to one another or transfer to other parties, and therefore are not great for market liquidity.

In Chapter 3, we reviewed the much-lamented track record of energy demand forecasters, but the poor performance extends beyond forecasting to many government-produced energy plans and the policies that flow from them.[43] The decision to sign a long-term contract is often the result of one of these policies or plans. When this occurs, and prices subsequently drop substantially, the buyer (usually a distribution utility) and the policies that prompted the contract are both sharply criticized for causing customers to "overpay" relative to current prices. Something akin to this happened when the Spanish government mandated distribution utilities to buy PV power at fixed prices in 2004. The government later viewed these contracts as too expensive and refused to pay, triggering a huge round of lawsuits.[44] Even larger problems of this type occurred in the 1990s, when the federal law known as PURPA required utilities to sign twenty-year purchase contracts with combined heat and power (cogeneration) and early renewables plants. These contracts helped finance the first generation of IPPs, which was PURPA's goal, but an ever-widening gap between contract and market prices created bitter opposition to this requirement from utilities and consumers.[45]

In some policy circles, this history has created a mental association between any sort of government planning, long-term contracts, and projects that flow from them. All of these are seen as inevitably flawed and to be avoided in favor of relying on competitive markets that don't lock in prices for so long or, put differently, transfer more risk to competitive sellers rather than utility buyers and captive ratepayers. Moreover, there are still many questions surrounding the most affordable and reliable mix of carbon-free resources for any one region. So, even the best of plans is certain to mispredict something important.

Some of these problems with contracts can be reduced by awarding them competitively, which is now the norm in almost all settings, and also by shortening their duration. Nonetheless, concerns with contracts have led to another mechanism for providing revenue guarantees sufficient to finance at least some new clean resources: *capacity markets*. The idea behind these markets is to create a standardized product much like one year's worth of payment under a long-term contract. Then, instead of signing a long-term contract, sellers bid the capacity from existing and planned new units into a centralized regional auction. Buyers who need capacity also submit their needs for capacity in coming years. The auction operator selects the cheapest supply bids for each future year that add up to the total demand requested by buyers, and the highest needed ("market-clearing") price is paid to all successful offerors. These markets, run by neutral agents such as regional transmission operators (in the United States) or the government (in the UK), typically buy for delivering three to five years into the future. There are three such markets in the United States, each covering large regions, and about fifteen in the EU, each centered on individual EU countries.[46]

Capacity markets sound simple, and as we'll see, they provide some real benefits. However, they are *extremely* complicated animals that present several thorny, hotly contested implementation issues. Most importantly, several of these issues involve a deep-seated conflict between the competitive goals of capacity markets and policies that promote decarbonization.

Like most markets, the basic purpose of capacity auctions is to buy from the suppliers whose prices (as recognized by the current market system) are cheapest. Unpriced externalities—now including carbon—don't count. You can't determine what's cheapest unless you standardize the attributes of the capacity offered in auctions, and since generic power capacity has been a pretty standard product in the industry for decades, it was the logical way to start. Then, if you really want to find what's cheapest, you let every type of capacity offer that meets the standard attributes participate in the market.

The first few generations of capacity markets, which are mainly what we're living with today, thus intentionally define capacity as pretty much anything that provides electricity when it is demanded—"a commitment by a power supplier resource to provide energy when needed" in the words of the world's largest capacity market operator, PJM.[47] "These resources include," PJM continues, "new and existing generators, upgrades for existing generators, demand response (consumers reducing electricity use in exchange for payment), energy efficiency and transmission upgrades."[48] In short, almost anything that can supply on-demand large future energy blocks, or guarantee large load reductions at any time during a year, can bid in the best capacity auctions.[49]

As one would expect, capacity markets open to all types of resources often find market solutions much cheaper than simply building new plants (of any type). DR and EE programs often win because they are cheaper than new plants, even if their contribution is guaranteed only for a few years. As these markets efficiently locate pockets of low-cost, short-term resources, they lower capacity prices and make it harder for new carbon-free capacity investments to compete.

In this kind of market, fossil-fueled generators are evaluated only on their offer price. Some existing fossil plants can profitably bid into capacity auctions at very low prices, meaning that they are very likely to be part of the winning group awarded capacity payments, extending their operating lives. Then there is a higher-priced group of fossil-fueled generators whose bids "clear" in the auction only if the market-clearing price turns out to be relatively high. Similarly, newer types of carbon-free resources may also be more expensive, since they are still developing in maturity and scale.

With capacity markets searching for the cheapest rather than the lowest carbon capacity, there's no reason why they should promote decarbonization. In fact, these markets have enabled demand-side measures to offset fossil generations. So, there are certainly some environmental benefits.[50] Nonetheless, a couple of inherent features of the current designs tend not to favor renewables. As we saw in Chapter 6, the two dominant sources of carbon-free power (wind and solar) are inherently variable. Since the definition of capacity requires that you deliver one unit of power whenever requested, a wind or solar plant with a 1 MW generator can't promise to provide 1 MW of capacity. Instead, you must build about 3–10 MW of these resources in different locations so that this collection of plants will always be able to supply at least 1 MW if asked.[51] From the standpoint of providing capacity, wind and solar are therefore not nearly as cheap as they are at providing "as-available" energy sold in the separate energy markets mentioned above. Put differently, the basic idea of buying plain-vanilla capacity plant by plant is a vestige of the old Chapter 6 design paradigm; in the new paradigm, resources are blended together to make reliable supplies, not purchased one plant at a time.

Other features of current capacity markets are also unhelpful.[52] For technical reasons, prices in capacity markets tend to be highly volatile and unpredictable, sometimes changing dramatically from one year to the next. In the largest U.S. auction, prices dropped 85 percent between 2011 and 2012, almost doubled the next year, and then rose almost 500 percent the year following.[53]

In addition, the farthest ahead you can sell your capacity in U.S. markets is usually five years, and often it is only two or three. These two features couldn't be *less* helpful to new, long-lived clean resources that are trying to obtain financing as cheaply as possible.

The results of many recent auctions bear all this out. The last PJM base auction in 2018 cleared roughly 5,000 and 30,000 MW of existing coal- and gas-fired plants, respectively—more than 75 percent of all winning supply bids. About 14,000 MW of DR and EE were also selected, but only 205 MW of PV plants won; wind did a little better at 1,300 MW.[54] Similarly, 85 percent of the winners in the UK's 2020–2021 capacity auction were existing coal, gas, and nuclear plants; the remaining winners included two more new gas plants and 500 MW of battery storage. Noting that twenty times as much existing coal capacity as new gas capacity won, one power executive said that the auction "signals a willingness to sweat these [old coal] assets until the bitter end."[55]

Fixing the Long-Term Markets

Horrified by these outcomes, climate-action proponents are pressing hard for changes and alternatives to capacity markets so that they somehow help force coal plants to retire and simultaneously accelerate new clean resources, balancing these goals against cost and reliability. Market designers have responded in two very different ways. The first is to recognize that the current standardized, old-school definition of capacity isn't correct for the future. As one U.S. expert group put it, "Capacity has never been a very well-defined term, which makes it subject to stakeholder influence, and the balance of stakeholder interests supports conventional resources rather than new technologies."[56]

To rectify this, the theory goes, the generic concept of capacity should be disaggregated into a number of more complex but appropriate products that provide services future system designers will use to plan new-paradigm systems. For example, capacity auctions don't reward plants that can adjust their output very quickly to balance wind and solar, but this "ramping" ability is an extremely valuable service that batteries and hydro can supply but coal plants cannot. Another idea is to define capacity units by season, allowing wind and solar to take advantage of their seasonal strengths.[57] By disaggregating capacity into many products and buying the required amounts of these more complex products—the theory goes—the capacity market will fund more new-paradigm resources.

In fact, as disaggregations continue, the bright line between energy and capacity markets will blur into a series of interconnected "power services markets" with exotic names like *synchronous inertial response* and *steady-state reactive power.*[58] The future price of the residual product called energy may continue to be driven low by wind and solar, but other new products will—if everything works as hoped—produce the revenues sufficient to prompt the construction of each new-paradigm power asset called for in the least-cost system.

By now, there is near-universal agreement that capacity and energy markets must move toward disaggregation, but it is a slow, halting process.[59] Every market for a newly defined product must have rules in place, including checks to make sure that the new market is competitive, pricing backstops in case it isn't, and supply backstops in case it doesn't produce adequate quantities. California's market operator concluded that productizing fast-adjusting resources was a good idea in 2011 and petitioned the U.S. federal regulator to start a market for this new product; it finally won approval to begin sales in November 2016.[60] Meanwhile, fossil-energy advocates often do everything they can to slow down any new product or market rule that hurts their prospects. At best, note rule changes move slowly; at worst, they act to reverse good climate policy. For the rapid progress we need, many climate-action proponents argue, capacity markets need a way to target carbon-emitting resources directly.

The response some market designers give to this admonishment is straightforward: if carbon free is the capacity resource attribute you want, put a value on it directly. Put a price on carbon that all carbon-emitting plants must pay, and this will automatically make these resources more expensive and less likely to win in auctions. Don't try to do indirectly what adding a carbon price and an efficient market will do directly and automatically: raise the price of fossil resources high enough and put these resources out of business. That way, you'll have both economic efficiency and zero carbon.

In fact, a surprising number of countries and "subnational units" have adopted carbon pricing in one form or another. The latest World Bank tally counts forty-five countries and twenty-five subnational units that have adopted, or are scheduled to adopt, some kind of carbon price. By 2020, the tally finds, about one fifth of global greenhouse gas emissions will be subject to a carbon price.[61] Thus far, however, only six countries have set prices at or above $40 per ton—a level that approaches the social cost of carbon and promotes rapid turnover. Several experts have also proposed clever ways to insert carbon charges automatically into energy and capacity market auctions, but so far these have not been adopted.[62]

As a result, carbon pricing provides little solace to climate-action proponents, especially in the United States. But the picture gets murkier still when the debate extends to consider the effect of dozens of other state and federal policies that affect the price at which each type of capacity seller might be willing to bid. Opponents of clean energy argue that capacity markets are already biased against them because solar and wind already are advantaged by tax credits, R&D subsidies, and other policies that enable these sources to bid lower than they would absent these supports. Furthermore, fossil interests argue, when a clean source gets a contract for part of its output thanks to something like a state RPS policy, the remainder of that facility's output is effectively subsidized and can bid a price that is unfairly low. Clean-energy proponents counter by noting that fossil fuels and nuclear power have received decades of state and federal subsidies of their own, and that fossil plants have many more unpriced negative externalities than renewables, but because fossil subsidies are less visible and direct, they are often not considered in these debates.

The final dimension of this meta controversy involves whether the sellers and buyers of capacity in an area are barred from, or penalized for, entering into bilateral contracts and bypassing the capacity market. It may seem odd that public policies would ever prevent a willing buyer and willing seller from voluntarily entering into an agreement that both of them believe is advantageous, and indeed such policies are generally rare. They are implemented in some U.S. regions with capacity markets out of a belief that capacity auction prices will not be efficient or fair unless everyone buys and sells in the same central market. Predictably, sellers and buyers who prefer the stability of long contracts regardless of price lock-in despise these requirements. With respect to decarbonization, they cut both ways, but on balance, they are probably unhelpful.[63]

Alongside the reality that different capacity sellers receive an uncountable number of diverse negative and positive subsidies, this creates a conundrum. On the one hand, most market designers prefer that everyone buys and sells in one market. Yet, various bidders have (or are alleged to have) subsidies of one form or another. U.S. regulators, which have approved these markets under the premise that their prices will be neither too high nor too low, have reacted by adopting rules designed to prevent sellers from using *identified* subsidies to lower their bid. In other words, they force some "subsidized" bidders to bid no lower than a minimum price regulators calculate to be fair to other "unsubsidized" bidders. And all of this must be done while regulators are also trying to ensure that the auction isn't subject to monopoly power exercised by buyers and sellers who are too large, insufficient numbers of buy or sell offers, or other sources of market power.[64]

In both the United States and the EU, all of this has proven to be a recipe for bitter disputes and litigation, much of it pitting fossil-fuel interests against the clean-energy community. There is always a long-standing underlying disagreement over which form of energy has been most subsidized over time, but it is particularly damaging to clean energy in the United States when its RPS requirements and tax credits are singled out as the primary subsidies that must be remedied. For example, the FERC recently ordered the PJM capacity market to factor in *all* renewable subsidies in its next auction as part of its policy to promote coal, prompting an immediate outcry and appeal by clean-energy advocates.[65] Meanwhile, in Europe, the European Commission is debating proposals that would limit capacity bids from some or all coal plants after 2025.[66]

This interconnected array of issues and disputes has created four main schools of thought on how to reconcile faster climate progress with Big Grid capacity markets. The arguments cut along two primary dimensions illustrated as the two axes in Figure 8-1. The vertical axis divides between those who argue that *significant* carbon pricing is essential—there is simply no successful decarbonization pathway without it—and those who think that it is unlikely and therefore can't be treated as essential—other policies and markets must play a large role. The horizontal axis divides between those who believe capacity markets can be made to work better than long-term contracts via product disaggregation and those who don't.

A series of specific long-term market reform proposals, recently issued by an umbrella group called America's Power Plan, map into the bottom half of the quadrant (i.e., don't *require* carbon pricing,

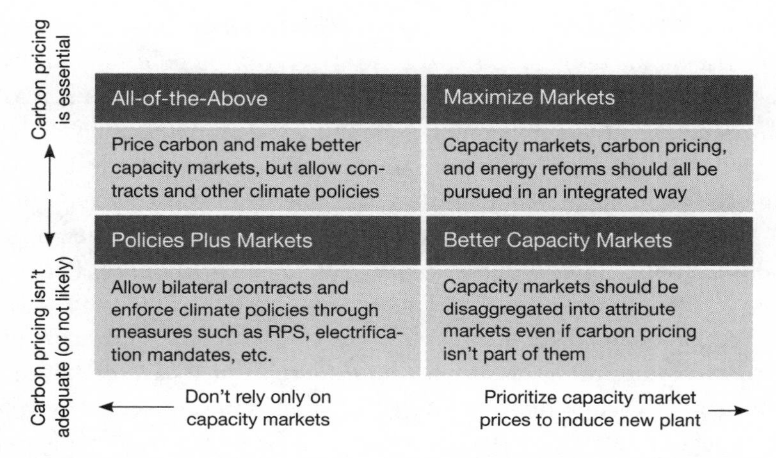

Figure 8-1 The Four Schools of Thought on Reconciling Strong Climate Policies and Capacity Markets.

although they fully support it).[67] One proposal, by market experts Rob Gramlich and Michael Hogan, avoids centralized capacity markets entirely and instead proposes a decentralized bilateral contracting marketplace.[68] Each vertically integrated utility or deregulated electric retailer negotiates and signs contracts for supplies that conform to local, state, and federal climate and other policies. One state might require its power sellers to offer only carbon-free energy by 2040, while another may engage in periodic energy or integrated resource planning that yields a specification for required least-cost purchases: *x* megawatts of renewable energy, *y* megawatts of storage, and *z* megawatts of DR, for example. Brendan Pierpont makes a similar proposal in the same series, adding a more centralized contract award process. Both of these proposals belong in the bottom-left quadrant of Figure 8-1.

Contrasting proposals in this series that belong in the bottom-right quadrant of Figure 8-1 come from experts Steve Corneli and Eric Gimon.[69] In both these proposals, there are centralized regional procurements for long-term (capacity) resources. To decide how much of what to buy, the independent regional auction operators use a state-of-the-art planning and grid simulation process that factors in all of the policy mandates the electric sector must meet. The process for selecting the resources to buy is much like the transmission planning process recommended in Chapter 7—as it should, since transmission is simply one (albeit unique) resource needed to create a least-cost, zero-carbon grid.

Perhaps someday it will be possible for experts to demonstrate empirically that one of these proposals works better in all markets, but I doubt it. As is often the case, the different proposals have their pros and cons. The reality is that there is no perfect way to engender the financing of carbon-free grid resources. Furthermore, there are political economy dimensions of each of these approaches that often make one or more of them off limits or extremely slow. However, where capacity markets are used, a set of best practices is steadily evolving. These best practices include allowing DR and EE to offer into these markets, allowing for bilateral contracting outside the market, allowing imported power to participate where it offers comparable security, treating variable renewable capacity bidders fairly, separately recognizing and rewarding environmental attributes, and extending the auctions far enough into the future to enable new resource financing.[70]

One important reason why no one long-term market approach will dominate everywhere is that each region's approach to financing Big Grid resources must interlock effectively with the distribution and sales portion of the industry. After all, it is the latter that must ultimately collect the customer revenues that pay for all of this. In other words, the structure, governance, and business models in the upstream part of the industry must work as seamlessly as possible—politically, economically, and technically—with the distribution utilities and retailers that make up the downstream part of the business. Given the urgency of reducing emissions, the imperative is simply to pick an approach that eliminates emissions as fast as possible at an affordable price—the same goal as system designers used in stage 1.

The Big Grid's Future

Unlike the Small Grid utilities we're about to examine, no one is predicting a huge disruption in the business models of Big Grid generators or transmission owners. The Big Grid may shrink in a few places that are rich in local generation and organize themselves accordingly, but as we saw in Chapters 2 and 3, Big Grid power is destined to grow in most places for decades to come. Big Grid business models will inevitably become more sophisticated, but at their core, they will involve revenues earned from contracts, spot markets, and / or regulated rate basing. It would not surprise me if many generation firms that are active now around the world remain leading players in 2050. This relatively secure growth scenario explains why the Big Grid is a perennially attractive sector for long-horizon investors.

Nonetheless, the Big Grid must go through its own physical and institutional transformation. It quickly has to retire or retrofit thousands of coal- and gas-fired plants and just as rapidly replace them with carbon-free generators and balancing resources. The transmission systems that host these plants need to adapt to this rapid shift with much more ambitious and integrated planning. Each region must also reshape its long-term capacity financing mechanisms to enable financing for a fully decarbonized system. These changes are not new business models for generation companies per se, but rather changes in the institutions and processes that shape their revenues, risks, and ultimate investability. These institutions and processes are inevitably tied to the new business and governance models we are about to examine in Part III. So, while Big Grid firms and institutions grapple with their own large changes, they are also on the receiving end of even greater challenges and disruptions downstream—challenges and opportunities to which we now turn our full attention.

PART III

Running and Regulating Post-Carbon Utilities

9 The Utility Business in Three Dimensions

In the United States, there are 174 investor-owned electric distribution utilities (IOUs); approximately 2,000 distribution systems are owned by municipalities and other government agencies, and 809 more are customer-owned cooperatives.[1] The layout of distributors in the United States makes a sprawling, beautiful mosaic of shapes and sizes (Figure 9-1). In the thirty-three EU countries, the landscape is nearly as diverse. "In some member states," one EU group writes, "there are hundreds of DSO's [distributors]; in other countries there may be only one or two."[2] Germany has 900, while the UK has only fourteen.[3] Australia has eighteen, and Japan and Korea have ten and one, respectively.[4]

If you're in charge of one of these, you have a plate filled with challenges. Sales or deliveries of power are flat or falling, while customers are increasingly making their own power and being enticed away by a new generation of nimble, digital competitors. So far, most of the competition is coming from lesser-known upstarts, but the likes of Amazon and Google and other large utilities are eyeing your territory as well. Meanwhile, policy makers are demanding that your delivered power come from clean sources and your system be more resilient against storms and cyberattacks; you simultaneously promote energy conservation and other public goods, enable new markets and innovation, and continue to provide universal service at the lowest possible rates. The grid is on the verge of splintering off microgrids here and there, and yet it requires millions of dollars of new investment. And this is before we've even considered

Figure 9-1 U.S. Electricity Distribution Utility Service Territories.
National Academies of Sciences, Engineering, and Medicine (2017) pp. 2–4. Used by permission. S&P Global Platts (2018).

sociopolitical trends, as we'll do a few chapters ahead. No surprise, then, when global consultant PwC reported that among utility executives, "concern about nearly all the major risks facing the . . . sector is rising."[5]

Today, it is still overwhelmingly the case that electric utilities make their money by either selling or delivering kilowatt-hours. Regulators set prices so that utilities typically recover all prudent out-of-pocket expenses and earn a fair return on invested capital or *rate base*. This gives rise to the oft-mentioned incentive for utilities to goose profits up by choosing capital-heavy solutions rather than cheaper approaches that involve more non-capital purchases—substituting capex for opex in industry parlance. Regulators have become quite skilled at counteracting this tendency, but regardless of how well they do, the real problem is this: what do you do when the *only* way to earn the profits that are built into prices is selling a product whose sales are flat to down?

Lest anyone panic, we should be clear that the *average* investor-owned utility is now on something of a roll. The cost of electricity as a percentage of average household budgets is 1.03 percent—its lowest share in thirty years. Average electricity bills haven't changed much, while the costs of most other household purchases have increased quite a bit. The cost of natural gas, the largest U.S. power-plant fuel, has actually *declined* substantially, dragging down the cost of the second largest fuel (coal), with renewables costs also dropping enormously. Alongside low interest rates and wage growth, all this has *lowered* average utility costs in real terms, creating an unusually large wedge between costs and revenues, even when the latter are flattish.

Utilities have mainly filled this wedge with investments in renewables, smart-grid technologies, and resilience. Utility capital budgets are at their highest point since the crash of 2008.[6] Since these capital expenditures boost rate base and therefore profits, utilities have enjoyed a golden moment of stable prices yet increasing profits. This description of a prominent New Jersey utility's plan by Wall Street analysts provides a recent example:

> This morning, PSE&G filed its $3.6B Clean Energy Future plan. This along with the $2.5B Energy Strong II plan is expected to drive the utility to the high-end of its 8–10% rate base growth target. The plan included $721M of "Energy Cloud" (AMI) spend that was incremental to the originally previewed $2.9B at the Analyst Day. The plan seems to have strong political support in alignment with Governor Murphy's push towards a cleaner New Jersey. Putting capital to work in the state is not a problem, with customer bills the primary restraint.[7]

Combined with the fact that many other asset classes have been much more volatile and less profitable, U.S. utility stocks have done extremely well lately. The average electric utility earned a very healthy 76 percent cumulative return on investment since 2012.[8]

The longer-term picture is far less stable. As we've seen, it is quite likely that customers will make a good fraction of their own power and the grid will splinter off many microgrids. These trends will make the system much more complicated and expensive to operate while kWh deliveries don't increase. In dollar terms, the challenge to utilities is to find new ways to grow revenues and their share price. As Chapters 10 and 11 will illustrate,

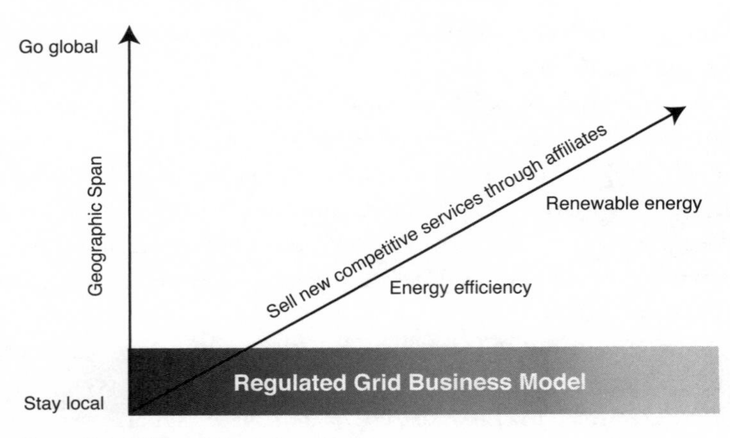

Figure 9-2 Utility-of-the-Future Strategy Space.

each new business model has its own method of offering new revenue prospects. But the challenge isn't only financial: the traditional business model, based on selling ever-more kWh, no longer describes the new multifaceted core mission of electric utilities.

In response, utilities are trying to chart a path away from their old business model to a new destination somewhere inside the strategy space illustrated in Figure 9-2. The three axes of this imaginary space are the three main directions along which a utility can adjust its business strategy.[9] The dimensions of Figure 9-2 cover the first fundamental questions a utility management must ask itself: What is my regulated business model (if any)? What unregulated lines of business will I pursue? And finally, where will I do all this?

The Business Model Rainbow

The first and most fundamental dimension of your strategy choice is your future as a regulated or publicly owned power distributor. Although microgrids and DG are making the grid less of an absolute natural monopoly, this business will remain regulated in some way for a long, long time. If you intend to stay in the business of running a distribution system, you need a business plan that meets all the policy goals that regulators require, along with adequate returns for your shareholders.

In *Smart Power,* I suggested two new business and regulatory models that might yield good future outcomes. I called the first one the Smart Integrator (SI) and the second one the Energy Service Utility (ESU). Both business models are designed explicitly to address the distribution system's role addressing the industry's main transformation challenge. The idea behind both business / regulatory models is that the company and the regulator stop viewing the main product that is regulated, priced, and sold as kilowatt-hours—what one company calls "commodity thinking."[10] The alternatives to selling kilowatt-hours in the two models are, however, quite different.

The animating vision on an industry of SIs is a thriving ecosystem of unregulated energy service firms that sell electricity itself along with energy efficiency, renewable energy, and other non-energy products such as home security. These competitive multiproduct retailers need a smart, controllable, information-rich distribution system to deliver their services. There may be several communications channels through which these firms can reach customers, but there will be only one set of power delivery wires. Many experts now propose utilities pursue versions of this model, naming them Network Orchestrators, Platform Utilities, Marketplace Advisor, among others.

This model addresses the industry's transformation challenge largely by leaving them to external policies and markets they facilitate in sophisticated ways. Decarbonization comes from mandates or fees on sellers and buyers, for example, while SIs create the sales platform. SIs must also facilitate greater resilience and the integration of distributed energy resources (DERs) through investments in their systems and the creation of localized markets and procurement.

The ESU is a business / regulatory model that is nearly the mirror image of the SI. Here, the utility *expands* its role as a provider of services to individual customers beyond the sale of invisible electrons. Its regulators allow it to sell energy efficiency, energy management, EV charging, energy-efficient equipment, and renewable energy to customers, all under approved terms and prices. The ESU must still operate a smart modern grid, much like the SI, but its primary revenue source isn't other companies using the grid, but rather individual customers buying retail energy services of one kind or another.

The ESU also addresses the main transformation challenges, but its role is more integral and less one of facilitation. It "owns" decarbonization of its supplies, resilience, and DER integration, and its regulatory and market incentives must be designed to reflect these added responsibilities.

The SI and the ESU were simplified regulated business models—almost caricatures really. A more refined spectrum of business models for distribution utilities of all types is illustrated in Figure 9-3. The key measure of change along the spectrum is the degree to which the distributor itself transacts with retail customers to provide services, often called *customer engagement.* As you move left along this spectrum, the utility's customer

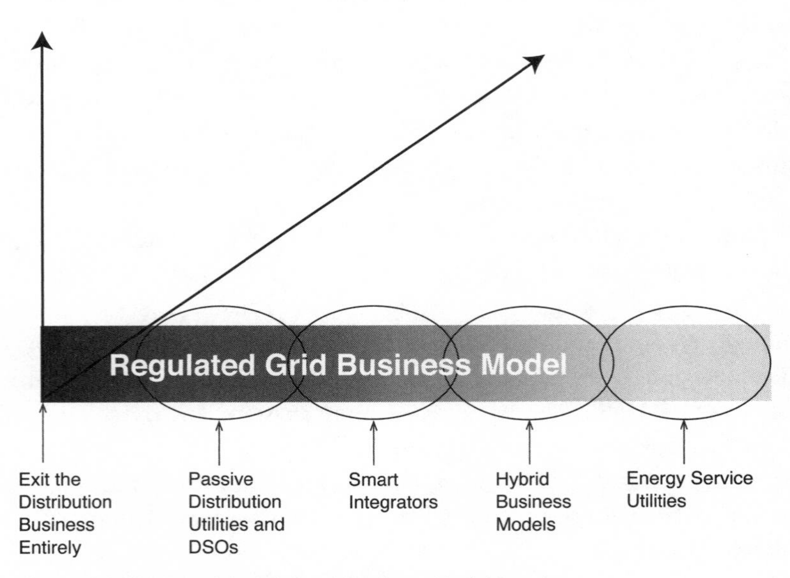

Figure 9-3 The Regulated Network Business Model Spectrum.

engagement declines until, at the end, it is out of the business entirely. Each business model along the spectrum is accompanied by a matching regulatory or oversight system and a somewhat different ecosystem of surrounding energy service companies, whether deregulated affiliates or fully independent third parties.

At the far right end of the rainbow, the ESU distributors just described choose (and are allowed by law) to sell the full range of energy services to customers. These services are likely to come mainly from closely associated partner companies, but the utility is seen as a business-to-consumer provider and, most importantly, as the brand. The ecosystem surrounding the ESU consists of two types of companies: those that are part of the utility's service delivery operation (but are not visible to consumers as a separate brand) and those that compete with the utility brand for whatever mix of services they offer.

The next stop along the spectrum is a utility that views itself not as the primary provider of services, but nonetheless as an integral, engaged partner in delivering energy services from other providers. This could be called something like the Customer-Direct or "strong partnership" SI, but I use the simple term *hybrid* for this model. Hybrids do not own or self-brand most of the services they provide, but they form strong partnerships with independent brands and sometimes co-brand with these "preferred providers." The ecosystem is similar to that surrounding ESUs, with a little less rigid utility integration.

The next portion of the spectrum is occupied by SIs. This somewhat obtuse name refers to utilities that continue to serve as platforms and network orchestrators but have stepped back from being the primary branded transactor with customers. Most energy consumers transact with independent, deregulated service providers for electricity purchases and all other related products [energy service companies (ESCOs)], and the utility delivers power and runs the network in the background, unseen by retail buyers. The utility transacts mainly with competitive third parties, so it is now a business-to-business provider or wholesale brand rather than a retail brand presence. This is by far the most radical shift in focus along this business model spectrum.

The next leftward step on the spectrum is to a passive ownership business model accompanied by an independent (nonprofit) Distribution System Operator (DSO). Here, the distribution utility owns but does not operate its own system, giving it virtually no engagement with retail customers at all.[11] This mimics the prevailing business / regulatory model for the Big Grid. In most of this grid, utilities are allowed to *own* both generation and transmission, but they are not allowed to *operate* their own transmission because they may discriminate in favor of affiliated generators. Without independent control, for example, a utility could operate its transmission system so that a rival power generator had a difficult time getting permission to sell its power to a buyer, while its own plant could easily transmit its juice to that same buyer.

The ability to discriminate could be policed by regulators, and indeed there are extensive, well-enforced rules that require "open access" transmission in most of the West. However, policy makers in some areas weren't comfortable relying only on oversight. So, they went a big step further by requiring transmission owners to hand over day-to-day control to a dedicated nonprofit system operator known by the acronyms TSO, ISO, RTO, or TO. Transmission owners continued to be paid their approved tariff rates by whoever the system operator scheduled to use their lines, but they had no control over who used the lines. Their job became maintaining or replacing their existing lines and planning and building new ones as needed—somewhat exaggeratedly described as *passive ownership.* Transmission-owning utilities didn't love losing day-to-day control of their power lines, but they agreed to it as an alternative to an outright ban on their simultaneous ownership of generation.

The proposal to create DSOs comes from the same line of reasoning applied to distribution systems and their surrounding markets.[12] Unregulated affiliates of distribution utilities will be active competitive sellers of energy services, while other utilities will offer these services under their own distributor brand. Analogously to generators on the Big Grid, a distributor could discriminate against a competitive rival selling, say, car chargers by programming its local grid to work poorly with its rival's chargers, thereby favoring its own charger brand. To prevent this, some experts have recommended that distribution systems cede control of their systems to an independent nonprofit operator directly analogous to the TSOs that manage transmission. This business model keeps utilities in the distribution business, but barely. Their role is reduced to cooperating with a DSO that has the full authority to run the system, maintaining their systems, and planning and building new capacity.

Thus far, policy makers have not yet adopted the passive distributor / DSO business model anywhere in the world. Much of their reasons have to do with concerns that each local utility has accumulated extremely specialized knowledge about how their system works, how to operate during storms and restore the system afterward, and many other detailed capabilities. Under a DSO model, it is first necessary to figure out how to divide up the moment-to-moment activities of a current utility into those handled by DSOs and those kept by utilities, and gaps in how these tasks are allocated or frictions between the two organizations could prove disastrous.[13] Many utility employees have long histories and deep knowledge working with parts of their systems that cannot be easily transferred. So, many utility employees would have to go to the DSO.

The cost of establishing and maintaining an independent DSO has also given policy makers some hesitation. To create a DSO, you must build a new, totally independent distribution system control center and then take over major control from each nearby utility's current control room, while the latter continues to provide some functions and thus remains operational. Beyond the mere cost of duplicating facilities, the communications and control systems necessary for this arrangement are quite complex, involving extensive backup systems and security protections. This same approach is used to create independent TSOs, and these organizations routinely cost hundreds of millions to build and operate.[14] No one is quite sure exactly how much DSOs will end up costing per customer.

Accountability is also an issue, especially for something as essential as reliable, resilient power. Under the current system, regulators and governors have pretty much a single point of accountability for electricity problems in any one area. In the wake of problems in a DSO structure, regulators would have to sort out what went wrong, who was to blame, and how to fix it—a difficult and highly contentious process. After extensive consideration, New York state regulators concluded that it was too risky to reliability, too disruptive, and potentially too expensive to adopt a DSO structure.[15] Similarly, the Australian Energy Market Operator concluded that "it is unclear whether introducing this level of complexity and associated costs to customers . . . would be warranted."[16]

Public Power and Cooperatives

Governmental and cooperative utilities confront most of the same forces of disruption as the IOUs and also recover their costs via kilowatt-hour sales. Although they may not feel quite as much pressure to evolve their business model to something along the ESU–SI axis, changes in the way IOUs interact with their customers will trigger similar customer demands in their own areas, and they too will need to evolve. They are also positively affected by new movements to support community-based power purchasing, though these movements center on new supply rather than distribution.[17]

A recent report by an elite group of nonprofit experts compared the factors affecting business model changes for municipals and co-ops with those affecting IOUs. On the negative side, the experts noted that munis and co-ops were sometimes constrained by their ownership of power plants or power contracts, political limits on their operations, scarce investment capital, and the short-term need to keep rates as low as possible. On the positive side, these types of utilities have no profit motive and therefore are free to move in any way that they think will make their customers better. Because they are nonprofit entities, they face no pressure to earn a return on equity, though this also gives them no cushion to fall back on. Even more so than IOUs, they are unlikely to be allowed to fail by their government owners or creditors. Nonetheless, they have debt to be serviced, new investment demands, and bills to pay. In addition, utilities in these categories usually do not need regulators' permission to change business models or operations. So, they can move faster than a typical IOU–regulator duo.[18]

To the extent that publics and co-ops have steered toward a new business model, they mostly favor the ESU approach. Two municipal utilities—Seattle City Light and the City of Fort Collins, Colorado—have already moved faster toward an ESU business model than almost any IOU in the United States. Fort Collins, for example, has city-wide energy policy that integrates energy efficiency and transportation reform activities directly with utility operators.[19] After surveying alternative business models, the association of U.S. municipal utilities, APPA, argues that "communities would be better served having utilities play a central part."[20] The U.S. trade association for electric cooperatives makes an even more forceful case for ESUs, which it calls a "consumer-centric utility."[21] Public power and cooperatives may also have other opportunities to expand in the coming transition—a topic we return to later.

Finally, none of these new business models are guaranteed to work every time. There are undoubtedly scenarios in which future regulators and stakeholders will not be able to balance competing interests. When this happens, that IOU's leader may well decide that being in the electric distribution business under any model isn't worth it. Going out of business and closing the doors isn't an option, but they might find another IOU willing to buy their system or transfer it to a government or cooperative utility.[22]

New Products and Horizons

Beyond distribution business / regulatory models, IOUs have two more strategic dimensions to consider. The first of these is diversification into unregulated products. Although it is not widely discussed outside utility circles, many IOUs have started to do this quite extensively—some at breakneck speed. According to GTM research, utilities in Europe and North America have invested more than $2.9 billion in more than 130 distributed energy companies since 2010, and both the pace and size of the deals are accelerating. The most active acquirer, France's utility conglomerate Engie, bought a startling fifteen separate companies, but the United States's Exelon is not far behind with thirteen buys. Forty-two utilities have made unregulated DER investments; ten of these have made five or more.[23] Regarding Engie, Accenture writes:

> Facing significant market transformation, ENGIE sought to review its retail operations and transform the digital experience for its business and residential customers. Its transformation includes reimagining the delivery of traditional commodity services, such as selling gas and electricity. It also includes designing new services to disrupt the market, challenge competitors and new entrants and, ultimately, position ENGIE to move into new markets and regions. Among the possibilities: servicing the new era of electric and self-driving vehicles, connecting the coming wave of home solutions in ways that delight customers, and helping customers in their energy transition projects.[24]

Utilities are not simply buying other companies—they are also starting their own in-house new business units, funding external incubators, and forming investment funds and coalitions. National Grid has an unregulated new product venture located right in Silicon Valley; Exelon has both an external investment fund and an in-house incubator; and many other utilities are following suit. Energy Impact Partners, which I work for, is a coalition of fourteen utilities that own shares in more than twenty-five new clean-energy startups, all directed toward providing benefit to their distribution system customers and new revenue opportunities.[25]

The third dimension in the strategy space is the geographic sphere of operation. Regulated utilities are tethered by law to a prescribed service area—generally a contiguous part of a state (in the United States) or country (in the EU and Asia). However, a utility-holding company might nowadays acquire many distribution grids in disparate locations, each answering to a local or national regulator. To cite just a few examples, Warren Buffett's Berkshire Hathaway owns four large utilities in the United States and one in the UK; U.S. utility Exelon owns six large distribution networks in different U.S. states; and Enel, the Italian mega-utility, owns majority stakes in more than twenty distributors in nine countries.[26] Since around 1990, the number of companies owning one or more electric distributor in the United States has shrunk from more than a hundred to forty-nine—sixteen in the past ten years alone.[27] Similar consolidation has occurred in Europe.

A utility's purchase of another distribution grid obviously brings it into a new geographic area. Similarly, a decision to enter an unregulated business greatly influences or even dictates geographic operations. The new venture needs to go wherever its strategy and opportunities take it, whether into a single market near the distribution area or global cyberspace. However, regulators may frown on the new business operating in the same service area as an affiliated regulated distributor, constraining geographic operations to everywhere the regulated affiliates aren't. This may have implications for harvesting synergies between the new ventures and the regulated wires—a topic we return to in Chapters 10 and 11.

Many utilities are already well into their journey toward one quadrant or another in this overall strategy space. Companies such as Wisconsin's Alliant Energy have declared their intent to move toward an ESU model, retain their geographic focus, and avoid competing with deregulated affiliates. E.On, Germany's largest utility, leans toward an SI, as does Chicago's Commonwealth Edison. AGL, an Australian "gentailer," sold off its regulated networks and sells competitive energy services on the *yz* plane in Figure 9-2. Avangrid separated itself into two large divisions: networks (regulated utilities) and renewables (unregulated). Many utilities have already expanded in their region or globally. "Over the last 20 years," two European experts write, "many European utilities have reached a high degree of internationalization. Even smaller players like Swiss utility Axpo are present in more than 20 markets (based on company data from 2015). Half of the 16 European players now earn over half their revenues outside their country of origin."[28] Conglomerate utilities are also organizing operations into regulated and unregulated business units, such as Enel. This utility, based in Italy, now has four global divisions: power generation, which owns and operates traditional power plants; renewable energy; infrastructure and networks, holding regulated distributors; and

Enel X, "intended to foster greater customer focus and digitization as accelerators of value."[29]

Toward Customer Love

Where is the electricity customer in all of this? The spectrum of business models shown in Figure 9-2 has been described in dry, business-school terms as if utility customers will be passive cogs in the wheels of these new firms. For customers, these models aren't abstract points in a strategy space; they are descriptions of the products they buy, who provides them, and their total customer experience as an energy consumer. What about this perspective? Under the ESU model, the customer sees the utility as a sort of one-stop shop for electricity itself, a variety of other related products such as EV charging and smart home management, and of course electricity delivery through the exclusive set of wires (top row of Figure 9-4).

Under alternative models, customers tend to buy all energy services from their ESU utilities instead of from a single ESCO (middle row) or from specialized providers of each of the main types of energy services

Customer Perspective Label	Electric Commodity	Local Solar Power	Electric Vehicle Charging	Smart Home of Building Energy Management	Electricity Delivery Service
Utility Is My One-Stop Shop for All Energy Services	HOMETOWN Energy	HOMETOWN Energy	HOMETOWN Energy	HOMETOWN Energy	HOMETOWN Energy
Independent "Chesco" Is My One-Stop Shop	CHESCO Total Energy Services	CHESCO Total Energy Services	CHESCO Total Energy Services	CHESCO Total Energy Services	HOMETOWN Energy
Multiple Independent Providers of Energy Services	HOMETOWN Energy	LOCAL Solar Power	LOCAL Electric Vehicle Charging	SMART Building Energy Management	HOMETOWN Energy

Figure 9-4 Utility Sector Business Models as the Customer Sees Them.

(bottom row). Under the latter model, they buy electricity from competitive electric retailers, solar panels from solar companies, smart thermostats from Ecobee or Nest, and so on.

As customers see it now, the utility landscape resembles the first or third row of Figure 9-4. If you live in an area with traditional regulated electric service, your only option is to buy the bundle of electric commodity and delivery from the local utility. If your utility expands to offer other services, it will become an ESU and begin to look like row 1. If you live where retail choice is allowed, your world looks like row 3: you may already choose a separate supplier for electricity itself and can also choose other services from competitive specialty retailers operating in your area. For utility strategists, the billion-dollar question is whether customers who live mainly in one of these two worlds today will be attracted to one with less regulation, buying a wider set of services from conglomerates or specialists. Alternatively, do they want to see their utility expand to offer more services itself?

We get hints at the answers to these questions from surveys of utility customer preferences and satisfaction. These surveys, which every utility conducts routinely, overwhelmingly find that the single most important determinant of satisfaction is low prices or, more precisely, low monthly bills. Regardless of whether your provider is a nonprofit city, regulated utility, or deregulated retailer, when rates go up, satisfaction goes down. However, we don't yet know whether any of the business models in Figure 9-2 will systematically yield the best combination of low cost and great service or whether the outcome will vary across utilities, states, and countries. There are far too many confounding unknowns to make confident predictions.

A few added hints can be gleaned from polling on customer satisfaction—or CSAT as it is known in the business. For decades, all types of electric utilities have used market research firms to collect CSAT scores that rank them against other types of customer-facing firms, other current utilities, and themselves over time. These statistics mainly show that utilities rank lower in customer satisfaction than most other industries. One 2017 ranking from J. D. Power shows utilities near the bottom of twenty-one consumer sectors (with a score of 720), below airlines (with a score of 760) and car dealers (with a score of 820).[30] Scores have been improving markedly in the past five years—probably because rates have not been increasing—but are still well below most other sectors.

Many strategists point to these numbers and suggest that utility customers would welcome supply alternatives. However, when customers are asked about their satisfaction with their present utility supplier, the aggregate results suggest a different narrative. Deloitte Consulting's survey found that 73 percent of electricity customers rated their utility excellent or very good; only 4 percent rated it not good or poor. The Smart Energy Consumer Collaborative (SECC) found about the same levels of "high and moderate satisfaction," with "low satisfaction" below 10 percent. Many utilities have watched their own unpublished CSAT scores creep up over the past few years. It is only when customers are segmented into marketing cohorts that one finds pockets of unhappiness, as in the segment the SECC calls "movers and shakers," characterized as "selectively engaged, college educated, and unemployed."[31] Interestingly, in these segmented surveys, millennials report almost the same rates of satisfaction as older customers.

Most consumer-facing businesses are now racing to create a customer experience that allows all types of business interactions to occur digitally at any time on any device, especially on mobile phones. The many ways businesses can communicate with customers and the vast amount of data they can now collect and analyze on individual interactions are leading to sophisticated ways of viewing and measuring customer experience. Accenture's latest advice to electricity companies suggests that they should use detailed measurable data [key performance indicators (KPIs)] to move customers not merely from satisfaction to the stronger feeling of brand affinity, but rather all the way to *love:*

> To become a customer-experience-driven organization, and drive customer retention and loyalty, energy providers need to adopt forward-looking customer experience KPIs in their scorecards. In the era of liquid

> expectations, consumers are benchmarking their experience with energy providers against those with other service providers like their retail bank or Uber car service. Energy providers are competing against customer experience leaders across all industries. It's no longer enough to create something that people like—energy providers need to craft experiences that people love.[32]

Accenture goes on to offer customers a Love Index that utilities can use to rate their service.

The lure of electric romance notwithstanding, perhaps the most relevant information comes from surveys that ask utility customers how interested they are in buying from other suppliers. The results are decidedly mixed. In one recent survey, 69 percent of all customers preferred their electric utility over other power vendors; in another, 73 percent said they would "consider a provider other than a utility / electricity provider, i.e. retailer, phone, cable provider, or online site." Asked in one survey for their preferred vendor of solar panels, energy storage, or wind power, customers voted for their utility over "independent energy service companies" by 57 to 45, 56 to 43, and 51 to 40, respectively.[33]

In a related exercise, the SECC asked customers whether they preferred to buy energy-related products and services from an online platform operated by their local utility versus buying them elsewhere. Only 23 percent strongly preferred the utility platform, leading the SECC to conclude that "energy providers will need to fight for their share of platform-enabled products and services." Accenture's especially detailed survey shows that utilities were scoring excellent or good about two-thirds of the time on providing key modern-day motivators, such as "seamless customer experience" and "products and services specific to my lifestyle."

If these imprecise results suggest anything, it is that customers are not monolithic, nor have they made up their minds. Our shopping experiences have evolved dramatically over decades, from neighborhood stores to department stores to malls to e-commerce. With accelerating digitization and bandwidth, the electricity buying experience is sure to change in ways we can't predict.[34] Right now, customers are clearly open to buying a wide range of energy-related products from their utilities, but they are also clearly open to alternatives. If regulated, public, and co-operative utilities can offer multiple services at competitive prices, and regulators allow it, utilities' generally good reputation positions them to make the ESU model an option. But the battle has hardly been joined, much less won.

10 The *Really* Smart Grid

Toward the center end of the business model spectrum, the utility acts as a Smart Integrator (SI). Utilities step back from retail customer transactions and hand over the market entirely to competitive providers of electricity and related services, which we're calling ESCOs.[1]

An industry future of SIs supporting ESCOs hinges on a widely held view that today's utilities won't be capable of providing customers with the cutting-edge, mass-customized products Accenture described in Chapter 9 nearly as well as a market served by a stable of hungry competitive rivals. There are three primary reasons.

The first is that utilities have spent a century or more providing one simple homogeneous service with almost no direct interaction with customers or product diversification. They did not need to build brand allegiance or learn marketing—two critical skills in the new energy service marketplace. They have neither the culture nor the capability to engage customers or offer product innovation, especially at the speed of private, digitally enabled firms. Meanwhile, modern electricity consumers have become much more varied and sophisticated in their preferences. As Accenture puts it:

> While price is still the bottom line when it comes to energy, [electricity] consumers are exhibiting shifting ideology. Increasingly, they want tailored products, services, and experiences that align with their personal values. Today's consumer is better informed than in the past, with unprecedented

access to information and an increasingly amplified voice through online and social channels . . . The possibilities for new energy-related value propositions are growing exponentially thanks to new energy management technologies, increased information on usage, microgeneration options and increased awareness of where energy is coming from. For energy providers, the challenge is to gain insight into how the values of specific groups of consumers are changing and to identify new ways to engage and capitalize on [these] shifting ideologies.[2]

A second major reason is that every product offered and price charged by regulated utilities typically goes through an approval process that is often lengthy and highly politicized.[3] Yet, any mass-customized SI marketplace will have hundreds of rapidly changing service options. Realistically, it is not possible for regulators to preapprove pricing and terms for every service bundle that customers might want in a fast-evolving market. Even if they somehow could, the time and resources required, and likely political compromises involving price or services, would make the results far inferior to letting the market sort things out by itself.

Finally, millions of electricity consumers are poised to become partial producers of power, or *prosumers*. The service bundles utilities offer now must include economic arrangements that allow customers to make some or most of their own power and to trade surplus power. Utilities and regulators may be okay at managing "one-sided" markets in which all customers buy from their utility, but that doesn't mean they can succeed managing markets consisting of thousands of sellers of surplus rooftop solar alongside millions of EV owners selling power back to the grid from their vehicles. As one group of experts wrote, "direct control by utilities is not practical nor desirable at this time for the thousands if not millions of DER systems in the field."[4]

For these reasons, the SI utility doesn't try to be the supplier of customized product bundles; it manages the distribution grid so that competitive ESCOs supply them well. The utility's role in this vision—also sometimes referred to as the transactive utility, platform utility, prosumer utility, network orchestrator, and other names—is illustrated in Figure 10-1.[5]

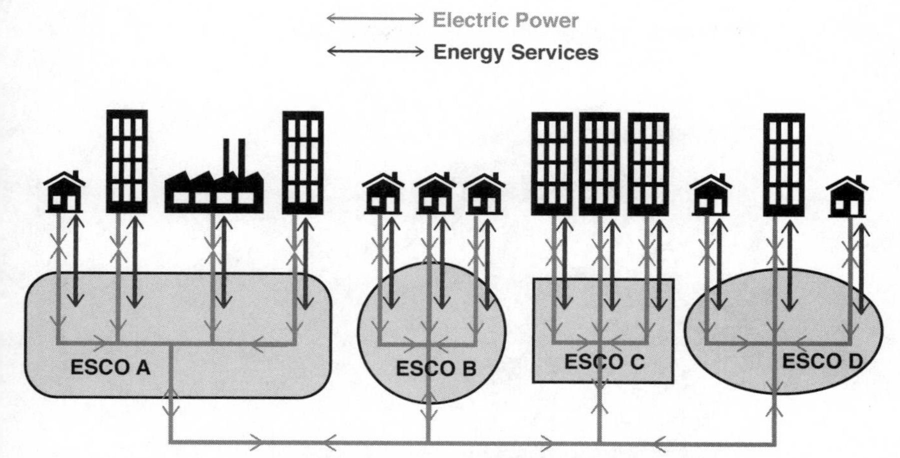

Figure 10-1 The Downstream Industry Structure in Smart Integrator (SI) Areas.

The home and business prosumers at the top of Figure 10-1 buy and sell their power through the ESCO, or trade with other prosumers directly (more on this below). These transactions are symbolized by the lines in the figure originating in the ESCOs and going to each customer. In theory, the pricing of all these services and all other trading between ESCOs and prosumers is unregulated and uncontrolled, allowing consumers and ESCOs to experiment and improve the customer experience.

Of course, all power bought and sold by ESCOs and prosumers has only one path along which to flow: the distribution grid in the areas it serves. Since ESCOs must use the local grid to move power around, they will have to pay for delivery services. This makes ESCOs the main, and perhaps only, source of revenue for the SI utility. ESCOs will have to pay the SI a regulated price for each of its uses of the grid—principally delivering the power transferred or purchased from each prosumer customer (black lines) and for delivering the rest of the power it supplies to customers, probably mostly from the Big Grid (gray lines). The regulators' main job in this is to set the prices ESCOs pay for each

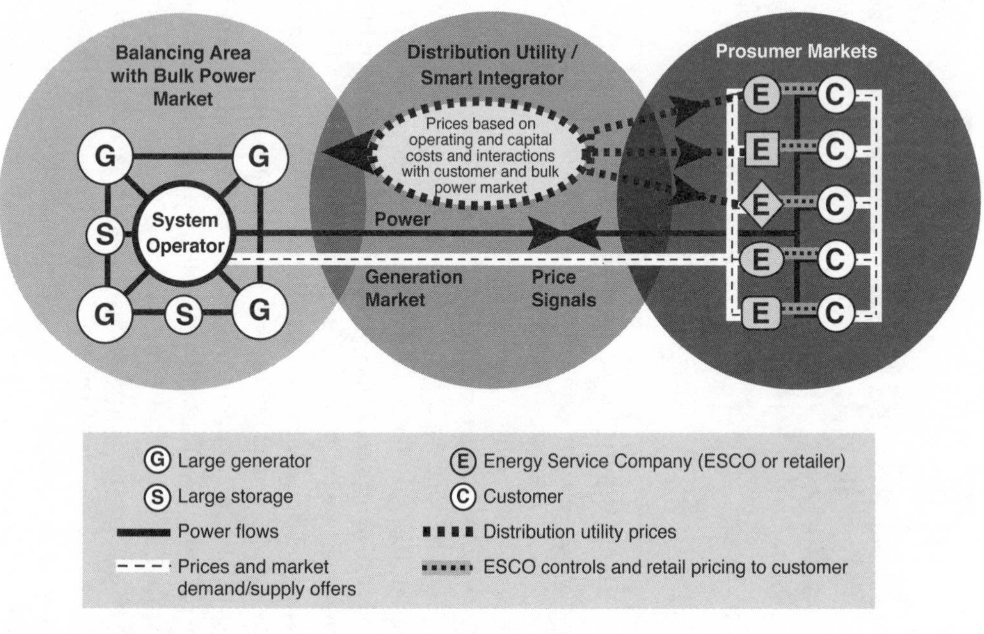

Figure 10-2 Three Overlapping Market Stages in the SI Structure.

different type of customer or use, and any other rules applying to how ESCOs can use the grid.

The SI industry in Figure 10-1 can be thought of as three interactive, interdependent transaction layers, as shown in Figure 10-2. The first layer is the marketplace between prosumers and ESCOs. The second is the interactions between the ESCO and the utility, managed and priced by regulators or other overseers. This layer includes customer and privately owned microgrids as well. The third layer is the large volume interactions between the distribution utility, the Big Grid, and ESCOs, which we're calling the Big Grid and others call the wholesale power markets.

In Chapter 8, we saw several competing ideas for how to organize long-term contract or centralized capacity markets. In a region with SI utilities, centralized capacity or contract markets are a more natural structure. The precursor to an SI structure is deregulated retail sales, and in the parts of Europe and the United States where this occurs, almost all already have regional centralized energy spot and capacity markets. A regional market for long-term capacity or contracts is the least disruptive upstream evolution because utilities and retailers are already accustomed to having an independent regional authority lead grid planning and operate all power-related markets. Bilateral contracts between ESCOs that sell retail power and Big Grid generators will also be common, but centralized competitive markets will enhance the competition between ESCOs that is the heart of the SI model.

The Prosumer ESCO Marketplace

We've already heard that future electricity customers will want customized bundles of electricity and related services; prosumers want even more choices. The customization of these bundles will involve all sorts of new ways to create and set prices for product mixtures. The ESCO will have to pay its own aggregate costs of purchasing and creating its products, including buying or making enough power to supply its net customers' demands. However, a fully deregulated ESCO will be free to set prices and raise revenues in ways that are new and unfamiliar to the industry today.

ESCOs will suss out and leverage the ways customers like to buy and pay for energy services, becoming, as one expert puts it, "the infrastructure that supports value-added personalization."[6] As discussed in Chapter 9, the most important things customers want is reliable on-demand service at the lowest possible cost. Most customers also want extreme simplicity in their transactions, rather than a complicated user experience, since energy is not something they want to spend a lot of time shopping for or thinking about. Although it is a somewhat contradictory impulse, many also want custom service bundles with customized pricing. Most also want energy that is green and carbon-free, but willingness to pay extra for this varies widely. Finally, not many customers care about energy conservation or efficiency for its own sake—they want it only to the extent it contributes to lower bills.

Most experts believe that ESCOs will seek to satisfy these tastes through pricing approaches very different from traditional utility per-kWh rates. One of the first examples of a radical pricing innovation, already adopted by many ESCOs in Texas, is free energy on nights and weekends, when the costs of buying energy to supply customers is very low.[7] We are starting to see many other innovations, from loyalty programs and gift cards, to solar and battery installation bargains, to discounts on energy for particular uses, such as car charging.[8]

One idea currently emerging in ESCO marketing is that customers will buy energy services under subscription-like arrangements, mimicking Netflix and Amazon. Within certain limits, customers can have all the energy services they want, customized using convenient online tools, for a simple monthly fee. Your subscription service would probably come with some automatic free energy-saving devices and sensors and options to buy other services and devices at a discount. You might give your ESCO the right to use all the data it collects to manage your energy use remotely, or rights to buy your surplus solar power. Curated new product offers, special sales, and energy tips would come at you regularly.[9]

A second new product / pricing idea that is rapidly gaining traction is *Energy as a Service* (EAAS). The idea underpinning EAAS is that consumers do not buy energy for its own sake; they buy it to get a particular type of service from that energy, such as heat or light or computing time.[10] We've almost always sold power to consumers as kilowatt-hours and then let the customer combine those kWh with a computer or furnace, which they had to buy and maintain, to get their services. The EAAS model takes over buying and maintaining the energy hardware, supplying the kWh, and combining them to create a stream of services. For example, instead of charging a customer for electricity and letting them buy their own light bulbs, an ESCO using the EAAS approach would sell the customer 250 lumens of light per square meter throughout their office for ten hours a day for a fixed annual fee. Even though the light fixtures reside in their customer's building, the ESCO would own and maintain them and guarantee that they worked as needed. Similarly, the ESCO could own and operate furnaces and air conditioners and provide a guaranteed temperature. EAAS pricing is still pretty new, and it won't be feasible everywhere, but it fits well with the needs of some customers who want to outsource all the work required for heating and cooling, managing EV fleet charging, operating data centers, and so on. Navigant Research, for example, sees the EAAS market rising to $221 billion by 2026.[11]

Market Optimum and Public Interest

The main conceptual argument in favor of an SI-ESCO industry harks back to the economic policy proposition that private good industries are most efficiently managed by markets that use prices to allocate resources. Because smart, distributed grids will involve hundreds of thousands of producers of power, millions of EVs with batteries, and billions of smart, connected devices, the only way to "optimize" this system is to let it self-optimize via markets.

Very few Western power experts reject this argument. However, an industry where prices are set by ESCOs will optimize around the maximization of their profits in service of aggregate consumer preferences. Left to their own devices, their version of an optimized grid will be the one catering predominantly to their wealthier customers, from whom they will earn most of their margins. Although the system may end up more energy efficient and lower in carbon than the current grid, it will not push these two public policies any farther than individual consumers will pay for. Other public goals, such as low-income customer protections and environmental justice, will not figure into this optimization much, if at all.

In 1961, Columbia University professor James C. Bonbright published what has become the definitive text on public utility ratemaking. Bonbright listed eight "desirable attributes" for rates charged by regulated utilities to fulfill their public service mission: simplicity and public acceptability, freedom from controversy, revenue sufficiency, revenue stability, stability of rates, fairness in apportionment of total costs, avoidance of undue rate discrimination, and encouragement of efficiency.[12] These attributes were obviously not invented to apply to unregulated ESCOs selling electricity at competitive prices—an idea Bonbright probably never even imagined. However, some of these attributes remain important whether electricity and related services are sold by regulated or unregulated firms. The important ones include an absence of undue discrimination, pricing so that each customer pays their fair share of the costs of creating and operating the entire system, and "efficiency in discouraging wasteful use" or what we would call energy efficiency (EE) today.

It is therefore a pretty fair bet that policy makers will try to keep the ESCO marketplace from straying too far from the particular goals of nondiscrimination, universal access, fair pricing, and EE, as well as the more recent critical goal of decarbonization. The question is *how*.

To inject social goals into the market optimization process, policy makers have two realistic choices. They can place obligations directly on the ESCOs, such as forbidding discrimination or setting required environmental metrics. This is already the case for "deregulated" retail power sellers in New York; these sellers now face several constraints on sales to low-income families.[13] Alternatively, they can leave the ESCOs alone and try to affect market-driven outcomes through the prices and policies of the SIs. Since every ESCO will require the delivery and reliability services of their local SIs, policy makers have an indirect channel by which they can guide the ESCO-consumer market optimum to include public-interest goals. Both approaches pose challenges that we will return to in a moment.

One key intervention regulators will have to consider is whether to continue to allow prosumers to transact directly with their utility. This same question arose when states began allowing retail choice. In twenty-five of the twenty-six states that adopted choice, and thirteen of the fourteen that now continue it, state legislators were not willing to impose a complete ban on customers buying power directly from their local utility, even if many competitive options were available.[14] It will take quite a change of heart for policy makers to remove this option under an SI structure, but allowing direct utility kWh sales to customers moves away from the structure's core paradigm.

Though still early in the process, the essential goal of increasing carbon-free power has meshed relatively well with the current competition framework. Fourteen U.S. states and all of the EU and Australia allow individual customers to buy power from price-unregulated retail suppliers.[15] In just about every case, these retailers are required to purchase a mandatory percentage of carbon-free power and blend it with the rest of the power they buy and resell to customers. U.S. utilities that provide last-resort service, and sometimes municipal utilities and cooperatives, face the same requirement.[16] As long as this mandate continues to apply equally to all power retailers, and the bulk power system itself transitions steadily to zero carbon, there is no conflict between decarbonization and this industry structure.

Looking farther out, however, neither of these two conditions can be taken for granted. Most of the mandated percentages apply to years between now and 2030, and many mandated levels are far less than 100 percent. Delaware, for example, requires only a minimum blend of 25 percent renewable power by 2026. Some states have considered eliminating the mandates altogether or have not updated them in recent years. Countries and states that have combined retail competition with relatively strong renewable and other climate policies are demonstrating that sound climate policies need not be a determining factor in the organization of the distribution segment of the industry. Nonetheless, there have been times in the past when policy makers perceived a choice between expanding markets and competition, and sound environmental policies and the environment lost out. If the SI-ESCO model finds itself facing political or economic barriers to fully carbon-free power by 2050 or sooner, it must get fixed or come off the table.

Grid Pricing and "Optimizing" the System

In consumer goods industries, delivery costs are usually a tiny fraction of the purchase price and therefore not much of an influence on purchase behavior. In electricity markets, it is not unusual for the local delivery portion of a customer's monthly bill to hit 30–40 percent. As a result, the prices that SIs charge ESCOs will greatly influence market outcomes and thus the degree to which the system delivers lowest total costs.

Economic theory teaches that maximum short-term efficiency is achieved when the regulated rate for delivering each ESCO's power to each consumer equals the SI's marginal cost of that delivery. Until recently, it was technically infeasible to measure customer-specific differences in power delivery costs on a utility distribution system. The closest you could come to assessing delivery cost differences was to separate customers into a handful of broad rate categories, such as residential large and small, commercial large and small, and so on. For this reason, nearly every delivery rate for power today is still a single tariff for each "rate class."

With much more widespread sensors and computing power, experts are now approaching the point where they can measure the marginal costs of delivering power hour by hour at hundreds of locations on a distribution system. This super-granular pricing, known as *nodal pricing* or *locational marginal pricing* (LMP), has been used for many years in Big Grid markets to measure the marginal cost of transmitting large amounts of power. When it is applied to measuring the costs of power delivery on a distribution system, it is called DLMP.

Setting rates based on DLMPs is not like calculating rates based on the costs of a utility as reported in its financial statements. To do the latter, you mainly need some relatively uncomplicated accounting calculations, in the nature of dividing total costs by sales to get an average price. But DLMPs are not a cost-based rate. Instead, they are a price that is set by the highest marginal cost or supply offer price needed to satisfy demand at each point on the grid. Computing DLMPs works by taking bids for energy from each producer and consumer at each spot on the distribution system, and computing prices as if these bids were in a competitive auction to provide power to everyone using power at that location.

As you might expect, this requires a super-sophisticated computer program that models the flow of power on each part of the distribution system and conducts the equivalent of simultaneous micro-auctions at hundreds of spots on the grid. No utility distribution anywhere has yet come close to collecting all of the necessary data and applying the necessary software to estimate DLMPs on a widespread basis. The process will almost certainly be simplified when it is ultimately used on real distribution systems, but it will remain a sophisticated, assumption-laden calculation[17] that is unlikely to be widely used for rate making for quite a long time.[18]

Along with other pricing elements, DLMPs are part of the most economically efficient delivery system price signals that could be charged to ESCOs. The more efficient the price signal paid by ESCOs to their SI, the more likely the retail markets they create will optimize the total system for lowest cost. For example, it may cost the SI utility less to deliver power to a customer located on a large, high-capacity power line than it costs to deliver power to a customer on a small, constrained line. If the first customer has energy systems that limit the amount of their deliveries or shape them by time of day, the cost of delivery may be lower still. Although the cost of *generating* power for these two customers may be identical, the efficient delivery charge to the first customer should be much higher than to the second in order to discourage the first customer from overusing their scarce, expensive delivery capacity.[19]

In a moment, we'll consider practical aspects of DLMPs that will temper their efficiency signal. From the optimization standpoint, however, regulators probably can't require that ESCOs pass through each customer-specific delivery charge to each of their retail customers. In fact, most ESCOs will want to do just the opposite.[20] Rather than showing each of their customers a complex basket of varying prices, they will want to charge the customer a simple price that is high enough to include all of their delivery charges, however much they vary by location and month. But that means that the distribution price signal the system intends to send to the retail customer—who, in the end, is the entity that is actually supposed to be responding to price signals—is perhaps entirely muted by the intermediary ESCO.

In this industry structure, it is the ESCO that pays and should therefore respond to the SI's delivery price signals. The ESCO in effect tries to induce its customers to adjust their energy use so as to optimize the system without sending them the specific prices that are designed to make consumers do just that. Market forces are still at work, but they are more complex and subtle. The ESCO must become the controller and the agent of the ultimate prosumer, and optimize the system through its contract with customers, even if the retail contract never shows DLMP price signals to the ultimate consumer.

How might this work? In theory, a modern prosumer could turn over control of most of its energy system to their ESCO, saying in effect, "ESCO, I am giving you control over my thermostats, vehicle chargers, ventilation systems, and other things, with the proviso that you keep all my services available within the limits I set. If you can operate my equipment, or get me to change my energy use behavior, so that my energy costs go down, and you pass most of those savings on to me, I will be a happy camper. Either way, keep me warm and safe and my bill low." The ability of clever, competitive ESCOs to induce their customers into saving money and thereby doing their part to "optimize" the total system—while still catering to their various pricing and product preferences—is a fundamental tenet of the SI vision.

Three more complexities add to this multilayered optimization. The first of these is a divided distribution grid. In addition to becoming prosumers, some large customers will undoubtedly start to own their own microgrids, while others will become part of microgrids operated by their ESCO or someone else. This adds another layer of constraints onto the market's search for optimization. As we saw in Chapter 5, microgrids will be subject to requirements for working with their surrounding distribution systems. In a perfect world, they will communicate and operate seamlessly with the rest of the distribution system, in which case price signals can be computed and propagated everywhere. In the real world, each microgrid will create new opportunities, but also pose some barriers to information and transaction flows across its boundaries.

The second complexity is bulk power markets. As noted earlier, large-scale power markets are already in use throughout the developed world, setting market-determined prices for power by the hour, day, year, or decade.[21] As its name suggests, transaction volumes in the bulk power market are thousands of times larger than most retail prosumers will ever generate. If this were all there was to the story, the downstream retail markets would view the bulk power price as a given. The outcome is much the same as typical retail stock market investors, who own so little of any one stock that they look at the exchange trading price and understand that this is the only price available to them.

However, ESCOs are starting to combine (*aggregate*) the surplus generation from hundreds or thousands of prosumers for the purpose of selling into the bulk power market. If the ESCO can assemble a large enough block of power to meet the size requirements of the bulk market, their offer to sell looks no different to the market operator than a supply offer from a big power generator (which is why this arrangement is sometimes called a *virtual power plant*). Smaller amounts of power with a high degree of controllability by bulk power market operators can also be sold. This creates two-way trading between prosumers, ESCOs, and the bulk market, which should in theory lead to greater market-wide efficiency. But it requires another layer of fairly strict aggregation procedures from both the bulk power market and the distribution system that ESCOs and prosumers must follow, and also more complex systems for measuring generation, billing and settlement, cybersecurity, and so on. Because this linkage of the bulk market and downstream prosumer markets is going to happen regardless of the distribution sector's form of organization, the industry is already hard at work trying to put these arrangements into place.[22]

The third complexity, discussed briefly in Chapter 9, is that prosumers are starting to want to buy and sell their power directly to each other, regardless of the services they receive from their ESCO. Under this industry structure, and perhaps under others as well, policy makers will probably not block the ability for peer-to-peer trading. However, if two prosumers agree to exchange their surplus power, the rest of the system must optimize around that trade. ESCOs can't touch the power that's been traded for any other use, including controlling it for optimizing customer use, optimizing the system, or sale back to the bulk power market. In other words, like microgrid transactions costs, peer-to-peer trades become constraints that the entire rest of the system must work around. If the entire market optimization process works without friction, the market in which peers trade will be a natural part of the entire chain of markets that together optimize the system.

In summary, the seemingly simple use of markets to optimize the distributed smart grid that animates the SI business model becomes complex, indirect, and constrained when it is implemented. The flow of information and price signals travels across several layers, as illustrated in Figure 10-2. Starting at the left, trading in the bulk power market including aggregated downstream DG will continue to set the main market price, which in turn will be transmitted to ESCOs and their customers. At the downstream retail end (right side), prosumers will buy service bundles from ESCOs and sometimes trade their power with peers. The distribution system transmits power between customers and the bulk market, providing balancing services, reliability, and resilience as it works with the microgrids in its area.

None of the deregulated electricity markets we have implemented to date are anywhere near this sophisticated, but in several places, all three layers of complexity are slowly becoming part of the marketplace. Competitive retailers can see and trade at time-varying wholesale prices, and in some places they can sell aggregated power. However, the distribution system is now largely a passive, inert lump between the retailers and the bulk market. Simple time-varying retail generation prices are only now becoming common, while sophisticated distribution rates or DLMP prices are still largely in the labs. And so far, very few electricity retailers use time-varying bulk power prices to optimize their customers' retail use.

The SI business model relies on price signals and market rules to optimize a massively complex network that stretches from power plants to microdevices. With enough computational resources and proper rules, this yields an elegant vision of a self-optimizing system. But the vision is about to get even more complex, intriguing, and uncertain in a rather big way.

Retail Choice's Next Act

During the 1990s, about half the United States and all of Europe and Australia adopted retail electric competition.[23] This created an industry structure with deregulated retail electricity retailers that transact directly with customers, competing against other retailers by offering lower prices and sometimes other services.

When retail choice began, almost no customers had solar systems or any other DG, so retailers didn't offer many products other than cheap electricity. Since almost no customers had the ability to self-generate, their business model was to purchase 100 percent of each customer's needs on the wholesale market, pay the local utility to deliver it over the wires, and collect an unregulated retail price from their customer. Under retail choice, all of the profits earned by distribution utilities in retail choice markets come from *delivering* someone else's kWh, not from selling it.

One would have thought that this arrangement would have caused retail electricity customers to stop thinking of their local utility as their power supplier. After all, none of us think that UPS makes the clothes or electronics we buy online, and our customer allegiance belongs with the clothing and computer brands. In fact, we have no allegiance at all to the delivery method—in most cases, we'd just as soon have whoever is cheapest or quickest deliver our stuff.

But in most of the world, retail choice hasn't played out this way. Most households still think of their local delivery utility as their supplier. For the few who can name their supplier, they continue to think of their utility as the most important part of their service. In areas with retail choice, utilities that could be "out of sight and out of mind" are instead quite deeply involved in the retail customer experience. There are several reasons for this somewhat unforeseen outcome.

First, electricity supply is homogeneous and invisible. The laws of physics prevent you from identifying who or where your power is actually coming from, and human laws require that all the electrical attributes of the power you buy (such as the voltage) must be the same. You can specify that you want power only from green sources, but the grid mixes everyone's power into a big pool and no one can trace whose electrons go where. Your competitive electricity retailer doesn't so much purchase title to specific electrons as purchase the rights to suck the power you need out of a big amorphous pool. If you had to pick the worst possible product to try and create a competitive market for, it would be something that can't be seen, touched, evaluated, individually purchased, or traced back to a specific supplier.

The one aspect of their power supply that customers do observe and experience is its delivery—the one function left to the utility. More than 90 percent of the outages or other power problems customers experience are due to electrical or physical problems with the distribution system in their area, not the availability of sufficient power plants.[24] When there's a problem with power, the website and text messages that post updates and the crew that shows up to fix it come from the utility. Depending on their experience, customers may love or hate their utility, but they often feel a visceral connection to a company that makes their home and community livable and safe, not to mention the other benefits of power service.

Utilities' unsevered bond to customers has also been helped by legislated policies that require utilities to continue to offer power to any customer who does not choose a competitive retailer. This was intended to be a transitional measure so that your lights would not be turned off if you had not chosen a competitive provider by the time retail choice began. However, many customers continue to like Provider of Last Resort (POLR) service or switch back and forth between competitive POLR service, depending on who is cheapest at the moment. The future of POLR is a topic of perennial industry debate, but there are few signs that it will change in the near term.[25]

The final reason why retail choice hasn't dislodged utilities is regulators' approach to preserving social and environmental goals. Earlier in the chapter, we noted that in the SI-ESCO business model, regulators had two main routes to shaping market outcomes to preserve public goals: impose those goals on ESCOs or impose them on the SIs, even if the latter were no longer in the retail sales business. Facing a very similar decision at the onset of retail choice, nearly all policy makers chose to continue to place public-interest responsibilities on the utilities. While utilities in retail choice areas don't sell power (except as a last resort), they continue to operate EE and DR programs, promote and purchase renewable energy and storage, support low-income energy assistance, and many other civically oriented tasks.

Because these public-interest activities are more tangible and often more important to customers than boring, invisible power service, the fact that regulators gave these responsibilities to utilities has helped the latter maintain their bond with customers. It was a little messy to separate the company that sells power (your retailer) from the company that helps you save money by reducing use of that power (the utility), but the approach has worked acceptably well. Utilities now spend almost $9 billion a year on customer EE programs, saving a little under 19 percent of electricity use each year.

However, this approach faces new challenges in an SI-ESCO industry. The original electricity retailers did not view themselves as being in the business of selling anything other than power. So, the utility's provision of EE or DR wasn't competing with their product line. In contrast, the underlying premise of the SI model is that ESCOs should be the customers' agent for optimizing their use of EE, DR, renewable power, storage, and so on. If regulators continue to place responsibilities for running programs in these areas on utilities, they will be impinging on ESCOs' market opportunities, if not competing head-to-head. This is something policy makers have not yet sorted out. If it's any consolation, alternative business models we'll examine next have a similar, equally serious problem.

If twenty-five years of retail deregulation haven't convinced customers to care about who supplies their power, why do we think that an industry of SI utilities and ESCOs will be any different? Current thinking holds that oncoming changes in the marketplace are going to make competition much more vibrant. Instead of offering a single undistinguishable product, ESCOs will sell customized bundles of sophisticated services that *are* visible, such as solar energy, DR, or outage protection. The range of services allowed by the smart grid goes far beyond simple invisible electricity service, and so will the varieties of pricing and service options. Prosumers can observe and judge the quality of these kinds of options and choose brands accordingly. Prosumers who see others getting innovative, valuable service bundles will start buying from ESCOs.

Another factor complicating this picture is the emergence of nonprofit mass buying organizations known as Community Choice Aggregators (CCAs) or municipal aggregators. These organizations undo the individual retail shopping choices in deregulated retail markets and instead buy all retail power competitively on the wholesale market and resell it to all retail customers in a service area. This maintains the distributor's role as an SI while it pushes for-profit (unregulated) electricity retailers out of selling energy itself. This constrains the product mix of these retailers, forcing them to omit a key product, and theirs diminishes their commercial prospects. But while CCAs shift the landscape of the prosumer marketplace—the third circle in Figure 10-2—they continue to place the utility in the role of an SI.

These changes and considerations are naturally forcing currently deregulated retail markets to evolve toward an SI-ESCO structure. In regions with retail choice, further regulatory changes are not *essential* to allow the SI-ESCO structure to emerge, although many are advisable. Wherever retail choice exists, any company is free to enter the market and start selling customers whatever smart, customized services it chooses along with power. This is precisely the strategy space on the *z* axis in Figure 9-2, and many non-utilities and utility affiliate multiservice ESCOs are busy selling their wares. In this sense, the SI-ESCO structure can be viewed as the final act in the addition of competition to the power sector. Contradictory as it may be, this expansion of competition will be accomplished by quite a lot of complicated regulation.

Machine over Market

Over the next decade, we are likely to see artificial intelligence (AI) continue to spread quickly into business processes. Over the same period, an estimated 125 billion devices will become connected to the so-called Internet of Things (IOT), and 5G networks with ten times the current mobile bandwidth will be deployed. While there will undoubtedly be some twists and turns, these giant strides in technology seem likely to change the fundamental character of grid optimization and what it means for ESCOs to compete. Ultimately, these changes seem likely to advance the SI-ESCO industry model and perhaps lead it to dominance, though not necessarily greater competition.

As we've just discussed, optimization of an SI-ESCO power system where humans are the decision makers involves three layers of interaction. The ESCO has to offer a human customer a package of technologies, control options, and pricing that appeals to its human objective function, whatever that may be. ESCOs will compete for customers by offering them more efficient (hence cheaper) or more responsive service, and technologies that deliver the required level of energy service needs most cheaply. The ESCO's costs and service offers depend on its interaction with the other two layers—the distribution utilities and the bulk power market.

It is widely believed that AI will soon surpass human intelligence for many tasks, undoubtedly including building energy management. "Artificial intelligence will bring an unprecedented level of efficiency to energy production and distribution," Intel confidently predicts.[26] At this point, what ESCOs will offer customers is a choice of algorithms that will take the customer's preferences and learn how to run the building to meet those preferences most cheaply. In fact, saving energy costs may not even be a main part of what the ESCO markets to customers—the expressed goal may be having a smarter home that is more fun to occupy or a more adaptable or secure commercial building. In this case, ESCOs' control over energy for lowering costs or carbon will be secondary to other targets set by customers.

Incidentally, the idea of turning the operation of a huge energy facility over to machines has already been demonstrated by putting neural networks on large power plants. These machine-learning programs teach themselves how to run the power plant while obeying all facility rules and delivering the exact amounts of power and other services specified by the plant's customers. A unit of General Electric, which installs these networks, claims that its machines improve the efficiency of power plants by 1–2 percent—a significant sum for a generator paying millions of dollars a month for fuel.[27]

When AI becomes universal, ESCO-to-ESCO competition will be over the best algorithms and how easily humans can interact with them. ESCO 1 will brag to customers that its algorithms save customers more money than the algorithms offered by ESCO 2. ESCO 2 will respond that its algorithms save just as much money as those of ESCO 1 and are much easier to adjust by simple voice commands ("Alexa, workers on the fifth floor are feeling cold in the early morning hours, please fix this"). ESCO 1 responds that its new system scrapes internal emails to learn about new energy needs as soon as employees talk about them online, and then Alexa asks you if you want to take action. The human / machine interface and customer service will be the most visible realms of differentiation, as it will take enormous expertise to compare the performance of energy algorithms that operate invisibly in the background.

In theory, these algorithms will be able to interact directly with the separate algorithms the local utility will be using to manage the local grid and send out price and control signals. Whereas no person I know relishes the thought of keeping track of delivery prices that change every few minutes, the algorithms won't care how often prices change and will simply recalibrate their settings, within their human-set limits, when they get new prices. Similarly, the algorithms that run the bulk power system will in theory interface naturally with aggregated ESCO offers to buy and sell large blocks of power.

Once we've given the ESCO our preferences, its own algorithms, the local grid's algorithms, and the bulk power market's algorithm will together try to solve for a solution—control settings on the grid that deliver what I want at the lowest calculated cost. As the system solves for my preferences, it must simultaneously solve for a solution that also serves every other customer according to their instructions. This is essentially solving, in parallel and near or actual real time, a multi-million-term simultaneous equation with millions of local and global power system constraints. Since most vehicles will be electric and connected at this point, this super-giant optimization will include electricity for charging, communicating with, and controlling electric and connected autonomous vehicles.

The computational and prudential issues raised by an optimization challenge this large boggle the mind. The power system must be managed in real time for near-perfect balance. So, this algorithmic control will have to check for solutions that involve thousands if not millions of customers and check these solutions against system constraints, all the while looking for a solution that is cheapest. As mentioned in Chapter 5, the expert committee assembled by the U.S. National Academy of Sciences concluded that there were many aspects of future smart grid controls for which there were no known computational solutions.

It is possible to imagine that computers will eventually be able to solve these problems, but it isn't possible to imagine this occurring without extensive regulatory oversight. First off, it isn't clear what least cost even means when there are many customers, each with different preferences and technologies that lead to different bills. Do we tell the system to weigh the importance of each customer's cost equally? What if one customer's preferences are so extreme or unusual that serving them at lowest cost raises everyone else's costs quite a lot? What if a part fails in my building that raises costs for everyone on my part of the distribution system—and the algorithms automatically adjust and pass the costs on to everyone?

In addition to many questions surrounding how to define and allocate costs, cybersecurity will be a major concern in this system. In all IOT systems, the opportunities for inserting malicious code are enormous, but the potential for damage to large parts of the power system will ensure that it remains a favored target of cyber warriors and terrorists. In addition, regulators will have to protect against ESCOs and customers using software to game the system by finding ways to evade or transfer cost responsibility. California began to set up its first-of-a-kind power marketplace, which went live in 1998, in early 1997. Before the market was even operating, at least one software vendor figured out how to game the new market and began offering to sell this information for a fee to future market traders. Enron also discovered these games but kept them for their own use.

Finally, the algorithms that run the local grids and the bulk power grids will all be explicitly subject to regulation, just as the physical wires are regulated today. All ESCO systems will have to communicate seamlessly with these systems, and this will unquestionably involve licensure, training, and other such processes.

This could be a technologically amazing future, but it does not look like one that is highly deregulated. The prices ESCOs charge customers for deploying their technologies and algorithms may never be set by regulators, but it seems quite possible that greater reliance on AI and machine-to-machine control will require more and more complex regulation, not less. One can imagine an oligopoly of AI-enabled, federally licensed ESCOs competing by offering customers the latest new options for setting preferences, or rewarding repeat customers with old-fashioned loyalty programs. That is certainly competition of a sort, and it may be quite vibrant as customers and ESCOs experiment with the epochal shift from human to machine control. However, it seems quite unlikely that regulators and regulated distributors will fade into a small, no-growth background role in this future. It will take years, if not decades, to get the rules for machine-to-machine interaction in the three layers of this industry structure set in a way that gives regulators the visibility and control they need to feel secure in their responsibilities. Rather than deregulating the industry, we may end up creating a new kind of quasi-regulated firm, or perhaps even reintegrating the ESCOs into a fully regulated network. Perhaps, in the end, the machines that run everything will view atomistic competition as a quaint twentieth-century notion that is not needed when they can simply solve for the lowest-cost solution and make it happen.

11 Governing a Really Smart Grid

At its very core, the economic regulation of utilities consists of two big tasks. The first is to cause the regulated firm to carry out its public duties as efficiently as possible. The second is to set the utility's products and rates consistent with its public duties, but also so as to yield revenues that allow the utility to raise capital, repay investors, and function properly. The first task could be called prudential oversight; the second, rate setting for multiple objectives.

In the SI-ESCO structure, the distribution utility's high-level duties are extensive.[1] It must provide nondiscriminatory universal service, which has come to include special protections for low-income customers. It must provide reliable, resilient, cybersecure delivery services for power in its area, including areas served by microgrids, as well as buy generation for last-resort customers. In addition, it must help lubricate and facilitate the competitive market for energy services. This means it must technically integrate and compensate the energy services that ESCOs or prosumers sell back into the grid. It must do all of this on a moment-to-moment basis as it operates its system while continuously searching for cost-reducing improvements. And it must plan and invest to make improvements in all these functions in the future, anticipating and adapting to changes in its surrounding ESCOs and microgrids. Anyone who surveys this list of duties and thinks that the SI business model will be boring or easy to regulate is certifiably nuts.

The second big task is setting prices that are consistent with these duties and which yield adequate revenue. In the SI model, the utility must charge ESCOs and / or prosumers for its services, but it must also pay for the services it receives. Because the services the SI utility provides and receives are no longer as simple as delivered kilowatt-hours, regulators must define and set prices for many entirely new products. As we've already discussed, the prices charged by utilities should have the effect of helping the entire system self-optimize, but these prices face many practical constraints, the largest of which is that they need to generate proper net utility revenues and yet be viewed as acceptably fair.

It would be hard enough to do this if the grid was such a strong natural monopoly that every electricity prosumer had to remain a customer of the utility one way or another. As we've seen, however, some customers will have the ability to defect entirely, and millions more will be able to reduce their use of grid power. This places utilities and regulators in the conundrum famously labeled the utility death spiral: raise delivery rates to customers who can defect; lose more of them; need even more revenues from remaining customers; raise prices to them; and prompt yet more customers to leave, restarting the cycle.

Most experts (including me) no longer think that this simple story is likely to occur. In most places, not enough individual customers are likely to defect to create a full death spiral. A universal, reliable electricity service is just too important, and any utility experiencing a revenue shortfall will undoubtedly get help one way or another. Nonetheless, setting rates for purchases and sales by the SI remains by far the toughest challenge of the SI business model.

Setting Regulation's Goals

Simply put, the purpose of regulation is to give a private firm an incentive to do something its profit motive doesn't already motivate it to do. Traditional utility "cost plus" regulation is often maligned as giving utilities a blank check, but it was designed specifically to give utilities the incentive to build out a grid that served every customer with ample, reliable power at a time when power service was incomplete, expensive, and unreliable. In its day, it did a very good job accomplishing this mission.

In the emerging SI industry, the mission has changed in important ways. Developed-world grids already serve everyone with highly reliable power. The SI's goal is not to expand the delivery capacity of the system using its own capital, but rather to induce and implement a system that is more efficient by integrating hardware owned by thousands of others with its own system. Along with its many other objectives and constraints, its objective is to blend together others' distributed generation, storage, and other resources to yield lowest-cost carbon-free power.

Why don't regulators simply tell utilities this is their new goal, measure their performance against the goal, and then set their profits accordingly? That's the idea behind performance-based regulation (PBR). Elements of PBR have been used in the industry for decades, but it has become widely apparent that PBR is the logical way to change regulation into its new role as overseer of utilities with the business objectives of the other SI or the ESU. The legislature of the State of Hawaii illustrated this beautifully when it enacted a new public utility bill requiring that

> the public utilities commission shall establish performance incentives and penalty mechanism that directly tie . . . an electric utility revenues to that utility's achievement on performance metrics and break the direct link between allowed revenues and investment levels . . . In developing performance incentive and penalty mechanisms, the public utilities commission's review of electric utility performance shall consider, but not be limited to, the following: (1) The economic incentives and cost-recovery mechanisms described in section 269–6(d); (2) Volatility and affordability of electric rates and customer electric bills; (3) Electric service reliability; (4) Customer engagement and satisfaction, including customer options for managing electricity costs; (5) Access to utility system information, including but not limited to public access to electric system planning data and aggregated customer energy use data and individual access to granular information about an individual customer's own energy use data; (6) Rapid integration of renewable energy sources, including quality interconnection of customer-sited resources; and (7) Timely execution of competitive procurement, third party interconnection, and other business processes.[2]

PBR can't be implemented in the abstract by dreaming up a set of goals and incentives. To keep the utility functioning, you have to start with its current physical system, revenue, and rates, and use incentives to move all these in the right direction without disrupting the utility's ability to serve. This places an important constraint on the totality of the incentive mechanisms that are adopted. PBR can inflict financial pain by reducing profits, but unless regulators are prepared to bankrupt the utility, they can't starve it to death. Conversely, regulators can reward utilities that do a very good job with somewhat higher profits, but they can't get *overly* generous, as this is sure to attract public opposition. As a result, the power of any PBR mechanism to affect utility performance is naturally constrained to operate within a profit band with upper and lower limits centered somewhere around a reasonable return on invested capital.[3] Though it is still quite early, there is a widespread hope (if not confidence) that the usable range of incentives is wide enough to work well while keeping the patient in basically good health.

There are several leading approaches to electricity distribution PBR. The first and most ambitious, known as RIIO, was adopted by the UK's regulator Ofgem in 2013.[4] Under RIIO, utilities submit a detailed, eight-year business plan that contains budgets for capital and operating outlays and specific performance targets for a number of objectives or "outputs." The outputs are grouped into five categories: reliability, environmental performance, customer service, social obligations, safety, and how well the utility has integrated DERs into its network ("connections").[5] These outputs map nicely into the updated obligations and goals of SI utilities, but notice that none of these goals force decarbonization of the overall power supply—this is left to other UK policies that intercede directly in the generation sector.

According to each utility's plan, each of the output areas has several metrics by which its performance is measured. If performance exceeds the target level, the utility earns a little extra profit, and vice versa. There is also a major incentive to limit total expenditures (the sum of capital and operating costs, or "totex"); if a utility underspends its approved plan outlays, it gets to keep part of the underspending as added profit. There are also incentives for utilities to add innovations to their service offering.

Ofgem's most recent (2016–2017) report card on the UK electricity distribution utilities subject to RIIO-style PBR found pretty good performance against most official metrics. All fourteen had exceeded their customer service and safety levels, and thirteen had beaten their reliability targets. Eight had failed to meet their targets for the time required to connect DG solar systems to their network. The utilities underspent their plan budgets by more than £1.2 billion, about 5 percent of the total, and earned £220 million pounds in incentive profits for this year (2016–2017).[6]

While all of this is encouraging, there are other aspects of this approach that provide sobering reminders of the challenges of using PBR on distribution utilities whose systems and surrounding ecosystems are destined to change dramatically. Any forecast of utility spending eight years in the future is heroic; the plans and financial information submitted to set performance targets run to a thousand pages or more for each utility. As AI-enabled systems, EVs, and local storage become the norm, these forecasts are likely get exponentially more difficult. The UK regulator devotes considerable expert resources to analyzing and negotiating these plans, in part to ensure that utilities don't game them by proposing targets they know they can easily exceed. This, too, will become far more difficult as future conditions become more unfamiliar.[7]

Of the many electricity regulators that are shifting toward PBR, most are using less sweeping, more incremental approaches. These measures tack specific, measurable goals on to the current framework of cost-based regulation, leaving the majority of the utility's objectives, rates, and revenues unchanged.[8] Nonetheless, an increasing number of these mechanisms are aimed at changing at least one aspect of the utility's objective toward a new business model. California encourages utilities to use the aforementioned non-wires alternatives in place of its own more expensive capital outlays by allowing them to make a 4 percent pure profit on non-wires alternative (NWA) expenses—thus departing from the principle that profits are based on rate-based capital.[9] Similarly, in 2017, Illinois enacted legislation that added a crucial new feature to traditional regulation: utilities could treat some outlays to pay for customers' solar systems or energy efficiency as if they were capital outlays—in other words, they could earn profits from these not-really-capital expenditures. "With this bill," one Illinois utility executive said after the bill was passed, "we can invest in community solar or EE or poles and wires and have the same value stream."[10]

Outside of Hawaii, the most ambitious U.S. use of PBR for the explicit purpose of resetting the utility business model has occurred in the state of New York under a program it calls Reforming the Energy Vision. One element of its PBR scheme again illustrates the regulatory care necessary to manage objective-driven regulation. As in Hawaii, the UK, and elsewhere, New York saw the importance of giving utilities new incentives to promote DERs in places where customers and the rest of the system both benefited. Rather than adopt a simple measure of DER encouragement such as the number of days required to obtain a DG grid connection, New York wanted a measure that reflected utility efforts more broadly. The New York Public Service Commission ruled in 2014 that a broad survey-based instrument be designed, the results of which would determine utilities' performance and thus their profit bonus or penalty. After three years of further debate over what the survey should look like, and other issues, the commission's staff recently suggested eliminating the incentive. In brief, staff felt that the utility's new planning mandates and other measures had adequately addressed the need.

There's no harm in a commission changing course on whether a particular incentive is needed; if anything, we should worry more about inflexibility in the face of changing evidence. More broadly, there is no doubt whatsoever that performance-based incentives are the only real tool we have to shape regulation into its new role overseeing a repurposed utility sector of one form or another. The point is that we have to be prepared for a new, higher level of effort and sophistication in regulation, in keeping with the sophistication and complexity in the SI business model itself.

Pricing Grid Services

The triple challenge of setting prices for efficiency, fairness, and revenue adequacy begins with the inconvenient fact that most of the costs of owning and operating the grid are fixed. To a first approximation, the total necessary revenues needed by an SI utility over the course of a year or two don't vary a lot with the number of kWh delivered, the number of solar systems connected, or the number of EVs charging each night. Of course, the long run is very, very different. It will be a second giant challenge for SI utilities and regulators to agree on future plans and capital budgets, but put this aside for a moment.

While the near-term *costs* of operating the grid are mainly fixed, the *services* provided by the SI are variable, that is, they can be separately defined, measured, controlled, and priced. The services provided by the grid can be denominated as kilowatt-hour delivery, transfer of self-generated power back to another user somewhere else on the grid, maintenance of proper voltage and frequency, backup power in the event of an outage, cybersecurity, and other services. There is also an overlapping set of products that distributed generators can provide back to the SI, the largest of which is surplus kilowatt-hours sent back into the grid. For example, many rooftop solar systems have equipment that the local utility could use (by remote control) to manage the voltage on nearby wires.

For each of these services, SI regulators will have to set prices or more accurately agree on algorithms that set these prices. For a variety of reasons, this is going to be a mind-numbing and contentious exercise.[11]

First, in the world of utilities, there isn't a single conceptual basis for price. In competitive markets, the price is where supply and demand intersect, and that's that. For a regulated firm, the rate can be the short- or long-run cost of making an additional unit (marginal cost), the short- or long-run cost of not making another unit (avoided cost), or the short- or long-run average actual or projected cost. In cases where there are the right conditions to create a liquid market, there is also the option of using a market price. Each of these concepts has its pros and cons and can be the subject of lengthy debates between policy makers, stakeholders, and experts. Each also comes with a small army of lawyers and consultants highly skilled at fighting over the many details of implementing any one of these concepts.

One dimension of these pricing standards—the aforementioned ability to compute prices that vary by the hour and location—casts a giant implementation shadow. As explained in Chapter 10, distribution locational marginal pricing systems (DLMPs) are miniature competitive auctions conducted by automated software using detailed system information from the utility and offers from prosumers to sell their excess power or buy power at each location on the local grid. The software required to implement DLMPs is still largely in the laboratories, but the approach has been used for many years in the bulk power market. So, we have some idea of the massive implementation challenges. The software required to calculate DLMPs will be formidable, and will need to be checked and monitored continuously by regulators (which is how LMPs are treated by federal authorities today). State regulatory agencies will have to acquire extensive new expertise, and many stakeholders will find it nearly impossible to participate meaningfully in highly technical discussions of how well the software is working. In addition, the costs of developing and operating the software will be significant, and it will have to interface with ESCO algorithms that respond to DLMPs.

Computing and then charging ESCOs delivery prices that reflect DLMPs will be an especially challenging task for regulators. By their very design, DLMP delivery prices depend on the electrical attributes of the distribution wires between you and the nearest substation—something a normal customer can scarcely understand, much less control, as well as the number and type of distributed generators, storage contributors, and other distributed elements attached to your part of the local system. This, too, is outside the control of the customer or any one ESCO. Accordingly, your own delivery rates could change significantly after your neighbors install a solar system or buy an EV, or after dozens of other actions taken by your neighbors.

To illustrate this, imagine one distribution circuit that hosts a handful of rooftop solar systems. Only one of these systems has new technology that the utility can remotely control to stabilize voltage on that circuit. The utility pays the owner of that system $10 a month for voltage support using a price set by the regulator based on avoided cost. By the next year, five more solar systems have been installed on that circuit, all of which have hardware that the utility can use to support voltage. The utility only needs one system's worth of services. So, it needs to decide which system to use. The utility holds a competitive auction among the six systems to see which one will allow it to support voltage most cheaply. The lowest (winning) bidder offers a price of $1—90 percent less than last year's regulated price. This kind of price shift occurs routinely in these unique kinds of markets.

DLMPs won't just vary—they will vary continuously and somewhat unpredictably based on a separate calculation done for each hour of each day. You won't actually know them until each hour's calculation is finished. While this real-time variability will theoretically allow for greater efficiency, many customers won't have the time, skill, or interest to respond to their price signals. The signals will work only when customers (or more likely their ESCOs) install smart systems that automatically respond to DLMPs. In any case, if regulators allow the use of DLMPs at all, they will almost certainly constrain them within a range, so that your delivery rates are not too dramatically different from your neighbors and do not vary too much from hour to hour or from month to month. The more they are constrained, however, the more efficient price signals to the ESCO and customer are attenuated, defeating the use of markets and prices to optimize.[12]

No regulator has yet allowed DLMP-based pricing. So, we don't know the tolerance of policy makers and the public for varying delivery prices, but we can get an idea of the difficulty of changing these prices from efforts to incorporate just one of the two dimensions of DLMPs—time differentials—in traditional electricity rates. As far back as the 1970s, regulators and utilities knew that the cost of generating electricity varies greatly throughout the typical day. Traditionally, making a kilowatt-hour costs two to four cents or so in the late night but costs can quickly rise to twenty cents or more on a hot summer afternoon.

In the ensuing forty years, electricity pricing experts have argued passionately that the flat-rate generation tariffs used by regulators almost everywhere are neither fair to all customers nor economically efficient. Utilities have conducted more than sixty experiments with dynamic prices all over the world. The world's leading pricing guru, Ahmad Faruqui, has carefully catalogued the results of these pilots and reported that they increase efficiency, reduce bills to the majority of customers, and produce proportionally greater benefits for low-income customers.[13]

Despite this extensive record, regulators are only now starting to adopt dynamic prices. Even now, these rates are the required form of pricing in only small numbers of U.S. and foreign utilities.[14] As late as 2017, Faruqui concluded that "progress has been stymied because of the persistent fears about a customer backlash or a failure to realize expected benefits."[15]

The expense of creating DLMPs, or any other pricing approach that sends granular optimization signals, is also certain to set off a protracted regulatory debate over fair allocation of implementation costs. Advocates for low-income and elderly customers will argue that these groups are not going to install fancy systems in their homes that allow ESCOs to optimize their energy use, and so they will get no benefit from the systems that create optimization signals. This is a simplification, as any proper social optimum will provide some benefits to everyone, but it will also be true that the greater benefits will be realized by those in high-tech smart buildings. Regulators will have the unenviable job of sorting through these claims and modifying the rates calculated by pricing algorithms to yield acceptably fair results.

When all of this is finished, regulators will have a set of approved prices for approved products sold by and to the SI utility. Their next task will be to estimate how much of each of these products the SI is going to sell, and how much of each they'll buy from DERs. They will need all this to estimate the utility's net revenue after it buys and sells everything at these prices, but this forecasting exercise is going to be quite challenging at first because the markets for the new disaggregated utility and DER services are new and evolving.

Just above, we saw an example of where one solar system's payment dropped 90 percent in one year. Multiply this situation by many locations and services, and you have a pretty strong revenue forecasting challenge.

Because we haven't gotten far in implementing this industry structure, we don't know what the numbers will look like when SI utilities implement advanced pricing structures. Based on laboratory experiments, theoretical calculations, and experience with large-system LMPs, most experts think that the net revenues SI utilities will earn from the locational pricing of disaggregated services will fall short of the SI's annual costs.[16] If there is such a shortfall—and the SI is already collecting all the revenue it can from the sale of all identifiable products—we come to the classic regulatory question of how to recover a lump-sum estimated shortfall from ESCOs and their customers.

The Problem of Fixed Costs

Collecting the portion of a utility's costs needed to make a utility whole, but which don't vary with its levels of sales, has vexed regulators and utilities since the birth of regulation. Over time, regulators and electric utilities have settled on a variety of approaches to recoup these revenues. They all have their pros and cons, and they are all going to get more complicated as the SI industry unfolds.

The primordial way to recoup fixed costs has been to add a margin to the price of delivering each kilowatt-hour and devote the accumulated margins toward fixed costs. In many delivery systems, the vast majority of revenues are still priced by taking all the costs of the utility, fixed and variable, and dividing by the number of kilowatt-hours sold. Since nearly all the costs of delivering power are fixed in the short run, much of this average price is in recovering fixed costs. In a typical $100 power bill, roughly $35–60 goes to the variable cost of actually making the power; the remainder is the largely fixed costs of transmission and distribution.[17]

As customer self-supply and grid defection become more common, this approach is going to have to change. When a prosumer installs their own power system, and supplies, say, two-thirds of their own kilowatt-hours, they then buy and pay for delivery of only one third of their former level of purchases. At the same per-kWh delivery charge, the SI's delivery revenues have dropped by two thirds. But because the costs of serving that customer are fixed, the utility still needs the same level of revenue from that customer as it used to be earning to remain whole. In many places, the penetration of DG customers is low, and the revenue shortfall triggered by self-generation is small enough to patch over. In places such as California and Hawaii, where 5 and 30 percent of all homes already have solar, respectively, the revenue gap is too large to ignore.

There are a number of ways regulated prices can be changed to recapture the revenues lost when prosumers reduce their kWh use under volumetric rates. Regulators can add a fixed charge onto each customer's bill or require a minimum monthly payment regardless of use, add a charge based on highest momentary (peak) use, or change per-kWh rates so they vary by time period or with total monthly use. In addition, they can use a variety of processes to readjust per-kWh rates more frequently.

There are several complicated and often conflicting aspects of each of these approaches. First, their effectiveness in reversing revenue shortfalls is often uncertain, incomplete, and situation specific. They vary greatly in the way they try to make prosumers who have reduced their kWh use still pay their fair share of fixed costs. They also vary in the way they treat customers who use little power—generally (though not always) low-income and / or elderly households or small businesses—versus those who use a lot, and in the way they incentivize greater EE or DG use. Perhaps most importantly, each of these approaches is bitterly opposed by at least some stakeholder groups, forcing regulators to choose between several unpopular alternatives.

It could easily take another book to sort through the nuances of these approaches and their surrounding arguments. To give a flavor for what regulators are in for, Table 11-1 provides an overview of the arguments for and against the main alternatives.

The first column in the table is higher fixed or per-customer charges. This is the main approach already in use in the industry. Essentially every

Table 11-1 Recovering Grid Fixed Costs

Name	High Fixed Charge	Minimum Bill	Demand Charges	Tiered Rates	Time-Varying Rates / Dynamic Pricing
Description	Increase part of a monthly bill that is lump sum; charge varies by customer class	Monthly total minimum charge regardless of use; if bill exceeds minimum, charge using normal rates	In addition to a charge per kilowatt-hour used, a separate charge based on the highest level of use during the month	Rates per kilowatt-hour increase significantly at specified levels of monthly use	Per kilowatt-hour rates vary during certain hours / time periods
Current Status	Many utilities doing this now	Extremely rare	Much consideration and some adoption	Rare	Becoming the default rate in California
Perceived Fairness Toward Low-Use versus High-Use Customers	Highly unfair to low-use customers (often low income)	Same as above	Perceived as somewhat unfair to low-use / low-income customers	Generally favored by low-income advocates; shift costs to high-use customers	Research shows low-income customers benefit
Perceived Fairness Prosumers versus Non-Prosumers	Fair—both groups pay similar fixed costs	Similar to high fixed charges but more complex	Improved fairness	Unfair to non-prosumers	Improves equity, but does not necessarily address fixed-cost deficit

Directional Impact or EE	Reduces EE incentives	Complicated effects	Reduces EE incentives	Mixed-improved incentives for high-use customers, reduced for low-use	Positive
Directional Impact on DG Installs	Reduces DG	Reduces DG	Reduces DG	Increases DG	Increases DG
Directional Impact or Economic Efficiency	Improves efficiency	Does not improve efficiency	Mixed impacts	Generally negative	Positive
Relative Effectiveness at Increasing Fixed Revenues	High	High	High	Low	Low

Source: Adapted from Wood et al. (2016) and Faruqui and Aydin (2017).
EE, energy efficiency; DG, distributed generation.

monthly power bill has a lump-sum "customer charge" on it—typically $5–25 a month for residential customers. When solar prosumers started reducing their bills, several utilities set off firestorms of protest by proposing to double this charge, or more. Theoretically, this would solve the revenue problem, since solar prosumers would now have to pay significant power bills no matter how much of their own power they generated. However, this approach "raises profound equity and social justice concerns," in the words of one advocate, because it raises prices for low- and fixed-income customers who tend not to have DG and thus do nothing to cause the revenue gap.[18] Higher fixed costs also reduce customer incentives to save energy and adopt DG and induce customers to defect entirely, if they can, to escape the higher charges.

If we covered the remaining columns of the table in equal detail, we would learn that every other approach has equally varying degrees of unfairness vis-à-vis different customer groups, differential impacts on EE, differential impacts on DG customers, and different degrees of confidence in its workability.[19] Predictably, representatives of the utility industry favor the approaches that generate the most revenue, low-income advocates want the lowest rates for their group, EE advocates want high incentives to conserve, and renewable energy groups want the lowest possible charges on prosumers. No one solution is likely to work by itself, and SI regulators will simply have to choose a combination that represents a solution that is both politically acceptable and economically sufficient.

To be sure, the challenges of paying for a grid that is inalterably a shared public good is common to all future utility business and regulatory models. No matter where utilities operate on the business model spectrum, many of their customers will self-generate some of their power and others will splinter off onto microgrids. Revenues from kWh deliveries will continue to decline, and regulators will have to choose one or more of the approaches above. If this is so, is there any difference between the SI and the other business models that makes this problem any different or more difficult?

For somewhat obscure reasons, the answer might be yes. The SI utility has no other business other than providing grid services, which we think won't recover revenues equal to costs when we set prices to encourage market-wide optimization. As we've seen, setting regulated prices for all these services will be complicated to begin with; trying to adjust them or layering on unpopular mechanisms that raise any one of these prices will add to what is already a difficult price calculation and adoption process.

In the alternative business model, the ESU, the utility and the regulator have a little more leeway to solve this problem. As we'll soon see, grid service prices need not be quite so complex, perhaps yielding slightly more manageable regulatory proceedings. ESUs may also not need the same complex pricing, control, and communication systems, somewhat reducing the fixed costs to be recovered. Most differently, ESUs are allowed to sell products other than delivered kWh and related grid services. Like the margins on kWh sales in today's industry, the margins on these sales could produce a modest but very helpful revenue stream that can be applied to the likely fixed-cost gap.

Planning and Building the Distribution System

When the distribution system delivers power in one direction—from the Big Grid to the customer—expansion planning for the system isn't too difficult. Besides replacing system elements that wear out, you carefully monitor annual changes in power use and install larger transformers and wires in each neighborhood to stay ahead of the growth in power use in that locale.

This process gets a lot more complex when customers become self-generating prosumers. Not only must you anticipate customer demand from all uses (including EVs, which are large new users of power), you must also project the approximate pattern of each DG installation and the pattern of charging and discharging for all local storage installations on the system, including perhaps EVs. Utility forecasters will have to collect much more forward-looking information from customer/prosumers, but much more extensive forecasting and contingency planning will also be necessary. Utility planning and its regulatory oversight must become much more sophisticated about risk trade-offs and planning with multiple uncertainties.

Modern distribution planning is further complicated by a new ability to substitute resources that customers own and control in place of expansions of the distribution system. Suppose, for example, a wire leading into one neighborhood has flows close to its maximum capacity during hot hours in the summer (only) and is therefore scheduled for an upgrade. Instead, an ESCO that controlled a large storage unit in that neighborhood could sign a contract with the SI agreeing to charge up the storage before every hot summer's day and discharge when the line would normally be close to overload. The ESCO would have to manage its arrangements with customers to make this work, but this is certainly feasible, even today. If the price the ESCO offers to the utility for this customized storage use is less than the cost of upgrading the wire, total system costs are less under this so-called NWA. In effect, a NWA means that a utility has essentially outsourced the ownership and operation of a portion of its system to a contractor, almost as if an independent contractor operated a key portion of someone's factory.

Utilities and regulators are only starting to comprehend how different planning will be when utilities must consider NWAs. Utilities in California and New York must now publish data on each part of their system on which they plan to spend money. ESCOs use these data to figure out whether they can install, or get their customers to install, something that reduces the utility's planned outlay. This adds a major procedural loop into utility capital budgeting and regulating oversight. Thus far, the number of actual NWAs installed by utilities has been small. So, each one could be individually analyzed by utilities and overseen by regulators. No one yet is quite sure how the process will work if a utility must process hundreds of NWA proposals at one time.

Importantly, the fundamental elements of these planning challenges apply to every utility business model on the Figure 9-3 spectrum. All types of utilities now understand that regardless of how their business and regulatory models evolve, their planning process must change. In all of the business models, regulators or public/co-op owners must ensure that distribution utilities do excellent forecasting of DG, storage, and load and then work hard to find the lowest-cost system capital plan that maintains fair, high-quality service. The U.S. states that have the largest amounts of installed and predicted future DG and storage are, uncoincidentally, the states where utilities and regulators are changing planning processes most aggressively.

These planning processes are moving forward as each of these areas also makes initial forays toward adopting new business models for their distribution utilities.[20] In regions that adopt the SI-ESCO structure, the planning process becomes more complex because there are two sets of layered actors, plus multiple microgrids and owners, all of whose actions and constraints must be incorporated in good plans. In the SI-ESCO-microgrid world, you must forecast the market interactions and ultimate investment behavior of many interdependent actors and then plan and build power and IT systems that accommodate them.

This adds more uncertainties and risks to an already-difficult planning challenge. However, all future utilities will share many of these challenges. The SI world adds another layer of complexity, but it is much too early to know whether the added layer makes much difference in the quality of distribution plans and investments. All we can say is that planning, and planning oversight, is more demanding.

Of Elegance and Complexity

The vision of SI utilities serving an all-competitive retail sector is anchored in the sensible view that regulated distribution is too constrained and slow moving to create a competitive market for high-tech energy services and a vibrant, optimized power system. As the vision moves toward real-world implementation, however, the magnitude of the implementation challenges becomes apparent.

Optimizing a distributed system purely for efficiency (lowest total monetized cost) requires that the regulated prices for use of the grid be much more customer specific and changeable than they have ever been. Regulators face a huge challenge determining how they will allow variable grid prices so as to send efficiency signals, while also attempting to respect and preserve rate stability, simplicity, and nondiscrimination. On top of this trade-off, regulators must layer in policies and guard rails to protect social and environmental goals that the market won't naturally provide. Assuming all of this can be done, we hope that the resulting ESCO-led markets reach a decent level of cost efficiency, fairness, and reliability.

The SI-ESCO industry structure and its algorithm and IOT–driven, highly distributed prosumer grid seem much better suited to optimization via prices and agile, unregulated firms. Yet, the cruel hand of technology requires that these firms can't avoid relying on a regulated grid or electricity's public-interest attributes.[21] The realities insert a regulated entity with strict social goals squarely in the center of their high-tech marketplace. In the coming maelstrom of disaggregated, distributed services, AI systems, and new regulatory processes, will this arrangement be simply too complex to manage? We won't know until we try, and try we will. But there are some less complicated structures that might also work, and we ought to try those too.

12 The Business and Regulation of Energy Service Utilities

An Energy Service Utility (ESU) is a Smart Integrator (SI) turned inside out. Rather than running a platform for competitive energy service companies (ESCOs), the utility itself sells similar bundles of mass-customized services to retail customers. As in the SI-ESCO world, these bundles include electricity and its delivery, as well as many of the related products sold by ESCOs. Because the elements of these service bundles will vary, so will prices and the specific interactions between utility and customer.

The structure of the commercial interactions in an ESU industry is diagrammed in Figure 12-1. On the right-hand edge, prosumers have solar panels, controllable appliances, and other systems that can change the level and pattern of their power use. These options will be largely the same as those used by prosumers living where utilities are SIs: batteries, energy control systems, solar systems, and microgrids. The crucial difference will be in the commercial arrangements for the provision and operation of this hardware.

Utilities will provide service bundles by combining electricity purchased from the bulk power market (left side of diagram) with other hardware, software, and services. For example, one customer might be offered electricity that is priced differently by the hour, along with a control system for shifting washing machine, dishwasher, and EV charger use. A second customer may be offered electricity with just two price levels during the day and night, no load shifting hardware, but a package of efficiency improvements to their home.

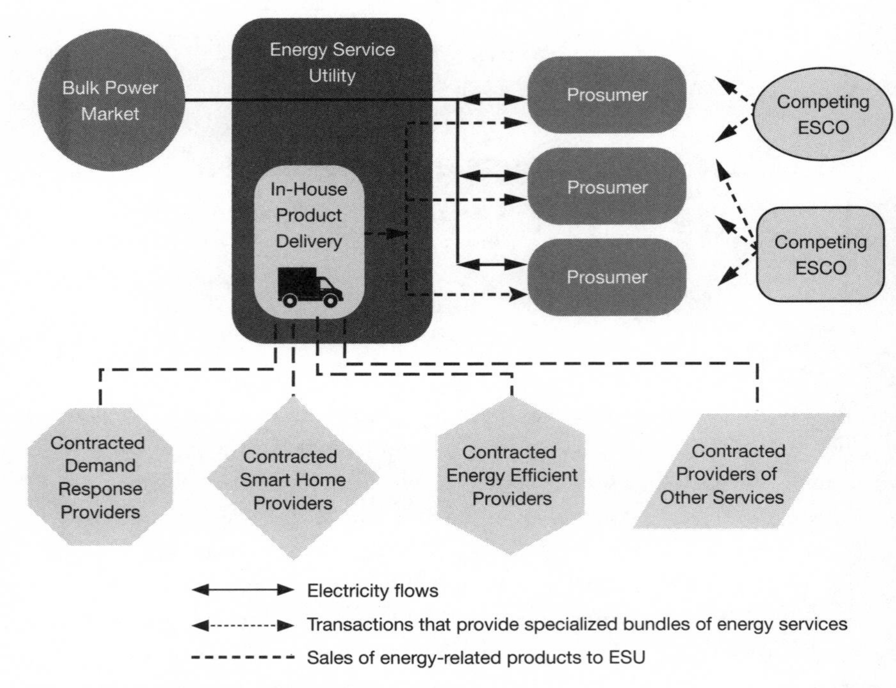

Figure 12-1 Transactions Map for the ESU Industry.

These other products may be created by the utility itself but more likely will be purchased from a network of preferred vendors (bottom of diagram). The ESU will not manufacture its own programmable thermostats offered to customers as part of many bundles; they will buy them from one or more partners that do. The key features of this model are that the ESU is seen by each prosumer as its trusted agent for providing the service bundle that is right for them. Simply put, the ESU is one large default in-house ESCO integrated into a regulated distributor of power.

Why on earth would anyone suggest replacing an industry full of agile, competitive ESCOs with one regulated ESCO that is part of the local utility? This contravenes a decades-long trend in Western economic policy, as well as the industry-specific reasoning behind the SI-ESCO structure discussed in Chapter 10. In the West, economists teach policy makers that firms should be economically (i.e., price) regulated only in the most extreme case: where competition *cannot* be made to work.[1] Some economists even think that economic regulation exists solely because the firms that are regulated want it to shield them from competition and earn excess profits.[2] The regulated firms then reward their regulators in subtle but important ways, such as future employment. These firms and their regulators will always want to expand regulated activities as much as they can get away with. From this, it follows that economic regulation should be limited strictly to the narrowest number of firms pursuing the smallest set of specific activities that do not work without regulation.

From the standpoint of utility business models, these views imply that the one type of company we are obliged to regulate—distribution utilities—should not be allowed to add businesses or sell things that are not natural monopoly products. If we allow an ESU to sell both natural monopoly and normal competitive products, we face two choices: either we regulate only electricity delivery and let firms sell the other services as they wish, or we have to set regulated prices for everything. Both of these options create serious headaches. Beyond these specific difficulties, however, what really nags at us is the heretical idea that we would ever want to willingly expand the scope of regulation—government's heavy hand in place of the market and all that. Crazier still, we are talking about doing this in a rapidly evolving, highly entrepreneurial space where the very description of products isn't stable, much less a regulated price for them. In conventional economic thinking, the ESU model approaches madness.

Sane or mad, a few operational dimensions of this idea are quickly apparent. First, the broad social and economic objectives of electricity service haven't changed simply because we're talking about a new business model. The central goal remains the "optimization" of the system to achieve the best possible combination of desired customer outcomes (lowest cost, reliability, convenience, etc.) and public-interest goals. In addition, many of the energy technologies, software systems, and other resources that customers and others must install and operate to achieve the optimum, such as energy control software and solar panels, will be equally available to ESUs and SI-ESCOs. New electronic, process, or marketing innovations will undoubtedly be more frequent in the freewheeling SI industry. However, one of the key differences between the "optimum" found by an industry of SI-ESCOs versus an industry of ESUs will derive more from differences in management processes and commercial arrangements, not from differences in available hardware or software.

It's also evident that regulation will have to change substantially under this model. Traditional utility regulation was designed to set prices for a handful of simple, slow-changing electricity products. Regulators have since gotten comfortable setting prices and terms on several other utility activities, and performance-based regulation allows for much greater price adaptability. In the ESU model, however, the ability to mix and match a wide range of services under very different pricing schemes is a giant leap in complexity. Regulation simply cannot individually preapprove all product and pricing bundles, as in the past, and expect to offer customers the same level of service as in the SI model.

Finally, it is an even greater heresy to think that policy makers will forbid any unregulated ESCOs from competing with ESUs to offer largely the same products and more. This means that ESUs will continue to have an exclusive franchise in electricity sales and delivery, but not in the rest of the products they sell. In the bigger picture, ESUs will be essentially a special competitor in the broader energy services marketplace, with unique advantages and disadvantages. These mixed blessings include the fact that it is regulated, it alone can bundle electricity itself into product offers, and it alone operates the local system. The unregulated ESCOs competing with the ESU are shown on the far right in Figure 12-1.

The Case for ESUs

On closer inspection, it may be that a full-service utility operating under new forms of oversight may be a viable path to a clean, optimized system. There is a plausible case to be made that in some circumstances, ESUs might work better than the alternatives.

Some competitive providers of some energy services may find some smaller geographic markets highly attractive. However, having to monitor the overall availability of competitive services continuously is arduous, and when it falls short, there is little to be done at that point other than to create a public alternative (i.e., an ESU). Risk-averse policy makers would rather not find themselves scrambling to create this alternative after the competition has come up short, even at the expense of foreclosing a competitive solution that *might* be there. Apart from size preventing strong competition, municipal utilities and co-ops may prefer the ESU model simply because it keeps them in closer contact with customers, which they feel is their mission, and enables them to continue to fulfill public-service responsibilities unique to their communities. In addition, municipals and cooperatives do not have shareholders who can absorb losses. Disruptions to their current business model are seen as risky by policy makers, who could be blamed if one of these utilities gets into financial trouble as a result of a policy mandate to change business models. For all these reasons and more, these types of utilities are often exempted from legislation that requires changes in utility regulation, such as Hawaii's 2018 SB2939.[3]

The first place this is likely to be true is in small towns or rural areas, where the market might not be large enough to attract enough ESCOs to create real competition. In fact, the U.S. associations of municipal utilities and rural cooperatives have both spoken out forcefully in favor of the ESU model—or the Consumer-Centric Utility in the language of the co-ops.[4] Small-town utilities and cooperatives exist because in the early 1900s, the provision of electricity itself did not create a large enough market to attract for-profit power sellers. It is quite possible that what was true of electricity then is true of new complementary energy services today, especially since demand for the latter is much newer and less uniform than the demand for power.[5]

A similar argument applies to areas with large populations of low-income customers. As with rural areas, the competitive marketplace may not see large enough revenue opportunities to offer service in these areas—not unlike large food chains, whose neglect of low-income neighborhoods has made them "food deserts" served mainly by convenience and dollar stores. Following a recent survey of U.S. utility regulators, two experts noted that regulators are open to "a different customer service model" in these communities.[6] Already, regulators in New York state have allowed utilities to install utility-owned solar systems in low-income areas; California has done the same thing for EV chargers.[7]

In large, healthy markets, the case for ESUs takes root within a corner of economic theory that asks why firms exist at all. What is the difference between a collection of individual workers who sign contracts with each other for the purpose of producing something jointly and workers and managers within a firm making that same product? What makes this difference valuable?

These questions were answered by the legendary economist Ronald Coase and then expanded by Nobel Laureate Oliver Williamson and others in the transaction-cost school.[8] Firms exist because it is more efficient to produce things in an ongoing enterprise with a management hierarchy and a highly varied and customized workforce that is hired for the long term. This allows for greater specialization of labor while also maintaining the close ongoing coordination needed to make complex products and services. Because firms commit to seeking profits in specific markets for the long term, they create valuable reputations and brands. They are then able to raise capital at scale, enabling cheaper production at large scale and investments in R&D and new products.

The concept of transaction costs is at the core of all of these firm-market differences. Transaction costs are the time, dollar expenses, and other psychic (non-monetary) costs required to evaluate, negotiate, and implement any sort of transaction. Firms exist because they can lower a variety of transaction costs in the making and selling of goods. The lower transaction barriers can reduce either the cost of producing groups of goods or the costs of marketing the goods—production or marketing synergies, in short. On the production side, lower transaction costs are monetized, that is, they usually translate into lower prices offered to consumers. Lower transaction costs on the sales side are sometimes reflected in prices, but they may also be realized in the form of time saved or a more satisfying and successful buying experience.

On the production side, it takes more time and money to negotiate contracts between many independent contractors, and then renegotiate them when small circumstances change, than it does to use a long-term employment agreement. The transaction costs of raising capital are lower for a firm with a reputation that can go to the capital market and sell billions of dollars of stocks and bonds in a single offering. And then of course there are the well-known economies of scale and scope that firms harness by expanding their production processes or adding new products. Every business day, untold legions of company workers and consultants search to see whether a firm can increase its profits by expanding new or existing product lines.

Brands also reduce non-monetary transaction costs by reducing the effort and time a customer needs for evaluating and effectuating a purchase, or *search costs.* Brands are a way to give a firm an easily remembered reputation that saves customers the trouble of evaluating the trustworthiness or quality of a purchase—they trust the brand. They also enable multiple purchases from one source, with assurance that the products will work together if they need to, as with Apple's computers and software. Amazon has succeeded in part because it has reduced the costs of shopping for millions of items far below the search costs of driving to the mall or a big box store and hunting for the right product at the right price.

Applying this line of thinking to utility business models, there is some chance that an ESU might act a little like an Amazon for utility customers—a single one-stop shop for energy services that reduces the transaction costs of finding your bundle of services and yields one that is about as good (or better) than would have been found without the ESU's help in a purely competitive world.

There is at least a little evidence supporting the proposition that customers might like their local utility to become the hub of their energy services provision. One way to interpret the mixed record of retail choice is that many customers prefer having their utility bundle electricity supply and delivery rather than having to shop for the two.[9] The polling data reviewed in Chapter 10 show that customers are open to buying services beyond electricity from utilities, and generally trust their advice on what to buy.

Shopping separately for each of five or six different energy services from an equal number of vendors is bound to be unattractive to all but the largest and most sophisticated customers. In this case, the transaction cost savings offered by an ESU comes from not having to select between competing ESCOs offering service packages that may be unfamiliar on pricing terms that may be quite variable and creative. One ESCO might offer five types of EE programs but only one demand response (DR) program and energy prices that vary every hour, while a second might offer only two EE technologies, three types of DR, and prices that are fixed for most hours but free at weekends. A numerical comparison of these two offers would be well beyond the reach of typical consumers. In the words of two experts, customers might be "more comfortable with the bundle of DER and grid-based services offered by their local utility than they would have been with any package they could assemble themselves."[10] For this reason, ESCOs will try to offer multiservice packages backed by some sort of performance guarantee that everything will work together.

Some experts speculate that practical engineering and administrative concerns will make it easier to optimize the future hyper-connected, multidirectional, AI–assisted power-grid argument if utilities own and control at least some of the systems on the customer side of the meter. According to this theory, if utilities own or at least control much of the smart grid, its ability to optimize the system for multiple objectives will be better than a system that relies only on prices. To be clear, price signals will unquestionably play a major and essential role; the question is the extent to which the system will need or benefit from supplementary utility-run control systems that include downstream devices. Some experts, such as National Regulatory Research Institute director Carl Pechman, speculate that this might be the case.[11]

Meanwhile, a regulated ESU will face great challenges trying to offer packages of services that are as broad, customized, and creative as a vibrant ESCO marketplace will offer. They will have to learn how to analyze customers' needs efficiently and formulate attractive customized service bundles; they will have to be able to offer a sufficient range of popular services so as not to seem content deficient; they will have to continue to honor social obligations, such as offering lower prices to low-income customers (unlike truly deregulated ESCOs); and they must purchase, deliver, and service these outside technologies and services at prices that are comparable to what ESCOs charge without losing too much money. In large urban markets, ESCOs will offer stiff, highly visible competition. If customers see the service bundles that ESUs offer as clunky and expensive compared to what they can get from the hungry ESCOs knocking on their doors, the ESU business model will fail.

Incidentally, the Big Grid capacity markets that most naturally fit with ESUs are the more decentralized, bilateral structures on the left side of Figure 8-1. ESUs are the exclusive electricity retailers in their areas and therefore the drivers of their own capacity needs. It is less natural for them to purchase from a centralized market for long-term resources than to hold their own competitive procurements for the resources their market research and engineers tell them they need. They could buy those same resources from a competitive central market—and probably more cheaply—but this means creating a joint institution, aggregating the demands of all ESUs to determine the total purchase, and standardizing product terms and purchase cycles.[12] From the political standpoint, utilities and regulators have less control. All of these factors increase the likelihood of left-side outcomes on the figure.

Changing Utilities' Cultural Stripes

It wasn't so long ago that the words *utility* and *culture change* didn't belong in the same sentence. Utilities were universally derided as stodgy bureaucracies filled with graying engineers and lawyers who instinctively resisted change and had no interest in interacting with customers. One of the true pioneers of utility customer engagement, O Power's cofounder Alex Laskey, recalled in a later interview that when he began in 2007, most utilities didn't have a senior officer whose job it was to think about customers. The typical so-called customer relations position, Laskey noted, "reported to [someone in] operations, or maybe to the COO"—never to the CEO.[13]

In the twelve years since, Western utilities have almost all awakened to the need to start behaving like unregulated companies that work obsessively to understand and engage their customers. The signs of this new mind-set are everywhere. Utilities are hiring change-management consultants and establishing innovation as a core corporate value and metric for employee performance evaluation. The UK's performance-based regulation scheme, RIIO, has an explicit element for rating and rewarding utility innovation. In 2015, Julia Hamm, head of the Smart Electric Power Association, said, "Utilities are coming at [their business] from a different perspective. They are thinking about all of the things available to the customer more holistically . . . versus siloing different product teams and offerings."[14]

Management scientists have also started to look at the process of utility management and culture change, with many interesting results. Last year, utility expert Steve Kihm used a technique called Construct Validation Computer-Aided Text Analysis (CATA) to compare the language used in annual letters to shareholders by the CEO from a Wisconsin electric utility and Nike, the shoe manufacturer. The CATA technique works by classifying the words in a text as being highly associated with different concepts and then counting how often these words appear in a particular text. Steve used the technique to find a set of words associated with an entrepreneurship, such as "novel" and "envision," in shareholder communications from a utility and a non-utility CEO. Not surprisingly, these entrepreneurial words appeared about twice as often in the Nike CEO's letter as in the utility CEO's letter—but the utility had increased its word frequency steadily over the last ten years.[15]

Many utilities now have a chief customer officer, reporting to the CEO, whose background centers on marketing, innovation, and change management. Utilities are studying and adopting the same marketing techniques and customer analytics software that non-utilities use to study their customers' preferences and improve customer service. Among the fourteen utilities represented in Energy Impact Partners, better customer engagement consistently ranks as one of the two or three highest priorities for new investments and research.

One good example of how utilities are migrating customer engagement products into their daily operations comes from a new software company called Innowatts.[16] This software was first embraced by deregulated retailers (REPs) in the Texas electricity market, which is notorious for its high levels of customer price sensitivity and customer churn. The software helped these retailers do what many wireless and cable companies routinely do: evaluate each customer's use to see if there is a pricing plan that will save that customer money or provide greater value. Since REPs are unconstrained and have many fast-changing pricing plans, proactively reaching out to customers to urge them to save money by shifting their plan is a smart customer-retention strategy.

Use of this kind of software wouldn't have made any sense in an electric utility that had only one tariff for each rate class and regulator-imposed definitions of who is allowed in each class. In an ESU future, rate structures will expand to include several time-varying rate options, energy efficiency (EE) programs, special rates for solar prosumers and EV owners, and much more, and use of software such as Innowatts will be a natural part of ESU operations.

One U.S. regulated utility has already adopted Innowatts on a pilot basis to show each of its customers how they can take advantage of better rate options and other products to lower their bill, even if it might mean lower sales revenue in the short run. "I'd like to have something like this on every screen in our call center," said the utility's director of digital transformation, "so that when a customer calls for any reason our representative can say, 'Oh, by the way, I see that your current rate package isn't your lowest cost option. May I suggest that you sign up for a new package of services?'" This is precisely the mass customization that ESUs must master if they are to compete with the SI-ESCO model.

It comes as no surprise that the investor-owned utility that is closest to a full ESU, Green Mountain Power (GMP), has also undergone a massive operational and cultural retooling. GMP's pioneering CEO, Mary Powell, was a former liberal arts major who spent most of her career in finance and small business, including two startups. The first thing she did when she was named CEO was to eliminate half the company's executive positions, keeping leaders who were most committed to radical change. She then led a full-scale reorganization, reducing the employee headcount from 345 to 191 and introducing service quality metrics and scorecards. Her mantras of team-based approaches, customer obsession, and dramatically flattened org charts are standard operating procedure in most competitive energy firms. These and other changes prompted the tech journal *Fast Company* to recognize GMP as one of the top firms in its sector three years in a row.[17]

Like most utilities, GMP had a corporate headquarters building that Mary described as "massive and park-like" and scattered smaller offices where lower-level employees ran service operations and met customers. She closed down the headquarters and moved everyone into a building she describes as a "Costco-like feel," with no private offices and no separation between the rarely-see-customers and always-see-customers departments. As someone who has visited dozens of other utility headquarters over the years, I can vouch for the fact that none I've ever seen remotely resembles a Costco.

GMP's success certainly does not imply that most or even many utilities will change their cultures successfully. GMP is comparatively quite small and fortunate to operate in a single, very progressive state.[18] However, GMP's changes suggest that it is not unimaginable that utilities could evolve to the point where their responsiveness and ability to mass customize would allow them to compete successfully with competitive ESCOs and DER providers.[19] During our interview, I asked Mary whether she thought her company could now keep up with these fast-moving upstarts. She didn't merely say yes—she said that some of the competitive DER firms that they encountered nowadays had become much larger and bureaucratic in their culture than GMP.

Regulating an ESU

There are two major processes that make up the economic regulation of utilities. The first is determining the costs and revenues that enable the utility to function properly under the chosen utility business model. The revenue-setting framework inevitably includes incentives and disincentives of one form or another, as well as additional dimensions such as planning approvals. The second major process is the way regulated rates are structured and set so as to hit the revenue target. In regulatory lingo, the objectives of these two processes are the *revenue requirement* and the *rate structure.*

As noted in Chapter 11, IOU regulators are moving away from pure cost-of-service regulation toward performance-based regulation (PBR). This move is at least as relevant for ESUs as it is for the SI business model. If ESU utilities earn profits based mainly on the amount of capital invested, they will have the incentive to try and sell customers additional services that require as much utility-owned capital as possible. All other things being equal, an ESU would much prefer selling power from a solar plant it owned rather than reselling solar power it purchased from a third party with no utility capital outlay, even if the latter saved customers money.

The IOUs that seem to be heading in an ESU direction have mostly not yet made it very far, but there is every reason to believe that ESU regulation will trend toward performance-based mechanisms. The U.S. IOU that is now closest to an ESU is Vermont's GMP, which is regulated by the Vermont Public Utility Commission (VPUC). The commission uses an "Alternative Rate Plan" that is much like a form of PBR. In the commission's own words, the plan "creates clear incentives for GMP to provide least-cost energy service to its customers" rather than rewarding higher capital investment, and also incentivizes GMP to innovate, improve service, and help the state achieve energy policy goals.[20]

The second major regulatory process is rate setting. When we think of utility regulation, we think of administrative proceedings that set electricity rates. A panel of regulators wearing judicial robes is trying to decide a fair rate based on sworn testimony by utility accountants and the "opposing parties." Lawyers and experts argue back and forth through voluminous written filings until the commission issues a formal written decision.

In reality, this represents a tiny fraction of utility regulation today. Electricity regulation mainly involves proceedings about aspects of utility service other than the level of rates. There are planning approvals, advance approvals to make large investments, and rules regarding special programs for EE or low-income policies or any number of other topics. A glance at the agenda for any regulatory commission will show the number of rate-setting dockets is dwarfed by planning and operational matters.

The process of utility regulation has also evolved. Utilities nowadays are in a constant, free-flowing dialogue with their regulators and other stakeholders, and almost every regulatory outcome is now negotiated rather than litigated to the point of a commission ruling. Whether the proceeding is setting utility-wide rates or a limited policy on an obscure issue, the chances are about ten to one that the utility, all opposing parties, and the commission itself will agree on a negotiated, collectively written document ("settlement") that the commission then turns into its official decision. Often, the process is an even less formal series of "workshops" where everyone tries to reach a consensus and the regulator writes up the result as official. In the rest of the developed world, the regulatory process is equally informal.

In a world where regulatory commissions litigated every action and rate proposal to a conclusion, there would be no hope for the ESU model. There would be too many service bundle options evolving too quickly to take months to agree on a price, after which it could not be changed until there was another similar process. The evolution toward negotiated outcomes creates some chance that utilities will be given enough latitude to make the model work, but this is very much an open question. In the case of GMP, the Vermont commission allows it to lease whole-house batteries, sell solar power to low-income customers, finance heat pumps and water heaters, sell energy audits, and give away EV chargers and electronic device controls. It also pays customers who install their own large batteries in exchange for controlling some of the battery's charging and discharging. This is about as close as an IOU comes to offering a product suite comparable to a full-scale ESCO, though not yet in separately priced bundles.

The rates GMP charges for electricity are set by annual proceedings. The VPUC gives GMP a bit more latitude than usual over the prices of non-electrical products, which it calls Innovative Pilots, but there is still a fair amount of oversight. The utility must notify the state fifteen days before starting a new product sale, with

> a narrative explanation of the Innovative Pilot and how it is consistent with the eligibility requirements, the number of customers it will be made available to and how those eligible customers were selected, expected costs and revenues, why the proposal does not conflict with work performed by Efficiency Vermont, a certification that GMP has collaborated with Efficiency Vermont regarding the proposal in advance of the filing, and the frequency by which GMP shall provide status reports to the Commission and Department on the Innovative Pilot's progress which shall not be less than six months.[21]

In this rule, the requirement that GMP not duplicate the work of Efficiency Vermont is a proxy for the turf conflicts IOUs will have with ESCOs and other market participants in other states. Efficiency Vermont is a somewhat unique public agency that has a statutory duty to help energy customers in the state to improve their EE. The passage above effectively bars GMP from usurping the EE role from the agency—a constraint no free-market ESCO would even want to put up with.

Cross-Subsidies and the Space for Political Bargains

The ESU model wields a double-edged sword that cuts for and against two long-standing problems in public utility regulation. We can't yet tell whether either edge of the sword will deliver fatal blows, but the challenges for policy makers are readily apparent, and we'll need to keep a first-aid kit nearby.

For as long as price regulation has existed, economists and practitioners have warned against allowing utilities to sell competitive and regulated products under one roof. Regulation guarantees that utilities that do a good job get prices for their *regulated* products high enough to make a profit. This gives them an incentive to sell their *competitive* products at prices below those of their competitive-sector rivals, even if they lose money on the sale. They don't mind losing money because regulators will allow them to recoup their losses by raising prices on regulated products, for which "captive" customers have no alternatives.

This phenomenon, known as cross-subsidization, is doubly harmful. The competitive market is hurt by one seller who can sell below cost and force efficient, competitive rivals out of the market. Competition itself is harmed, along with all of its benefits to customers. Meanwhile, customers of the regulated products are paying prices higher than they should because they are making up for the losses of the competitive products division. For these reasons, many economists, antitrust practitioners, and industry stakeholders warn against expanding the scope of regulated firm activities.[22] The U.S. Congress was so concerned with this problem that it initially enacted legislation during the 1980s that prevented utilities from financing or installing home-efficiency measures—a ban it quickly lifted.[23] More recently, the EU enacted a "Third Package" of continent-wide electricity regulations that included strict separation between the activities of regulated network operators and affiliated energy supply companies, including restrictions on the common use of the utility brand. Like the U.S. Congress decades earlier, this led the Netherlands to outlaw any ownership in any other business by distribution utilities.[24]

Utility regulators are acutely aware of their responsibility to keep regulated rates as low as possible and are therefore often highly skeptical of allowing utilities to sell added competitive services. If they allow it, they must oversee the prices offered for competitive products to make sure utilities are not selling below cost. But once they are doing this, the so-called competitive products sold by the utility are no longer sold at deregulated prices; they are back to being regulated. This puts a large new burden on regulators to examine costs and set rates for services they have little experience of regulating while they make sure the regulated firm isn't unfairly preying on adjacent competitive markets. "The more DSOs [distribution utilities] are involved in non-core activities," wrote the Council of European Energy Regulators, "the greater the need for regulatory oversight."[25]

If utilities start separate deregulated subsidiaries, which is the most common way utilities approach adding new products, protecting against excessive cross-subsidies becomes a little easier because each company keeps its own financial records. The only way a separate but affiliated company could get subsidized by an ESU company is to sell it something at high prices no normal company would pay. The overpriced purchase is recorded by the regulated company as a cost of operations, and if regulators don't notice it, then those costs are built into the price of the regulated products. For this reason, when utilities buy or sell anything between two separate subsidiaries, the price of the transaction is carefully scrutinized.

Ironically, the oncoming disruptions in the power sector may create their own inherent protection against competitive product cross-subsidies. Cross-subsidization is feasible only when there are captive customers who have no choice but to pay for regulated service, whatever the price. When power service from the utility's grid was the only option, electricity customers were certainly captive—electricity is essential, and customers will pay much higher costs if necessary to keep their lights on. However, from here on out, many types of customers will have some opportunity to make their own power or get it from nearby non-utility sources.[26] This is not true of all, but for those customers who will have alternatives, utilities cannot charge them higher prices without losing sales. If ESUs try to cross-subsidize, they will have to do it in a much more surgical way, avoiding even larger losses of regulated sales in order to gain less profitable or even unprofitable competitive sales. Cross-subsidization will become much more visible too. Utilities will have to explain to regulators why they want to charge some customers who might defect to lower regulated rates, while charging those who are truly captive higher prices.

It is hardly likely that regulators will accept the idea of setting regulated rate differences solely on the basis of the ability of customers to defect. This is even less likely if the reason for the rate difference comes from the need to cover losses on competitive products. As a rule, the customers who will be most able to defect partially or fully will be wealthy homeowners with ample roof space or adjacent land and well-financed businesses, while the customers who are most captive will be low-income renters and small shopkeepers. Woe to the regulator that chooses to raise rates for the latter solely to keep a utility in the money-losing business of selling competitive products.[27]

In fact, the political difficulty in covering losses may well end up being an inherent impediment to utility adoption of the ESU model rather than a means of squelching competition. Many competitive ventures run losses for a fair period of time before they become established and profitable; Amazon took six years to report a profit after going public, and Tesla has taken nine.[28] The challenge of building very new in-house capabilities to do mass-customized marketing and delivery as good as competitive rivals, when faced with the risks that regulators are unlikely to cover losses, will undoubtedly create much hesitation and soul-searching among the leaders of utilities.

The second edge of the double-edged sword involves the ability of regulators and ESUs to reach agreement on how the ESU will carry out its social and environmental obligations. ESUs' business activities will cover a much larger space than the activities of SIs, naturally including EE, renewable energy, and the sale of electricity itself. In most states, regulators want utilities to carry out programs that promote renewables, increase EE, protect low-income customers from shutoffs, promote local economic development, and carry out other public-interest tasks. In Chapter 11, we noted that regulators will be challenged to find ways to get SI utilities to carry out these tasks because under that business model, the utility should stick to a single core activity—running a smart transactional grid.

With ESUs, there are many more natural ways in which regulators and utilities can devise programs that address public-interest goals. For example, any ESU will want to offer EE and DR products to its customers, regardless of their state's policies on promoting efficiency. If regulators want to bolster or shape this program, they will be making incremental adjustments to something that is already part of the utility's core strategy. In the case of SIs, regulators will have to invent a workaround by which the utility gets partly into the business of efficiency or they try to influence or partly regulate ESCOs.

Time and again, I am reminded of the fact that most of the socioeconomic accomplishments of the power industry to date, including most of the progress on decarbonization, have been anchored by the existence of a franchise and the implicit government guarantees that come with it. This creates a financially stable power distributor that can be directed to test new technologies, buy clean energy, and implement new approaches at scale. Without state policies that almost entirely relied on the balance sheets and franchise guarantee of utilities, wind and solar power would be deployed at a fraction of their current scale.

The operational latitude to devise programs that combine the business strategy of the utility with public-interest needs of policy makers might be an odd reason to favor one utility business model over another. For those opposed to the public-interest goals themselves—for example, opposed to action to prevent climate change, or favoring only the use of carbon tax for this purpose rather than using utilities—the ESU is an invitation for political collusion. In their view, under the ESU model, regulators get their pet programs, utilities get protected profits, and electricity customers pay the price for both. We are back to the arguments in favor of the SI business model, relying as much as possible on competitive markets and other tools for achieving public-interest goals. However, throughout these processes, utilities and regulators will be disciplined by competitive ESCOs in large markets and by the existence of alternative business models in other states and countries that could work as well or better.

In the end, we are brought back to the optimization question. Which of the business / regulatory models will do best at optimizing the power system of the future against the key customer and public objectives? The summary case for ESUs is that utilities can do a good enough job at offering multiple services to keep the competitive market at bay, partly due to reduced time, money, and transaction costs. In addition, it may be easier and less disruptive to address public-interest goals, and the complex mechanics of controlling the distribution system may be more feasible with a larger utility role. The summary case for SIs is that the purported transaction cost savings are not large enough, for either production or marketing, to offset the likely benefits of greater agility and creativity from ESCOs dealing directly with customers. The question of whether transaction costs make it more convenient for electricity customers to engage in one-stop shopping is, in a sense, the mirror image of the question of whether the agility and creativity of competitive markets will allow them to optimize energy services better than a regulated ESU.

13 Forces and Fault Lines beyond the Industry

The transformation of the electric power industry is hardly occurring in a social and political vacuum. In economic sectors far beyond electric power, technological change is colliding with deep cultural and political crosscurrents, and technology's role in helping or harming the public is hotly debated.

Many of these trends are in the very early stages and are likely to play out over decades. It is difficult to determine their overall trajectory, much less their impact on the electric power sector. Nonetheless, our limited ability to predict the impacts of these trends on the power sector hardly guarantees that these impacts will be negligible. It is more likely that these large forces will indeed influence utilities, and we would be remiss in ignoring them, even if they raise uncertainties we cannot comfortably settle.

Big Tech and Monopoly Power

One major trend already influencing the future of utilities is the rise of so-called Big Tech. Four tech titans—Apple, Google, Facebook, and Amazon—have come to dominate the American economy in unprecedented ways. As Scott Galloway notes in his book *The Four,* between 2013 and 2017, the value of these four firms increased by $1.3 trillion—roughly the gross domestic product of Russia. Today, they stand at $2.7 trillion.[1] The size and reach of these firms is triggering calls for a reevaluation of global competition policies, which have so far done little to halt these firms' expansion. As the website Econofacts puts it:

> In recent years, a number of analysts have increasingly railed against large firms, particularly modern technological giants such as Amazon, Google, and Facebook, calling for such firms to be regulated . . . or perhaps broken up as was famously done with Standard Oil over a century ago . . . Very specifically, these commentators call for a change in antitrust policy from one that focuses primarily on consumer welfare . . . to one that views all large firms with sizable market shares as problematic. This invokes an American populist tradition that is suspicious of concentrated economic and political power.[2]

Stanford professor Tim Wu, one of the leading proponents of this revival of antitrust policy, calls the movement the New Brandeis School of economic thought. In his monograph *The Curse of Bigness,* Wu writes:

> Antitrust has fallen into hibernation before as ideologies have shifted, only to come roaring back to meet the needs both to return to its broader goals and upgrade its capacities. It needs better tools to access new forms of market power, to assess macroeconomic arguments, and to take seriously the link between industrial concentration and political influence. It needs to take advantage of all that economics and other social sciences have to offer. It needs stronger remedies, including a return to breakups, that are designed with the broader goals of antitrust in mind. Finally, it needs to put courts back in the business of policing what Brandeis termed as conduct meant to "suppress or even destroy competition.[3]

In the distinguished *Journal of European Competition Law and Practice,* editor Lina Khan noted that this viewpoint has attracted support in the United States from President Obama's Council of Economic Advisors and senators on both sides of the aisle. "In some ways," she writes, "the renewed attention in the USA echoes conversations in Europe, where the antitrust community is debating whether and to what degree competition law should embody values of fairness."[4]

Any sustained resurgence of interest in curbing economic concentration is certain to find its way into the utility sector. When it does, its effect on the emerging business models could break any of several ways. On the one hand, regulated utilities will face strong scrutiny regarding their ability to squelch emerging ESCO competition by leveraging their regulated status. This concern is not new, but as we saw in Chapter 12, it is especially important in the ESU model. On the other hand, the strength and span of large ESCOs could trigger antitrust scrutiny, especially if any of the tech giants move further into energy than they already are. The advantage here probably belongs to the SI model, but local market circumstances will be at least as important as broad policy directions.

Privacy and the Smart Grid

The rise of Big Tech is also associated with the business models that monetize data collected from consumers during their online activities. The profusion of ways online firms have devised to monetize customer data is steadily expanding to include energy behavior data collected by utilities and ESCOs. To data experts, this energy device information is a rich new addition to the already-large stockpiles of data they use to understand and target consumers. No one in the industry was surprised when Google acquired Nest, a company that made a connected thermostat better at capturing customer information than prior models.

However, a backlash against tech dominance has been slowly building for some time. Facebook and Twitter are facing enormous criticism for their role in election misinformation; Facebook has also come under strong attacks for misrepresenting and misusing the information it collects. An entire branch of policy literature critical of the power of Big Tech companies has emerged, beginning at least as far back as Jaron Lanier's 2011 *You Are Not a Gadget*, a self-described manifesto that mourned the loss of humanity on the worldwide web.[5] By early 2019, one of the most distinguished scholars of economic culture, Harvard professor Shoshana Zuboff, adopted the term *surveillance capitalism* to describe "a new economic order that claims human experience as a free source of raw materials." Zuboff further calls this new order "a rogue mutation of capitalism" and "an overthrow of the people's sovereignty."[6]

Political leaders across the ideological spectrum have taken notice. In 2016, Europe adopted the General Data Protection Regulation—new privacy rules that require consent for customer data to be collected and affirm consumers' rights to erase their data held by a commercial enterprise: the "right to be forgotten."[7] In 2018, the California Consumer Privacy Act was introduced, a law that gives customers the right to know what data are being collected about them, the right to opt out of collection, and the right to have their data deleted when they stop taking a service from the collecting firm.[8] Writing in the *Harvard Business Review,* digital expert Dipayan Ghosh noted the potentially vast impact of the law:

> Perhaps the primary issue that firms are contending with is that the law's requirements could threaten established business models throughout the digital sector. For instance, companies that generate revenue from targeted advertising over internet platforms—such as Facebook, Twitter, and Google—must, as the law is currently written, allow California residents to delete their data or bring it with them to alternative service providers . . . These measures might significantly cut into the profits these firms currently enjoy, or force adjustments to their revenue-growth strategies. They could also further impact any businesses that advertise on digital platforms, as the service they are purchasing—highly targeted advertising—might become less precise as a result of the new protections afforded to individual consumers.[9]

Broad concerns over the protection of privacy translate immediately into specific concerns over electricity-based data. The installation of smart meters led to isolated protests over utilities' ability to "spy on" their customers, but utilities' collection and use of meter data quickly became protected in several states and countries by measures that go beyond those applying to unregulated parts of the new electric ecosystem. The concerns are typified by the actions of township supervisors in Smithfield, PA, who responded to the introduction of smart meters by asking utility regulators to compel their utility to disclose how it was using and protecting their data.[10] As of 2015, three state regulators had passed regulations governing the use of smart meter data, six more were considering it, and four states enacted similar limits via legislation.[11] Both the U.S. Department of Energy and the UK's Department of Energy and Climate Change have passed codes of conduct applying to regulated utilities' smart-meter data.

As the analysis of electricity data is moving toward an ability to track intimate, moment-by-moment activities within a home or other space, some experts and activists are strongly criticizing elements of the smart grid. For example, three academic critics of the emerging system write that "The 'Smartening' of the energy grid, we find, is both an ideological construct and a technological rationalization for facilitating capital accumulation through data collection, analysis, segmentation of consumers, and variable electricity pricing schemes to standardize social practices within and outside of home"[12] and "Not only does the 'smartening' of the grid entail greater consumption data collection for analytic capabilities and profit-making activities, it also enables deeper surveillance of home life and its everyday activities. This invokes concern about whether the anonymous observation of mundane (vacuum cleaning), personal (what time one sleeps), or even intimate activities (heating a waterbed) porously violates rights of privacy."[13]

The authors go on to argue that involuntary smart grid data collection violates the fourth amendment of the U.S. constitution.

Calls to increase the regulation of the Internet and expand privacy protections naturally bolster the case for continued regulation of energy utilities, or for the regulation of tech companies as a new kind of utility. On the other hand, if the power and business model of data-driven Big Tech remains strong, this favors the market-centric SI model of the industry, with utilities managing the wires but little more. Whether Big Tech buys the wires, an unfettered private ESCO sector will ride this trend toward automated customer solutions with strong scale and network effects, emulating similar platforms in other consumer sectors. ESCOs in the nascent areas of SI utilities already aspire to become the energy-sector equivalents of Google or Amazon, serving the largest possible set of customer needs via an individually curated customer experience. It is largely for this reason that most industry insiders believe that the odds favor private utilities ending up closer to SIs than any other business model. The sheer execution capabilities of these new technologies, backed by the political power of Big Tech and its huge global investor base, are likely eventually to make the power marketplace resemble the other service sectors of the 5G economy. Indeed, many cutting-edge energy and tech marketers think that energy sales and controls will be just one more set of apps running on smart home platforms provided by Google, Amazon, Apple, and others. In this world, utilities simply run the last-mile delivery service.

To the extent that the privacy backlash leads to greater oversight of Tech and greater protections, it will surely extend into the energy sector and thereby encumber the SI-ESCO model. Although ESU utilities will not escape unscathed, their history of public service may give them greater latitude to collect and manage energy data within approved limits. This could favor the ESU business model or close hybrids, where the utility's central role naturally enables it to limit and police the use of data. This is the case made by Stanford researcher Stephen Comello, who thinks that utilities may become "information fiduciaries" with the formal public duty to protect and use energy data.[14]

Energy Democracy

Since the earliest days of the power industry, many small and a few large distribution utilities have been publicly owned. Especially in the United States, where government ownership of the "means of production" has been consistently out of the mainstream, these utilities have thrived by stressing the benefits of local control: accountability in governance, financial support for local governments, efficient operations, sensitivity to local concerns, and the "value of ownership."[15]

Beyond this historic business model, a new movement has arisen that generally goes by the name of Energy Democracy (ED). Although there is no one official definition, the concept has four consistent elements: the predominant, if not exclusive, use of local, small-scale energy sources; environmental stewardship through the use of clean energy; public rather than private ownership; and a strong emphasis on participatory decision making.[16] From the organizational perspective, the movement combines elements of the original antinuclear and pro-solar groups; groups favoring local, public, and worker ownership; traditional environmentalists; green libertarians; the environmental justice community; and climate-change activists.[17]

Interestingly, the ED movement looks at the very same disruptions that spark the industry's transformation and draws very different conclusions about the best future form of utilities. The ED movement concludes that the improved ability to generate affordable power locally and the new, grid control technologies that allow for sectionalizing and microgrids mean that electric utilities no longer need to be large or privately owned. Instead, Szulecki explains:

> This new prosumer-citizen is characterized by a set of virtues reminiscent of the Tocquevillian citizen in the nineteenth century . . . The prosumer gains political power through ownership of means of production (of energy). The element is strongly emphasized in all activist accounts of energy democracy, and not simply because it has a Marxist ring to it. Like coal miners in Mitchell's account (2011), increasing numbers of prosumers are not only independent from the energy oligopolies, which would create a libertarian, atomized system; rather, in a dispersed but interconnected and interdependent system, prosumers become a vital component without which the system cannot work.[18]

While some of its proponents emphasize the libertarian aspects of ED, others focus on the importance of public ownership. Northeastern professors Mathew Burke and Jennie Stephens write:

> Central to an energy democracy agenda is a shift of power through democratic public and social ownership of the energy sector and a reversal of privatization and corporate control. Energy democracy seeks

> to shift control over all stages of the energy sector, from production to distribution, and extending to infrastructure, finance, technology and knowledge while reducing the concentration of political and economic power of the energy sector, particularly within the electricity industry.[19]

In most areas served by IOUs, local ED proponents are far from realizing their vision of an all-local, all-public electricity system. Transferring ownership of the physical distribution system to a new public entity, or *municipalization,* is a complex, expensive process that has a low record of success. To acquire an IOU system, a local group must raise millions of dollars to purchase the system, overcome legal disputes with their IOU over valuation and other issues, and engage in a lengthy process of planning and implementing changes that allow one formerly seamless grid to operate as two semi-independent, separately owned networks. Then, the local government must create a new organization able to perform all utility functions from day 1 forward.

For all these reasons, municipalization has been pretty rare so far. The City of Winter Park, Florida, acquired its system in 2005 after a five-year legal battle with its former utility—the first and only new city utility in Florida since the 1940s.[20] At least one German city also municipalized via referendum during the Energiewende, and the City of Boulder, Colorado, is in the midst of a conversion process that began in 2010 but will not be operational until perhaps 2023 or later.[21]

While some ED proponents still strongly favor outright grid ownership, others have come to the view that full possession is not necessary to achieve many important social goals. The former mayor of Boulder, Will Toon, who began the municipalization effort, no longer favors this approach. "We need to be acting on climate change now," Toon says. "I'm fine with changing utility business models, but not if it delays climate action." Noting that both pro-coal and green activists favor less control over local utilities by state authorities, "you could end up with a right-left alliance that could paralyze action on clean energy while we argued about the details [of how to organize the industry]."[22]

Recognizing the difficulties of forming a grid-owning public utility, many ED proponents have instead turned to an easier-to-achieve structure called Community Choice Aggregation (CCA).[23] Under this approach, the sale of electricity itself is separated from the delivery of electricity by the local distribution utility, exactly as occurs under retail deregulation. However, instead of creating a competitive market of electricity retailers and letting customers choose their supplier from among them, a new quasi-governmental nonprofit agency purchases electricity on behalf of all citizens in a city. The CCA thus becomes the default power supplier for all city residents, taking the place of both deregulated retailers and the "last resort" service offered by distribution utilities in the usual retail choice setting.

CCAs have taken off surprisingly quickly in the United States. By 2016, CCAs in seven states were serving about 3.3 million customers.[24] As of now, eight U.S. states allow CCAs, and legislation is pending in four more.[25] In California, where the idea originally took root, current projections are that community choice aggregators will serve 80 percent or more of the former customers of IOUs by 2025.

CCAs are a pragmatic marriage of two ideologically opposed ideas. The concept of getting the local utility out of the business of supplying electricity itself comes from the finding that the generation function is no longer a natural monopoly. So, a competitive market will supply electricity more cheaply while the local distribution utility still delivers all power. The idea of forming a CCA is that it is cheaper, and advances climate goals faster, if all buyers band together and buy from a single nonprofit retailer that purchases generation from the wholesale market on everyone's behalf. There is no longer *retail* competition; there is simply a nonprofit wholesale buying agent for everyone. This nonprofit buying agent purchases by soliciting bids from wholesale suppliers. So, the locus of competition is the wholesale market.

Due to its varied forms, ED does not map easily into the industry's new business and regulatory models. New grid-owning municipal utilities, while probably remaining rare, will likely follow their existing brethren and trend toward the integrated, full-service ESU model. In Winter Park, for example, the clearest benefit of their municipalization was greater control over distribution reliability, not lower power bills.[26]

Since they do not own their local grid, CCAs will have to interact with customers and utilities the way competitive ESCOs work in SI markets. In fact, CCAs are simply a single nonprofit community-wide ESCO. While most current CCAs sell only electric power, they may evolve into multiservice providers like private competitive ESCOs. In summary, one form of energy democracy begets new local ESUs, while the other creates a new variant of the SI-ESCO industry structure with all its attendant complexities.

One way or another, the ED movement will be a force in the industry for local control, public ownership, and strong climate policies. It will create an impactful counterview to an entirely private, marketized, and globalized industry that will directly or indirectly confer some advantage on the ESU model. However, any such advantage will undoubtedly depend heavily on place and circumstance.

Political Fault Lines

Beyond the immediate policy debates over Big Tech, privacy, and Energy Democracy, the tectonic political forces at work in the world today will undoubtedly shape the industry's future path. They are likely to be important, if not decisive.

We appear to be witnessing the end of an era when privatization, globalization, and market capitalism were the unquestioned preference of Western government leaders. Instead, we are seeing the rise of right-wing populism and, in response, a tug of war between mainstream center-left parties and stronger left-wing policies. As this book goes to press, many European countries divided along these party lines are locked in struggles to form governments, and the United States and many other countries are bitterly polarized.

Along with much larger impacts, the evolution of this deeply unsettled political picture will undoubtedly influence future utility structures. The populist right is generally anti-government and views regulated utilities as instruments of the state. Bill McKibben quotes an arch-conservative activist named Jason Rose who says, "Solar should be our issue. Obamacare is bad because it diminishes healthcare choices. Public education is bad because it diminishes school choice. You'd think it would apply as well to energy."[27] Rose's advice has been followed by a group of green energy libertarians like Erik Curren, the author of a manifesto called the Solar Patriot. In it, Curren likens American citizens' "right to" solar energy to their freedom from government tyranny and interference. To Erik, that tyranny comes in the form of government and utilities conspiring to squash self-generation of solar energy.[28] "In many cases," he writes, "the government puts up barriers to solar that make it more difficult for families who want solar to afford it. Why? . . . Such roadblocks to solar merely act to protect the profits of monopoly electric utilities." This school of thought meshes nicely with the zeitgeist of anti-utility disruption emanating from Silicon Valley. Indeed, a widely reported survey of Silicon Valley's leaders found them left of center on nearly every major issue except industry regulation, where their views strongly coincided with Republican activists.[29]

Conversely, a revitalized center left will harness the power of the state along with the private sector to expand and "democratize" renewable energy. The latest big idea to emerge from this constituency in the United States is a "Green New Deal" (GND)—an ambitious proposal with the goal of 100 percent renewable power throughout the United States by 2030. The proposal is too new to signal much thinking about the specific role of utilities in the program and the resulting implications for industry structure. Practically speaking, however, shifting the power generation base as quickly as GND proponents favor will require extensive engagement by current utilities and will probably push the industry in the direction of ESUs. As Will Toon observed in the context of Boulder's changes, rapidity of climate action tends to favor existing or little-changed business models.

It is probably worth noting that a variation of the ESU model that integrates competitive markets represents roles for government and industry close to the center of the current left–right political spectrum in the super-polarized West. Utilities following this model are well to the left of the libertarian vision of minimized regulation and unfettered markets, but also well to the right of complete public ownership. This point is made in an interesting context by Jerry Oppenheimer and Theo MacGregor, two former utility regulators who argue that U.S.-style regulation represents an effective means of achieving centrist outcomes in a democratic setting.[30]

Oppenheimer and MacGregor argue that U.S. electricity prices are lower than almost any other country served by IOUs, while U.S. service is universal and highly reliable (though not world leading). At the same time, U.S. utility workers are paid comparatively high wages and good benefits and are comparatively high in unionization. They conclude:

> What America has is the toughest, strictest, most elaborate system for regulating private utility corporations found anywhere in the world (with the possible exception of Canada). This may come as a surprise. America, after all, has sent out an army of consultants to every corner of the Earth to extol the virtues of deregulation, free markets and less government. But this is an export-only philosophy, not applied within the US itself, and for good reason. In the land of free enterprise, a century of practical experience has led Americans to adopt as faith the idea that public services—especially those owned by stockholder corporations—are unique monopolies which the government and public, not markets, must tightly control.
>
> . . . It is an extraordinary exercise in democracy, and it works.[31]

From this standpoint, asking about the future organization of the utility industry is essentially asking whether the center will hold in the political economy of utility regulation. Where the Silicon Valley and libertarian market are dominant, the SI model is likely; if there is a shift toward the left, the ESU model and public ownership will gain much more traction. And if, for whatever reason, the center holds, we will see business and regulatory models that combine utilities, regulation, and markets in pragmatic, adaptable forms we can call the *hybrids.*

14 Money Talks

In the financial capitals of the world utilities have a place of their own. Utilities have their own stock and bond indexes, index funds, and mutual funds, as well as their own categories in Standard and Poor's (S&P) Global Industry Classification Standard. Renewable energy firms, independent power producers, and energy traders need not apply. Utilities also have their own society of financial analysts, their own specialized financial conferences, and an investor base that includes many sophisticated, long-horizon investors such as pension funds and university endowments.

The attributes that distinguish utility securities are well known to every active investor and are closely tied to utilities' regulated business model. Electricity is an absolute necessity, yet takes a very small fraction of the family or most business budgets. So, electricity bills are almost always paid, and revenues are among the most stable of any industrial sector. "Utilities have historically been viewed as a safe haven," writes one Fidelity Investment expert, "providing ballast for a diversified portfolio when markets turn downward . . . They are primarily U.S.-based companies, and are less economically sensitive, as most consumers pay their utility bills during both strong markets and recessions."[1] One writer for Barron's calls them "bond substitutes."[2]

The second main financial attribute of utilities is the fact that most U.S. utility stocks pay a relatively high cash dividend—twice the S&P 500 average right now.[3] This means regulators are generally setting rates high enough for utilities to earn revenues that provide the opex and capex necessary for good service and still have cash left over for dividends. As we've noted several times already, regulation and public ownership also create a presumption that this strong oversight lessens the chance of serious financial underperformance or bankruptcy.

In recent decades, utilities have also been viewed as slow-growing firms without a high potential for large share-price appreciation. Up until ten years ago, sales were growing at several percent a year with stable or rising prices. So, revenues and capital outlays naturally grew steadily at a modest pace. During the ten-year sales flattening we've just finished, most utilities managed to keep growing via mergers and due to the lucky fact that fuel and bulk power prices (their single largest operating cost) declined. Many have also grown by getting into renewable power generation—more about that in a moment.

Even so, most investor-owned utilities can't take growth for granted and aren't standing still. The forces of change are motivating them to search for growth in new regions of the three-dimensional strategy space of Figure 9-2. As seen in the last several chapters, utilities are adding deregulated affiliates selling new energy services, expanding geographically, and moving toward new regulated business models. The whole purpose of these strategy moves is to increase actual and financial performance. In financial terms, the shifts within the Figure 9-2 space should increase the growth rate of earnings while they improve overall service and performance without adding risks large enough to offset these gains.

Some movements within the Figure 9-2 space have risk / return tradeoffs that are easy to understand, though not always easy to manage. Acquisitions that expand current services into new geographic areas, by purchasing distribution utilities, have been very common all over the United States, Europe, and Latin America. This offers automatic increases in size and earnings balanced by idiosyncratic operating and management risks. Many expansions are adjacent. So, the new territory and utility are from nearby polities and business cultures that are somewhat familiar. Along these lines, Enel (Italy) acquired Endesa (Spain) in 2009, and Ameren (headquartered in St. Louis, Missouri) acquired two more Illinois utilities in 2003–2004. One common feature of these acquisitions is that they tend to occur more in areas that speak the same language, as when the UK's National Grid came into the U.S. Northeast or Enel / Endesa expanded into Latin America.

The utility sector also has a long record of creating deregulated affiliates selling related services in competitive markets. The largest expansion by far has been into large renewable generation operating in bulk power markets. This particular move offers a risk / reward profile similar to, and occasionally even better than, regulated firms. As we learned in Chapter 8, nearly all renewable energy generation firms build plants and sell under long-term contracts with known prices for eight to twenty years. The counterparties to these contracts are mostly utilities or blue-chip corporations, both of whom are extremely unlikely to default on their payments. Renewables firms therefore may have even more stable long-term revenues than utilities and even less regulatory risk, once they are operating. And because renewable energy is gaining rapidly in market share, the growth of these firms is not constrained by the slow or negative growth of total power sales. All this applies to grid-scale storage as well.

Downstream retail competitive services are another matter entirely. As mentioned in the discussion on ESUs, the market for energy service offerings (demand response, building energy management, distributed generation) is likely to be highly competitive and yield highly variable cash flows. Utilities almost always buy existing companies rather than create them from scratch, but they still face the need to grow these firms into global competitors. As these firms become larger, they change the financial personality of the utility.

The risk / return proposition on these acquisitions will be the same for utility owners as it is for all other owners unless utilities find and harvest synergies between the regulated core companies and the acquisition. These are the same potential synergies we discussed in the context of an ESU, except that in the latter, these cross-service value additions are within one regulated firm. Absent strong synergies that lower the acquisition's cash-flow variance, its risk / reward proposition generally will not match the low-risk / slow-growth / high-dividend model of the regulated parts of the firm. With the possible exception of contract-anchored power and storage businesses, utilities that expand their deregulated operations are therefore creating, in the eyes of Wall Street, a hybrid or integrated electricity enterprise.

Wall Street and New Business Models

The financial community has spent decades watching utilities get into and out of new territories and new affiliated businesses. Financial experts are now quite accustomed to evaluating the upsides and downsides of these kinds of strategy moves on a case-by-case basis and pricing this into their valuations.

In contrast, new regulated utility business models such as the SI and ESU are a somewhat rare development. Wall Street experts and financial analysts have written almost nothing about how these new models could affect utility valuations and stock prices. I have long been curious about why this was the case, so I interviewed several well-regarded Wall Street utility analysts, asking them specifically about how they view the emerging new models.

I was quite surprised by their views. First off, they understood the new utility business models but were not especially excited by either one. "Investors like nothing in the way of exotic," one analyst told me, speaking of course about the conservative utility investor base. Wall Street's time horizon rarely extends beyond five years, and looking back over the last five years or longer, the current utility business model has performed extremely well. To improve financial performance, which is what moves The Street, there is already a very high bar. The experts I spoke with did not expect that financial performance was likely to get much better than it is now under any new model.

The most interesting point I took from the interviews was that Wall Street did not have an inherent preference for any one of the private ownership models on the Figure 9-2 spectrum. Instead, it has a strong preference for any one of them that has lasting buy-in from its regulators and surrounding stakeholders. "The most important part of any new deal is the regulatory compact," one analyst told me, "not the specifics of the new models." This view squares with observations in utility finance theory that political and regulatory risks are the most damaging to utilities because they cannot be diversified away or hedged.[4]

Along these same lines, my interviewees were not very enamored with performance-based regulation (PBR). Part of this is undoubtedly sheer caution toward anything new. In addition, some analysts mentioned a lack of confidence in the ability to set and then measure performance metrics without lots of added risk. There is not a long industry history of setting very specific performance targets and then allowing profits to fluctuate according to progress against them.

I pressed the experts for specific views on the risk / return trade-off differences between SIs and ESUs. Broadly speaking, most of them agreed that the SI business model was closer to current natural monopoly regulation and therefore less likely to be exposed to regulatory and political shocks. The SIs are utilities sticking to their own knitting and staying away from other services, sold either through the regulated firm or deregulated affiliates. In this case, risks should be low and revenues stable. However, the growth prospects for this kind of utility are more limited, and depend on getting favorable PBR treatment or expanding services into precisely the areas analysts don't like.

ESUs or conglomerate utilities provide a different risk / reward proposition. There are more prospects for growth, but also many more regulatory and political risks and a wider political bargaining space, as explained in Chapter 12. While this may suit the public interest in the long run, a larger latitude to introduce new products, buy new companies, and make political bargains inherently means more scope for missteps that reduce earnings one way or another.

What about utilities that have already tried to step out and become the leading avatars of new business models or deregulated service diversification? The analysts were keenly aware of these efforts, but so far saw them as too small to affect overall financial performance. "They are all science experiments," one analyst told me. Another told me of an unnamed utility that told its investors that these "science experiments" were expected to reduce earnings by something like 2 percent, or about five cents a share. The response from investors was overwhelmingly negative, and the company scaled back its plans. As noted, the one area in which many utilities have diversified into "hybrids" with growth and good performance is in owning and operating large generators, especially zero-carbon fleets. Even so, one analyst pointed out that hybrid companies commanded a lower valuation, even if they arguably had higher growth prospects. The hybrid company Exelon, he noted, trades for fifteen times earnings, while an undiversified Wisconsin utility he was watching traded at twenty times earnings.

These views were surprisingly similar across different financial experts at different companies. Assuming my sample wasn't somehow tainted, utilities are going to face a huge challenge moving very far in *any* direction in the Figure 9-2 strategy space. Wall Street's view is dead simple: *anything* that increases volatility for this kind of stock is unwelcome. They see it as exceedingly unlikely that regulators, policy makers, and market forces will come together to create a better risk / reward package than utilities have enjoyed in the last decade, and may continue to enjoy if the global economy slides sideways or backward over the next five to ten years.

As if we needed another reminder, these attitudes place regulators squarely at the center of utility business-model changes. It is not unusual to see Wall Street insisting that regulators make sure that the risks of investing in utilities are commensurate with the returns—this has been the core question in every rate proceeding from time immemorial. What's different this time is that changing a business model is much more structural than deciding a rate case. Whereas the original regulatory compact was a balance between only two groups—investors and customers—the updated compact has more constituencies and considerations to balance, not least of which is the health of the planet.

15 Power without Carbon

If anything is clear about our energy future, it is that the electric power system isn't going anywhere but up. The unsightly wires that bind us into a single unified grid may soon fragment into microgrids or traverse whole oceans and continents, but they aren't going away. The most essential of our sixteen essential infrastructures is going to become even more critical as it becomes the backbone of the Internet of Things, as well as a primary transport and industrial fuel. There is no realistic pathway out of our carbon emergency that does not involve larger electricity use.

Though it may sound counterintuitive, the pressing future demand for clean power only intensifies the importance of energy efficiency and clean nonelectric fuels. EE has long been an underutilized but popular resource with excellent economic and environmental co-benefits. In the fast decarbonization and electrification scenarios we need, however, efficiency moves from a money-saving virtue to a necessity. As we saw in the example of Boston's 2050 power demand, much stronger efficiency means the difference between an expensive, volatile local grid and one that is cheaper and easier to manage.

The world needs all the affordable and reliable carbon-free power it can get with acceptable environmental and social costs. In some instances, we will undoubtedly have to choose between distributed sources and the large-scale grid, but in most cases, it will be surprising if distributed energy resources displace rather than complement the Big Grid's growth. Where we must make explicit trade-offs, they should be based on a thorough, transparent evaluation of the public interest, not our preferences for one particular scale or animosity over comparative historic levels of policy support. And regardless of the ultimate balance, it is critical that we continue to improve market structures and regulations that efficiently integrate and compete Big- and Small-Grid resources.

Technologically speaking, many paths to power systems without carbon are becoming clear. We do not yet have a full suite of economical solutions at scale, but I would easily take a bet that technologists will find and improve our options faster than we can put them to work. However, winning this bet requires that we support and expand the clean-energy innovation chain using proven, cost-conscious R&D policies at multiple governmental levels.

Along with stronger innovation support, we need much better multi-fuel infrastructure-planning processes. Gone are the days when electricity regulators could issue permission to build a new line or plant based on simple forecasts of uncontrolled system load, but electricity planning still largely occurs within its own vertical and across areas too small to harvest all opportunities. Instead, multiregional and multi-fuel energy-planning processes should evaluate the trade-offs and synergies between the electric grid, new clean-fuel infrastructures, a carbon transport and storage system, heat storage and use, and much of the transportation system. This is a vastly more interconnected and complex planning challenge, but it lies at the heart of the carbon transition.

As we saw earlier, U.S. energy planning has an abysmal track record. Rather than serving as an excuse for avoiding plans, we should see our failure to act to stem climate change as another failure of planning and change our approach. Sound price signals are essential tools, but they cannot create highly capital-intensive new infrastructure sectors without plans and government support. By its very nature, energy infrastructure is always the slowest and most difficult element of the system to change, but it is also the most important enabling resource for solutions at scale.

There will always be a healthy tension between planning, regulation, and the use of markets, but all of them are essential. In the West, electricity has gone from a fully regulated commodity to one that is extensively governed by market forces at many, many points along the value chain. In both the Big and Small Grids, there is little chance that reliance on markets will diminish—it is now simply a part of the industry's DNA. If the history of twentieth-century deregulation teaches us anything, however, it is that modern-day markets in infrastructure sectors require more and smarter governance than the regulated industries of yore, not less.

Without a doubt, the most difficult transitions will occur on the Small Grid, where decarbonization is only one of a number of disruptions to all of the current business and regulatory models. Smart technology is enabling agile grids and microgrids offering different levels of reliability. Communications and controls are set to create feedback loops between supply and demand at the level of millions of individual devices, as well as aggregations and entire systems. The same technologies are enabling customer defection and vastly more complicated pricing for energy and distribution services and two-way markets between customers, energy service companies, and the bulk power grid. Meanwhile, the distribution system will remain an increasingly expensive fixed-cost asset whose efficient function is critical to everything else.

Distribution utilities must navigate these challenges by choosing a course through the three dimensions of strategy: geography, deregulated product lines, and a business model for their regulated or public networks. Geographic diversification is already extensive in the industry, but the expansion into deregulated products has so far extended only to similar lines of business such as renewable independent power producers, which do not change the risk / return profile of utilities very much. The relatively modest forays of utilities into much different unregulated electricity products so far has been met with resounding investor caution, if not outright skepticism.

Meanwhile, the tech giants and smart, well-funded competitive non-utilities are circling the industry, dead set on stealing the customer relationship, along with the data and brand value that come with it. To confront this threat, utilities and their regulators have a spectrum of new models, each with its own surrounding ecosystem and challenges. For all their complexity, the two business / regulatory models that have served as centerpieces for our discussion—the Smart Integrator and the Energy Service Utility—are somewhat simplistic exemplars. With surprisingly strong foresight and encouragement from regulators, many utilities are making progress, adopting pragmatic variations and hybrids of the models, and this is immensely encouraging. However, the changes to date have largely been incremental—too small to awaken the sleeping giant of Wall Street but also perhaps too small to declare a success. A few years back, Navigant's prestigious strategy team looked at utilities' efforts to create new business models and saw "no clear contenders," adding that "rarely in other industries has an existing player made the leap."[1]

Amidst all this change, the most underappreciated challenge to either new business model is not batteries, Amazon, or rooftop solar, but rather the technological achievement known as microgrids. A utility's ownership of a single unified distribution system and consequent role in operation and delivery is what anchors the regulatory compact in all the utility business models, old and new. "The creation of microgrids that offer customers the option to secede from the larger grid will, in some respects, be the functional equivalent of municipalizing those grid assets and taking them out of the hands of utilities," two experts explained five years ago. If the distribution grid becomes a series of small micro-utilities that interact according to complex rules, it is possible that the advantages of scale, optimization, and non-discriminatory treatment can be preserved. Right now, however, we have no roadmap or examples that give us confidence that this will be easy.

In earlier writings, I've likened the transformation of the utility business model to replacing the control systems of an airplane while it is in the air. We don't take the power grid down for maintenance or to change out its generators or software. With the advent of artificial intelligence, 5G, microgrids, new competitor business models and partnerships, and the proliferation of new generation and storage technologies, I wonder if it is more like rewiring a flying plane while the software programmers change to a new language, the new parts each come from an entirely different supplier, and the Federal Aviation Administration is changing the rules on where and how planes can fly.

"An electric utility is an odd beast," Bill McKibben once wrote, "neither public nor exactly private."[2] He's right. Electric utilities have been an extraordinarily long-lived and successful public/private partnership. Broadly speaking, the role for such partnerships in addressing the world's many social, economic, and environmental challenges has never been larger. The next few decades will tell us whether and how to use the utility model to meet the needs of a world gone mad with connectivity and AI, climate change, and unmet economic and demographic challenges. Used wisely, it is probably one of the best tools we have to heal our freighted future.

Appendix A

SUMMARY OF POLICY RECOMMENDATIONS

Comprehensive Energy and Climate Mitigation Planning. Each region should create long-term energy and climate mitigation plans with the goal of reducing net greenhouse-gas emissions from all sectors to zero by 2050 at the latest. The planning process should:

- Maximize the use of cost-effective distributed resources and energy efficiency (EE; see policies below) and ensure that utility business model and incentives are aligned with this goal;
- Integrate consideration of all end uses and fuels, and their carbon-free alternatives, to minimize the cost of decarbonizing all end uses of energy;
- Make transmission system and other energy infrastructure planning an integral part of the planning exercise (see below); and
- Include the development of politically realistic, actionable policies that implement the plan, including reduced or accelerated processes for facilities within the plan. These plans should extend beyond engineering calculations to address facility ownership, financing, and revenue recoupment.

In this context, infrastructure includes systems for the delivery of all carbon-free fuels and heat, as well as the infrastructure needed to charge electric transportation.

EE and Housing Costs. As part of the adoption of a new utility business and regulatory model, institutions should be developed within or outside utilities that:

- Give responsibility for EE either to utilities or to another qualified actor[1] that has access to adequate low-cost capital for EE measures, adequate operating funds, and operations data and is accountable to qualified overseers;
- Collect and use the best available data and a comprehensive framework to measure the dollar savings, carbon savings, economic development, and other benefits of EE versus supply-side alternatives;
- Adopt incentives, targets, and oversight policies consistent with allocated responsibilities so that the accountable party can profit from good performance;
- Structure new utility ecosystems [including those using artificial intelligence (AI) and the Internet of Things (IOT)] and electricity pricing policies to preserve or strengthen the incentive to adopt cost-effective EE;
- Make maximum use of codes and standards and similar measures to transition all structures to as close to net zero as possible;
- Ensure there is a designated agency / mechanism to deliver EE and low-carbon mobility programs to underserved markets and that all EE policies are shaped to expand housing affordability, not raise housing costs; and
- With the increased penetration of renewable resources, ensure that system operators have the capability to incentivize load shifts and dispatch load.

Distributed Generation, Pricing, and Access to the Grid

- Distributed generation (DG) owners should be compensated for the services they contribute to the grid at the fair value of these services, but not more; these services may include location-specific partial displacement of new distribution and transmission investments ("non-wires alternatives"). Automatic payment of full retail prices for DG power sent back to the grid, or a feed-in tariff higher than value, is not an accurate proxy for value.

- Policies that promote or mandate DG ownership (as distinct from paying for) should recognize the local economic development and community benefits of DG, should not reduce efficiency price signals, and should operate so as to make housing more affordable overall.
- Distribution system regulation should provide incentives and oversight to distribution utilities to host and integrate DG efficiently, but should also allow them to provide DG and other carbon-free energy supply options to all customers at fair prices in a manner consistent with the overall regulatory / business model.
- Distribution system planning should include accommodation and encouragement of new electrification loads emerging from climate / energy plans, as well as the potential for investment displacement from local supplies.
- Better coordination is essential between integrated resource planning and distribution system planning processes to ensure that the investments are co-optimized for both generation and distribution system purposes.

Pricing and Access on the Distribution System

- Distribution system and retail service pricing should adopt time- and location-based pricing as soon as is feasible, balancing the costs of the investments necessary for adoption against the long-term benefits. The benefits of time-based pricing have been thoroughly researched and demonstrated, and time-based pricing can be adopted where technical feasible without hesitation.
- Further research on and experimentation with effective and efficient locational retail pricing approaches and distribution system management technologies that enable increased electrification and improve system control are important.

Microgrids, Resilience, and Grid Defection

Working with technical experts and stakeholders, regulators should adopt microgrid rules that:

- Enable the resilience and decarbonization benefits of microgrids to be realized;
- Ensure ownership, communications, cybersecurity, and control protocols that allow the entire grid to continue to function fairly and efficiently; and
- Allocate the unique costs of the microgrid to its beneficiaries.

These rules should be revisited periodically, as experience with microgrids will inform their evolution.

Regulators should also closely monitor complete grid defection to ensure it does not jeopardize the financial or technical viability of universal service. Policy makers should ensure all customers continue to pay their properly determined share of the cost of maintaining a universal grid, and of the cost of services intended to be funded by universal grid charges.

Choosing Utility Business and Regulatory Models

Chapter 9 and Figure 9-3 describe the spectrum of business and regulatory models that distribution utilities can adopt. Each nation, state, and utility should consciously adopt one form of these models that:

- Continues to meet all the public-interest objectives of electricity service;
- Enables the fastest possible decarbonization via a just transition;
- Enables the distribution utility to be able to raise adequate capital at reasonable rates and otherwise function effectively; and
- Harnesses market forces and innovation via internal processes (such as competitive procurements) and external processes (such as opening a service to competition).

At this point, no model on the spectrum of options can be shown to dominate the others in terms of their effectiveness, especially when considering the current structure and evolutionary path of each utility. In general the natural transition for utilities operating in restructured or liberalized markets is toward the Smart Integrator (SI) model, while publicly owned and vertically integrated utilities will generally transition most naturally into Energy Service Utilities (ESUs). However, business models should not be rigidly applied, so that some hybridization will be the rule rather than the exception. The resulting business model should ensure that the customer is best served in terms of affordability, innovation, and the availability of green choices, and that reliability is not compromised.

Regulation of distribution, transmission, and / or vertically integrated utilities should migrate toward performance-based regulation (PBR). The precise performance objectives and incentive mechanisms will depend on the specific business model and circumstances. Regulators and utilities should take advantage of extensive experience and research with PBR.

Each utility should evolve its core competencies to match its business model. In general, SI utilities must learn how to manage their platform to achieve public-interest and business objectives together at the center of a fast-changing ecosystem. ESUs must learn how to deliver efficient mass-customized energy services by harnessing the external marketplace. All types of utilities must adopt modern customer engagement tools that allow them to understand and communicate with their customers.

Transmission System Planning

- As part of climate and energy planning, planners should consider the full value of transmission expansion to areas with low-cost, carbon-free power, as well as other grid enhancements and technologies that increase the grid's capabilities and support decarbonization, reliability, security, resilience, geographic diversification, and improved market competition.

- The energy / climate plans that give rise to transmission plans should first include EE, DG, and other resources that reduce the need for new transmission, but must also prioritize the fastest possible decarbonization of the energy system at reasonable costs.
- The environmental costs of new transmission lines should be mitigated to the maximum feasible extent, but should also be weighed against the costs of climate inaction.
- National governments (notably the United States) should provide national-level leadership and establish stakeholder mediation processes that enable new or expanded transmission projects emerging from these planning processes to be permitted, funded, and built without substantial delays.

Wholesale Power Market Reform

- Centralized energy spot markets should be adopted where they aren't yet in place, applying best practices learned from extensive worldwide experience.
- The scope of centralized markets should steadily be expanded to include the additional power services needed for a flexible, resilient, fully decarbonized grid, with proper market oversight and backstops.
- Long-term markets for energy and capacity should function alongside short-term markets.
- Forward capacity markets that use demonstrated best practices should be adopted wherever possible; they should extend as long as possible into the future and add unbundled capacity services that allow system planners and operators to use carbon-free technologies as efficiently as possible.
- Regardless of whether forward capacity markets exist, long-term contracts between willing buyers and sellers should not be prohibited or disadvantaged unless they disrupt a proper regional energy / climate plan or are otherwise contrary to the public interest. Whenever possible, long-term contracts should be competitively sourced.

Research and Development of Enabling Technologies

- International, national, and subnational efforts to fund R&D on new technologies that will enable or reduce the cost of decarbonizing all sectors of the economy, or otherwise lessen the effects of climate change, should be accelerated, consistent with the goals of Mission Innovation.

Increased Educational and Technical Resources for Energy Regulators

- Energy regulators and public-sector officials involved in electricity and energy policy and regulatory decisions should have access to high-quality, impartial advice and educational resources on the many new aspects of energy regulation; in the United States, this should include a national academy for regulators and energy policy makers.
- Energy regulatory agencies should have adequate funding, be able to hire highly qualified staff members, have access to advanced technical advice and resources, and be insulated against "regulatory capture" as much as possible.
- Regulators will need to expand skill sets in the areas of PBR, distribution system management technology, distribution and retail pricing, the role of AI and IOT in the grid, cybersecurity, transmission planning, and the monitoring, regulation, and oversight of price-deregulated markets for new distribution and retail services.

Carbon Pricing

- To help decarbonize the power system and the rest of the global economy, a carbon price signal in one form or another should be adopted; prices should be set at an effective level as close as possible to the social cost of carbon emissions.
- The revenues generated by carbon price signals should be recycled back into the local economies so as to make the overall effect of the

signal socially beneficial and equitable, including enabling a just transition for affected workers.

- On a case-by-case basis, the implementation of an adequate carbon price signal may remove the need for other sector- and technology-specific carbon mitigation policies, but where they are already in place and effective, these policies should not be discontinued until there is strong evidence that the carbon price signal is having an equally effective effect.

Appendix B

THE CHALLENGES TO ENERGY SPOT MARKETS WITH INCREASED WIND AND SOLAR GENERATION

Long before deregulation came along, traditional integrated utilities created markets in which they could trade small surplus amounts of the product now called energy. When one utility had a little more energy coming from its plants than it needed, and another nearby utility needed a little extra, exchanging the surplus for cash was an everyday occurrence. This trading was mainly bilateral, with prices set by regulator-approved formulas.

Restructured Big Grid markets continue to allow bilateral short-term energy trading, but the dominant form of a competitive wholesale energy market today is an auction-style market, run hourly (or an even shorter period) and daily, for energy delivered at a number of "trading hubs" within a region. As in the capacity markets section in Chapter 8, a single market-clearing price is paid by all buyers to all winning sellers. In this case, winning means that the market operator, which physically controls ("dispatches") the system, allows your plant to turn on and supply energy for the period you won and earn the associated revenue. The limits of the auction region, set by transmission constraints and jurisdictional boundaries, can be as large as four or more EU countries or U.S. states.

Because competitive energy markets were adapted from the original utility-to-utility energy versions, they were fashioned around the old industry design paradigm. MIT professor Paul Joskow writes:

> Perhaps ironically, the conceptual basis for the design of organized wholesale electricity markets in the U.S. during the late 1990s . . . can be traced directly to the mid-20th century economic-engineering literature on optimal dispatch of and optimal investment in dispatchable generating facilities . . . These models were developed to apply to pre-restructuring vertically integrated . . . monopolies subject to some kind of regulations, including government ownership.[1]

Under traditional regulation, the capital costs of power resources were automatically recovered through regulated rates once those resources were approved. This meant utilities never worried about earning enough revenue from energy sales to other utilities to cover the costs of their plant. Their objective was to cover the additional (incremental or marginal) costs of actually producing the power that they sold, plus a little profit to make the exercise worthwhile. A plant that could produce additional juice for three cents per kilowatt-hour, its per-unit cost of fuel and other variable costs, would want to charge a little bit more than this to make a profit on the sale. Charging less than three cents would be wrong, as the utility would lose money on the sale; charging far above three cents wouldn't be smart because the buyer might find another seller closer to three cents and you'd lose what would otherwise be a profitable sale.

When deregulated versions of these markets were created, the principle that every seller should offer to sell at just a little above its marginal cost transferred over. Economists have long agreed that in strongly competitive auction markets, the profit-maximizing sales strategy is to offer at your marginal cost. While this is the best of all bidding strategies, it doesn't guarantee that a plant owner will earn revenues sufficient to cover its capital costs. As explained in the main text, this is the reason why policy makers use contracts or capacity markets to assure "resource adequacy."

Competitive energy spot markets may not be great at inducing construction, but what they have been good at is making the most efficient use of old-paradigm system power resources that are already financed and constructed. Chapter 6 taught us that traditional power systems had baseload units that produced the cheapest energy (i.e., had the lowest marginal cost), followed by cycling units, followed by peaking plants. Put a system such as this into a restructured energy market, and each owner of each competitive power plant (i.e., each independent power producer) will rationally bid its marginal cost. The "offer curve" of bids will look like Figure B-1, with the first few bids from baseload plants, followed by cycling plant bids and then peakers. Each plant will have somewhat different fuel costs, operating efficiencies, and so on. So, the bids will ordinarily stack up gradually. The auction / market operator, whose job is always to maintain immediate, near-perfect balance between supply and demand, monitors the exact amount of power needed and marches up the "bid stack" until it reaches the level of supply that matches demand in that hour (see Figure B-1).

In old-paradigm systems, as long as sellers or buyers were not so large as to have market power, this approach worked extremely well. Existing plant operators were incentivized to find efficiencies that reduced operating costs and system operators nearly always selected as winners the group of old-school plants that together yielded the lowest total cost of meeting demand from the set of plants available to the system. This was, and is, the goal of energy markets—not promoting carbon-free resources or reducing carbon from existing sources. Like traditional capacity markets, they were all about greater economic efficiency for all costs that society charged energy producers.

In restructured areas, ever-increasing amounts of wind and photovoltaic solar plants have offered their surplus energy into these auctions.[2] Once built, these plants pay nothing for fuel and have almost no other marginal costs. So, their economically efficient bid is just a little above zero. Market operators put these energy bids into the bid stack first, since the stack always goes from lowest to highest. If the rest of the plants in the market are unchanged, this has the effect of shifting the bid stack to the right. As shown in Figure B-2, if demand hasn't changed the market-clearing price—the highest bid in the stack needed for cumulative bids

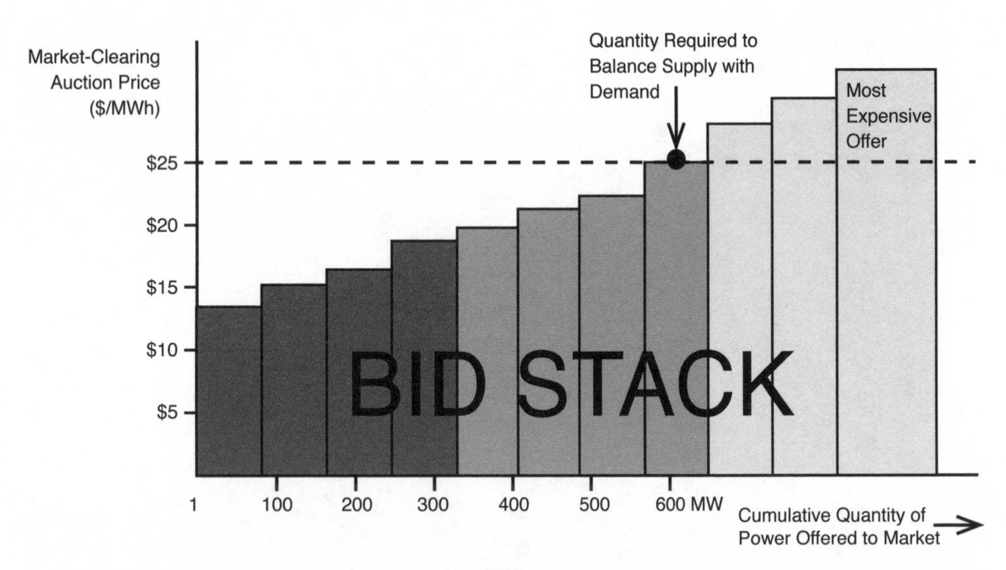

Figure B-1 Power Supply Bids into Spot Markets by Traditional Plants and Clearing Price for One Hour. To set the market-clearing price paid to all winning bidders, the system operator stacks up the bids from generators from lowest to highest and then finds the bid where cumulative offered megawatt-hours equals the system demand. D1 is the demand curve and VRE stands for variable renewable energy. Since there is none of this energy in this picture, both supply and demand are labeled "no VRE."

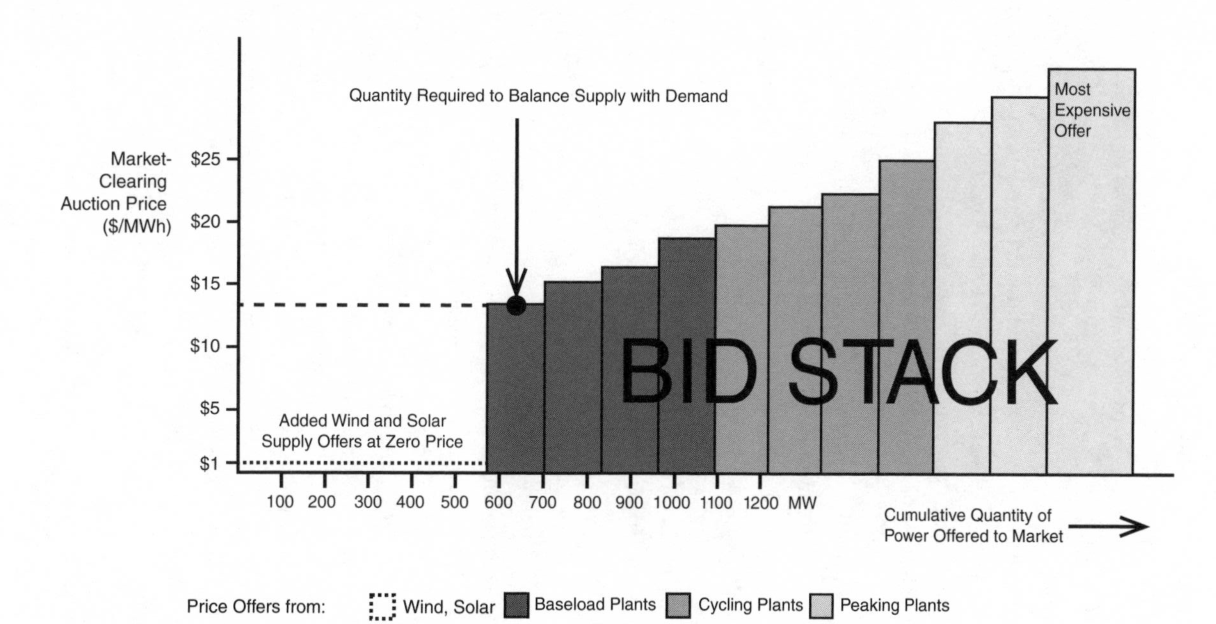

Figure B-2 Power Supply Bids in Market with Wind and Solar as Well as Traditional Plants.
When VRE in the form of wind and solar are added to the bid stack, and demand is unchanged, the market-clearing price drops considerably.

to equal demand—is much lower. Simply put, the addition of large amounts of near-zero marginal-cost bids from wind and solar into competitive energy markets causes market-clearing prices to plummet.

This effect has already been seen in nearly every energy market. Careful work by two University of California economists, James Bushnell and Kevin Novan, has isolated the effect of greater solar and wind generation in California. They find a "substantial decline" in daily average prices, with especially large effects during the midday hours when solar generation is largest. During the four-year period 2012–2016, average hourly California solar generation at 11:00 a.m. went from about 500 to 6,500 MWh; average hourly prices for that hour dropped 42 percent, from $35 to $20 per megawatt-hour.[3] Simulations of future systems with wind and solar together supplying about 37–47 percent of all energy conducted by the Lawrence Berkeley National Laboratory show even stronger effects, including regions with market-clearing prices equal to zero for 700–1,400 hours.[4] These results aren't a surprise, as the Texas and New England markets each had about fifty hours with zero prices in 2018; Germany had almost a hundred.[5]

There is a bit of irony in the fact that energy spot markets, which weren't remotely designed with the thought of favoring wind and solar, end up always making them winners—nearly the opposite of capacity market outcomes. Nevertheless, this isn't nearly as good as it looks for the giant, multifaceted Big Grid retooling needed to fully decarbonize. For starters, lower energy prices mean less revenue for every type of asset that sells energy, including important balancing plants, transmission lines, and storage facilities. Put differently, lower energy prices exacerbate all of the new-plant revenue deficiencies that contracts and capacity markets are trying to solve, placing even higher burdens on these mechanisms. They cut even more directly against scarcity pricing being a success, since this approach depends on very high occasional prices that happen far less often when the market is flooded with zero-bid energy.

One of the most important changes to energy markets is to allow customers to manage their electricity use in response to energy spot price changes in real time. This includes customers changing their actual pattern of use, choosing when to charge or discharge storage attached to the distribution systems downstream, choosing when to charge electric vehicles or send power back from vehicle to grid, and many other activities generally grouped under the rubric of demand response or price-responsive demand. This critical enabling element for Big Grid decarbonization requires changes to technologies, regulation, and pricing in the downstream distribution utility Small Grids. These changes will create a pricing and demand feedback loop between the Big Grid's giant auction markets and faraway individual customers. This presents us with a nice point of departure for the deep dive we are about to take into the downstream utility future.

In summary, competitive energy spot markets designed for old-paradigm systems simply don't make sense *by themselves* for new power systems that must blend zero-marginal-cost and high-marginal-cost resources. There is a rapidly emerging consensus that these markets are not so much wrong as incomplete. They must be combined with markets for other types of power service products, as well as the contracting and / or capacity market mechanisms discussed in the main text.

Appendix C

SOURCE NOTES FOR FIGURE 2-2

United States (1.67, 1.49, 1.18)—Electricity Consumption in 2050 as a Multiple of 2016 Consumption

The National Renewable Energy Laboratory's Electrification Futures Study was used to determine 2050 electricity consumption for three scenarios: reference, medium, and high. In the study, figure ES-3's historical and projected annual electricity consumption shows three scenarios evaluated for electricity consumption growth to 2050. Current consumption for 2017 is approximated to be 3,900 TWh.[1] The three scenarios of high, medium, and reference show electricity consumption values of 6,500, 5,800, and 4,600 TWh, respectively. Comparison of 2050 consumption values to the 2017 values yields ratios of 1.67, 1.49, and 1.18.

United States (1.73, 1.33, 1.77, 1.80)—Electricity Generation 2050 as a Multiple of Electricity Generation in 2005

The U.S. Mid-Century Strategy (MCS) was used to source 2005 value of 15 EJ (approximated from graph) as well as 2050 values for four scenarios: MCS Benchmark 26 EJ, Smart Growth 20 EJ, No CCUS 26.5 EJ, and Low Biomass 27 EJ.[2] Comparison of 2050 values to 2005 values yielded ratios of 1.73, 1.33, 1.77, and 1.80, respectively.

United States (1.32)—Electricity Use in 2050 as a Multiple of Electricity Use in 2018

The U.S. National Electrification Assessment published by the Electric Power Research Institute projects electricity use in the United States to grow by 32 percent.[3] The 32 percent growth was assumed to be relative to the approximate publishing time of the report, 2018.

Canada (2.15)—Electricity Demand in 2040 as a Multiple of 2016 Demand

Using the Statistics Canada website, total generation of electricity for 2017 was determined to be 650.2 TWh.[4] This total generation includes exports less imports. Demand for the year 2050 was determined using Pathways to Deep Decarbonization in Canada. Using the area chart titled "Energy Supply Pathways, by Resource," the 2050 demand value was estimated to be about 1,400 TWh.[5]

Germany (0.81, 1.00, 1.48)—2050 Electricity Demand as a Multiple of 2013 Electricity Demand

The current value of Germany's electricity demand was determined by estimating demand from the Germany Deep Decarbonization Pathway report as a value of 525 TWh. There are descriptions in the report describing the approximate values of three scenarios evaluated for Germany. The Government Target Scenario is about 100 TWh lower than the 2013 value. The second scenario—the Renewable Electrification Scenario—is described as about 250 TWh higher than the 2013 value (525 + 250 = 775 TWh). The third scenario is a 90% reduction in greenhouse-gas (GHG) generation, which estimates electricity demand about equal to Germany's 2013 level (~525 TWh).[6] Using 525 TWh for the 2013 current value and 425, 775, and 525 TWh for the three scenarios, the ratios of 2050 demand as a multiple of current electricity demand are 0.81 for the Government Target Scenario, 1.48 for the Renewable Electrification Scenario, and 1.00 for the 90% reduction in GHG scenario.

The UK (3.34, 3.35, 3.27, 5.10, 4.96, 4.82)—Electricity Demand in 2050 as a Multiple of 2017 Production

A range of values in the ratios is present due to the large number of studies and scenarios published in the UK. Starting with Pathways to Deep Decarbonization in the United Kingdom, figure 1, twenty-one different scenarios are represented from eight unique studies.[7] The eight studies' values presented were compared to the timeline of publication. No significant trend existed when comparing the studies. Two unique studies were chosen from those represented in the figure: The UK Energy System in 2050: Comparing Low Carbon Resilient Scenarios (UKERC 2050) and DECC 2050: 2050 Pathways Analysis. The two studies were chosen from the eight because their scenario values represent about the median values for power generation estimates across the total eight studies.

Starting with UKERC 2050, figure 2-1 has seven values unique from the reference value for 2050 energy demand in 2050. Of the scenarios presented in UKERC 2050, CAM (Ambition), CEA (Early Action), and CSAM (Super Ambition) were evaluated.[8] Values were estimated from the UKERC 2050's figure 2.1: Detailed results on Carbon Ambition Scenarios for CAM, CEA, CSAM: 4,250, 4,260, and 4,150 PJ, respectively.[9] Values were converted to terawatt-hours with the given conversion: 1 TWh = 3.6 PJ. Converted values were 1,181, 1,183, and 1,153 TWh.

In the DECC 2050 study, several trajectories and pathways were developed. Pathways alpha, beta, and gamma values for 2050 electricity demand are approximately 1,800, 1,750, and 1,700 TWh, respectively.[10] These three pathways represent scenarios and projections where the UK acts on climate change and technology policies in slightly different manners.

Current electricity demand was determined for the year 2017 using the National Statistics Digest of UK Energy Statistics (DUKES), which shows a 2017 total production value of 353 TWh.[11]

Mexico (2.59)—2050 Electricity Demand as a Multiple of 2015 Electricity Demand

Mexico's Climate Change Mid-Century Strategy, figure 29, was used to determine 2015 electricity generation and estimated 2050 generation. Current (2015) electricity generation is estimated to be 290 TWh, and in 2050, it is estimated to be 750 TWh.[12]

South Africa (1.88, 1.79)—Electricity Production in 2050 as a Multiple of 2010 Production

For South Africa's ratios, production was considered synonymous with demand for electricity. Two scenarios for electricity demand were evaluated in Pathways to Deep Decarbonization in South Africa, Economic Structure and High Skills.[13] From the chart titled "The total amount of electricity produced over time," current electricity demand in 2010 was estimated at 240 TWh. Within the same figure, economic structure in 2050 corresponds to a production of approximately 450 TWh, while the high-skills scenario corresponds to 430 TWh.

New England (1.31, 1.55)—2050 Electricity Consumption as a Multiple of 2017 Consumption

The Northeastern Regional Assessment of Strategic Electrification report was used to determine 2017 production of 259 TWh per year for the New England area. In addition, 2050 values were presented for two scenarios: max electric and plausibly optimistic. Values of each scenario for 2050 are 402 and 339 TWh, respectively.[14]

Portland, Oregon (1.72)—Electricity Sales in 2050 as a Multiple of Electricity Sales in 2017 for the Portland General Electric Service Territory

Values for electricity sales in the Portland General Electric area for 2017 and 2050 were gathered from Exploring Pathways to Deep Decarbonization. Electricity sales in 2017 were 2,242 MWh, and three scenarios—high electrification, low electrification, and high distributed energy resources—had electricity sales values of 3,866, 3,434, and 3,562 MWh, respectively.[15]

California (1.60)—Electricity Demand in 2050 as a Multiple of Electricity Demand in 2018

California Energy Commission's study Deep Decarbonization in a High Renewables Future describes an expected increase of 60% for California's electricity demand.[16]

Portugal (1.73, 1.36)—Electricity Demand in 2050 as a Multiple of Electricity Demand in 2014

A research report describing the decarbonization pathways for 2050 in Portugal models two scenarios: open system and closed system. Using figure 3 of the report, we are able to approximate the current Portuguese electricity generation to be 55,000 GWh for the year 2014. Figure 3 of the report also shows open- and closed-system scenarios, with 2050 electricity demand values of 95,000 and 75,000 GWh, respectively.[17]

Pacific Northwest (1.25, 1.60)—Electric Load in 2050 as a Multiple of Electric Load in 2019

Resource Adequacy in the Pacific Northwest discusses the 25 percent growth of electrical sector load with possible growth upward of 60 percent.[18] Growth was assumed to be relative to the current time of the published report, 2019.

Appendix D

SUPPLEMENT TO TABLE 6-1

Table D-1 Three Approaches to Identifying the Main Energy Elements in Power Systems

Type of Energy Needed	Obtaining This Energy in Traditional Systems	New Paradigm Systems as I Describe Them	Sepulveda et al. (2018) Description
Baseload—large steady blocks of around-the-clock power	Baseload plants, typically coal, nuclear, or hydro	Combine variable wind and solar resources over very large regions with long-term storage to create baseload energy blocks. Also, this category includes "firm low-carbon resources" (see right) where available	Use "firm low-carbon resources": hydro plants with high storage, fossil plants with carbon capture and storage, geothermal and biomass
Intermediate—large amounts of energy that vary considerably by season and time of the day	Cycling plants, most commonly natural gas, sometimes coal and hydro power	Same as above with more emphasis on seasonal storage (possibly including hydrogen-based storage)	Use "fuel-saving variable renewable energy sources"—these authors' name for wind and solar
Peaking—small but essential spikes of energy demand, usually during very hot or cold days	Peaking plants ("peakers"), typically oil or gas	Use electricity stored in chemical batteries, pumped storage hydro and storage technologies	Use "fast burst variable sources"—these authors' name for carbon-free peaking plants and storage

Notes

1. Les Jeux Sont Faits

1. Nordhaus (2013) pp. 1–2.

2. The U.S. Global Change Research Program's Fourth National Climate Assessment, issued in 2018, begins: "The impacts of climate change are already being felt in communities across the country. More frequent and intense extreme weather and climate-related events, as well as changes in average climate conditions, are expected to continue to damage infrastructure, ecosystems, and social systems that provide essential benefits to communities. Future climate change is expected to further disrupt many areas of life, exacerbating existing challenges to prosperity posed by aging and deteriorating infrastructure, stressed ecosystems, and economic inequality." Similar conclusions come from the most recent Intergovernmental Panel on Climate Change report (as of this writing; Hoegh-Guldberg et al., 2018).

3. According to a study led by Canadian scientist Katarzyna Tokarska, combusting the entire stock of known fossil-fuel resources would "transform Earth into a place where food is scarce, parts of the world are uninhabitable for humans, and many species of animals and plants are wiped out" (Tokarska et al., 2016). Another important source of climate change information is *The Lancet*'s countdown report—see Watts et al. (2018).

4. Department of Defense (2015).

5. Climate Home News (2012).

6. Wallace-Wells (2019) p. 1.

7. The World Bank. World Bank Development Indicators 2015, table 3.6, based on International Energy Agency data.

8. U.S. Energy Information Administration. Energy Transportation Use 2014, citing the International Energy Agency (2015c) Monthly Energy Review March 2015, table 2.6, and International Energy Agency (2015b), Key World Energy Statistics 2015.

9. International Energy Agency (2018).

10. There is a lively debate over the fastest pace at which a capital-intensive, highly interconnected energy system such as the power grid can change its generation sources. Vaclav Smil—one of the world's undisputed energy gurus—studied energy transitions and concluded that each one required many decades to achieve. Similar doubts are expressed in Temple (2018). Analysts such as Chris Nedler at Greentech Media disagree with Smil by noting that (1) wind and solar are growing at double-digit growth rates for sustained periods; (2) the world is much more connected, and both government and business processes are more flexible and agile; and (3) the policy push to decarbonize exerts a sustained, unique force. See Smil (2010) and Nedler (2013).

11. Rogers and Williams (2015).

2. The Future Is Electric

1. McKinsey's prediction is covered in the October 6, 1999, edition of *The Economist* (Cutting the Cord, 1999). Cell phones are tracked by the GSMA tracker (GSMA Intelligence, n.d.).

2. Smil (2008) p. 356.

3. International Energy Agency (2015a) p. 96.

4. Lamonica (2011).

5. Viribright (n.d.).

6. Alexander. Air Conditioners Really Are Getting Better (n.d.).

7. ISO New England (n.d.).

8. Numbers derived from St. Louis Federal Reserve Data (FRED, n.d.).

9. Electricity forecasters will recognize that there are other variables that greatly influence power use, notably indexes of economic activity, weather, and activity in especially electric-intensive industries such as manufacturing. Omitting them from most of the discussion is tantamount to assuming that their long-term effects are unremarkable, that is, that future long-term economic growth will occur at roughly the same average pace as it has over the past several decades and that the mix of demands for traditional end uses does

not change substantially. New end uses are a different story, and are discussed much more later in this chapter.

10. U.S. Energy Information Administration (2012).

11. International Energy Agency (2015c) p. 16.

12. See, for example, Allcott and Greenstone (2017), and especially Gerarden, Newell, and Stavins (2015).

13. See, for example, Gately (1993), Golove and Schipper (1997), and Schipper and Grubb (2000).

14. International Energy Agency (2015c) p. 8.

15. Laitner et al. (2012) p. 88.

16. The Dutch have pioneered a new retrofit process for standardized public housing retrofits known as the Energiesprong (energiespring.org) that shows some potential to lower housing retrofit costs substantially. The New Buildings Institute tracks progress in net-zero building additions in the United States. It reports that the number of certified net-zero structures has increased from 60 in 2012 to 500 in 2017 (New Buildings Institute, 2018).

17. Pyper (2018a).

18. Spector (2016).

19. This is 0.99 percent growth rate compounded over thirty-three years (U.S. Energy Information Adminstration, 2018).

20. A. Lovins, How Big Is the Energy Efficiency Resource? *Environmental Research Letters,* 13 (2018), 090401.

21. Nadel (2016) pp. 5–6.

22. Nadel (2016) pp. 5–6.

23. Email from Steve Nadel to the author, February 4, 2017 (available from the author).

24. For a recent update, see Schwartz et al. (2017) pp. 16–36, which finds many untapped opportunities for higher building and electronic equipment efficiency.

25. In addition, distributed solar was starting to become economical enough to gain a small but growing market share. If total power demand growth was 0.5 percent / year, and 0.6 percent of customers per year adopted rooftop solar, this alone would reduce sales growth to near-zero levels.

26. The White House Washington (2016) p. 30.

27. International Energy Agency (2017a) p. 4.

28. Zehong (2017) slide 11.

29. Williams et al. (2014).

30. The White House Washington (2016) p. 60; half of U.S. floor space is heated by fossil fuels.

31. See IRENA (2017), Northeast Energy Efficiency Partnership (NEEP) (2017), Lechtenböhmer et al. (2016), and Energy Transitions Commission (2017), among others, for useful studies of industrial electrification.

32. Jenkins and Thernstrom (2017) p. 2. A similar worldwide analysis by the Climate Policy Initiative and Copenhagen Economics for the Energy Transitions Commission in 2017 found that global electricity use would rise from its current level of about 17 percent of all global energy demand to roughly 25 percent of demand by 2040. Since electrification is far larger in the developed-country pathways studies, this suggests that electrification is less in the less-developed parts of the world. Nonetheless, in their scenarios, total world power use rises between 8,000 and 14,400 TWh by 2040—the equivalent of adding two United States in the low and three and a half in the high electrification case, respectively (Energy Transitions Commission, 2017, pp. 9, 13).

33. Jenkins and Thernstrom (2017).

34. Northeast Energy Efficiency Partnership (NEEP) (2017) table 10.

35. A team of EU researchers estimated that 100 percent electrification of all steel, glass, cement petrochemicals, ammonia, and chlorine production in the EU would increase industrial electricity use from 125 TWh in 2010 to 1,713 TWh in 2050, more than twelve times the power use in 2010. This is an upper bound, as much of the power demand is for electrolysis to create hydrogen, and this could be supplied by biofuels or other processes. Nonetheless, this exercise illustrates the huge growth potential from industrial electrification (Lechtenböhmer et al., 2016).

36. Most of the current pathways studies assume a significant role for both hydrogen and biofuels on top of the main strategies of efficiency and electrification. The key questions for our purpose are: (1) whether other low-carbon fuels could economically scale up for applications that forecasters now see as electricity growth areas, such as transportation and building heat; and (2) whether hydrogen will be produced from electricity by electrolysis.

With respect to the question, to the extent that hydrogen is made from power, its production becomes a potentially gigantic new end use. European hydrogen expert Ulf Bossel (2006, p. 1827) notes that twenty-five new power plants, operating continuously, would be required to create hydrogen sufficient to fuel the planes that leave Frankfurt's airport each day. Similarly, the IEA's 2015 Hydrogen Technology Roadmap notes that the electricity demand for hydrogen used to balance valuable renewables would be on the scale of "Gigawatt-hours to Terawatt-hours" (International Energy Agency, 2015d, table 6).

With respect to the displacement of electrifying end uses, even the forecasts of hydrogen planners indicate relatively small effects. The vision document, Hydrogen Scaling Up, shows hydrogen providing only about 10 percent of auto fuels, 25 percent of industry heat, and less than 20 percent of building heating and power by 2050. Obviously, this leaves 90 percent of auto fuels, 75 percent of industrial heat, and 80 percent of building energy for electric power.

The prospects for biofuels displacing electrification are even smaller. There is little growth in transport biofueling today, and the direct use of bioenergy in buildings or industry (the other two main electrification sectors) is also quite limited.

37. Germany's 2015 Deep Decarbonization Pathways Study provides an interesting contrast to the many studies that project high electricity growth. Two of its three 2050 scenarios have "very strong efficiency improvements" and do not electrify as much of the economy, instead opting to use biofuels and other nonelectric sources of no-carbon energy. These two scenarios have no power demand growth by 2050. A third scenario, with weaker efficiency policies and more hydrogen fuels made from electricity, shows 25 percent greater power demand (Hillebrandt et al., 2015, table 4 and figure 5).

38. Romm (2009).

39. Shehabi et al. (2016) p. ES-1.

40. de Vries (2018) p. 801.

41. Digiconomist (n.d.).

42. American Planning Association (n.d.).

43. See, for example, Wadud, MacKenzie, and Leiby (2016), Ross and Guhathakurta (2017), and Fox-Penner, Gorman, and Hatch (2018) and the many references cited in these papers. In my paper, for example, we conservatively estimated that vehicle miles traveled by car would increase 40 percent by 2050, which would only be offset by efficiency improvements once rightsizing kicked in.

44. World Economic Forum (2015).

45. Andrae (2017).

3. La Vida Local

1. Sandia National Laboratories (2006) p. 10.

2. The Atlantic (n.d.).

3. The Atlantic (n.d.).

4. EnergySage (n.d.).
5. Solar Energy Industries Association (2019).
6. Bronski et al. (2014) p. 5.
7. Byrd et al. (2014).
8. Pentland (2013); Chediak and Wells (2013); and Roberts (2013).
9. Magill (2015).
10. Distributed power production is often hidden as a deduction from total power use in U.S. Energy Information Administration data. So, for example, if residential total demand is 1,000 kWh, and self-generation is 500 kWh, national data and models simply report net residential demand of 500 kWh. Even the most recent U.S. national forecast, the Annual Energy Outlook (U.S. Energy Information Administration, 2018), simply reports renewable generation as a single line item—though interestingly it does not report gas-fired DG separately.
11. See figure 4.6 in the U.S. Mid-Century Strategy for Deep Decarbonization (The White House Washington, 2016), which shows about 50 percent of all generation coming from "solar" and wind, or figures 28–30 in the U.S. DDPP study (Williams et al., 2014).
12. Hillebrandt et al. (2015).
13. See note 5 for EIA and the IEA World Energy Outlook 2017, figure 1.3, as examples. Jenkins and Thernstrom's (2017) literature review of deep decarbonization studies does not discuss PV scales at all. There are, however, many exceptions, including Lovins's (2011) *Reinventing Fire* and Jacobson et al.'s (2015) model of a 100 percent renewable U.S. system by 2050. China and India have also recently produced long-term forecasts that track DG PV and large PV separately; see Sivaram (2018a, figure 1.1) for an example.
14. United Nations (2014). See also Wang et al. (2018).
15. Isolated rural homes and small buildings with ample nearby land will certainly self-make increasing amounts of power through 2050. However, the aggregate power use by these structures is not large. We return to the unique challenges facing rural utilities in Part II.
16. Miller et al. (2016), accompanying letters to Kammen and Sunter (2016).
17. Mapdwell (n.d.).
18. See Gagnon et al. (2016) and Margolis et al. (2017) for NREL's work and Byrne et al. (2015), Santos et al. (2014), Karteris, Slini, and Papadopoulos (2013), and Mainzer et al. (2014) for peer-reviewed studies of Korean, Portuguese, Greek, and German cities, respectively. Castellanos, Sunter, and Kammen (2017) compare the methods used for these estimates.

19. See Gagnon et al. (2016) and Margolis et al. (2017) for NREL's work and Byrne et al. (2015), Santos et al. (2014), Karteris, Slini, and Papadopoulos (2013), and Mainzer et al. (2014) for peer-reviewed studies of Korean, Portuguese, Greek, and German cities, respectively. Castellanos, Sunter, and Kammen (2017) compare the methods used for these estimates.

20. Gagnon et al. (2016).

21. Sivaram (2018b).

22. To clarify, we are discussing the solar-electric collection only in this chapter. The collection or reflection of solar energy for heating would reduce fossil-fuel or electric heating in the building and, for our purposes, is therefore an energy conservation measure.

23. County population estimates, domestic and international migration data (Governing, 2018).

24. Cities buildings database (SkyscraperPage.com, n.d.).

25. Phoenix is also interesting because a 2012 newspaper story by reporter Ray Stern concluded that powering the city with solar energy was uneconomical and would not happen soon.

26. NREL's research team did not have the same detailed aerial roof survey data for Phoenix that it had for the cities in Table 3-1. The estimates for Phoenix shown in Table 3-1 are based on NREL's extension of its detailed city results into estimated state total potential for the state of Arizona (Gagnon et al., 2016, p. 36), which is a current PV potential equal to 34.4 percent of current state electricity use. The figures for both roof space and PV electricity potential for Phoenix in Table 3-2 are based simply on Phoenix's share of the Arizona population (1,582,000 from 7,016,000) and Phoenix's 26,481 GWh electricity demand.

27. For any city, the long-term growth in roof surface can be decomposed into three important parameters: new population, which adds residential floor space; additions to the workforce, which drives changes in commercial floor space; and the proportion of each type of floor space created by going up rather than out. The complete decomposition of these driving factors would include changes in average residential space per inhabitant and average changes in workspace per city worker. As it happens, the national averages for these two parameters have been remarkably stable during the past few decades. Reyna and Chester (2015, figure 2) show the average U.S. residential area per person nearly flat at 400 square feet.

28. The most recent comprehensive study of urban development patterns by the U.S. Department of Transportation (2014) concludes that U.S. urban

density will continue to decline on average, despite the trend toward walkable urbanism—and this study does not include significant effects from autonomous cars. According to an extensive analysis by a distinguished team of U.S. urban researchers, the area covered by U.S. cities will, at a minimum, double by 2050, and is more likely to triple (Angel et al., 2011, pp. 93–95). As it happens, urban development trends suggest that these are fairly reasonable assumptions. For example, in spite of the growth in single-family home sizes, average U.S. living and working spaces have been stable for many years. As to density and vertical growth, there are competing forces at work. Cultural preferences among millennials and the increasing cost of low-density expansion are promoting greater density and "walkable urbanism," but other factors such as the onset of autonomous vehicles are likely to accelerate sprawl. Assuming that Phoenix remains at its current density and height mix probably understates the growth in roof space, meaning that our 2050 potential figures are more likely to be conservative (underestimates) than excessive.

29. Recent progress in solar cell efficiencies is charted in Green et al. (2018). Sivaram (2018b, figure 2-1) simplifies the usual tables and suggests that long-term cell improvements will come not from incremental silicon-based cell improvements, but rather from other cell materials or blends of other cells with silicon. This conclusion is echoed by another internationally renowned PV scientist, Dr. William Shafarman of the University of Delaware (a distant step relative; personal communication with William Shafarman, February 16, 2018).

30. One company, Onyx Solar, sells a wide array of glass building heating material and windows with embedded PV. Onyx claims to have won more than fifty awards for its products and that some of its materials have a two-year payback.

31. See Shukla, Sudhakar, and Baredar (2017) for a comparatively recent survey. A very interesting study of the potential for increasing distributed PV by rearranging building locations in London found that under a wide variety of conditions, facade PV potential was approximately one-ninth to two-ninths that of rooftop PV (Sarralde et al., 2015, comparing tables 5 and 6). In one of the few quantitative studies to date, two German researchers estimated the amount of building facade usable for PV generation relative to the amount of roof space in North Rhine-Westphalia, Germany, and extrapolated this to the entire EU. They concluded that usable roof space was roughly twice the size of usable facade space: 4,571 km^2 of roof space versus 2,411 km^2 of facade space. When combined with the fact that most facades are shaded by other structures

or greenery much of the time, and that this will increase if cities become more dense and vertical, it seems unlikely that facade PV will generate anything more than a fraction of rooftop power, even with much better technology (Lehmann and Peter, 2003).

32. Gardner (2011) in his Old Urbanist Blog from December 12, 2011. Some estimates are even higher, as in Ben-Joseph (2012, p. 2), where parking covers "more than a third of the land area." Most estimates center on a number about half as large; for example, Lawrence Berkeley National Labs estimates that, "[i]n a typical city, pavements account for 35 to 50% of the surface area, of which about half is comprised of streets and about 40% exposed parking lots" Chao (2012, p. 1).

33. Eucalitto, Portillo, and Gander (2018).

34. Email from C. Hoehne to author April 19, 2018.

35. Wind-power generation in urban areas inspires a lively discussion but no clean view of future potential. Kammen and Sunter's (2016) seminal review of urban DG potential surveys several sources claiming significant potential, especially from vertical-axis turbines mounted on the corners of tall buildings (see, e.g., Dabiri et al., 2015). Miller et al. (2015) and Miller et al. (2016) disagreed with these estimates, citing theoretical and practical limits on our urban wind generation. Ishugah et al. (2014) provide a useful survey of urban wind applications, indicating both pros and cons. Kamal and Saraswat (2014) describe several unique skyscrapers around the world that generate approximately 11–15 percent of their power use from attached wind. For perspective, Mithraratne (2009) claims that for vintage 2007 technologies, large New Zealand wind farms have seven to eleven times the generation potential of urban wind.

36. Tsuchida et al. (2015; updated with calculations available from the author). See also Hledik, Tsuchida, and Palfreyman (2018).

37. The advantages of new lighter PV technologies are described nicely in this passage from five University of Delaware solar researchers: "Compared with a conventional crystalline silicon PV ballasted system which has a distributed weight of 3–4 pounds per square foot (psf), a Lightweight PV (LPV) system, which uses fewer metal components, can weigh less than 2.5 pounds per square foot [1,2]. This lightweight character can reduce the roof load concern and help the LPV system to penetrate more existing membrane-based commercial rooftop and load-constrained industrial rooftop market [1,3,4]. Beyond the weight reduction advantage, the LPV system also provides several critical deployment advantages such as low Balance-of-System (BOS) hardware costs

due to fewer metallic components, lower transportation costs due to the integration of modules with racking hardware, and lower installation costs due to its lightweight character [1]. These advantages can help LPV systems to be deployed on more existing rooftops and speed up future PV market diffusion" (Chen et al., 2017).

38. California Energy Commission (2018).

39. In-building storage will probably always be the gold standard for outage avoidance, but local utilities will offer better and better alternatives that also prevent or mitigate outages.

40. See Kuffner (2018) p. 1.

41. Vlerick Business School (n.d.).

42. AEMO and Energy Networks Australia (2018) pp. 15, 17.

4. Why We Grid

1. Freeberg (2013).

2. Isolated Grid System from [D8221] Electric Light—Edison Co for Isolated Lighting. Courtesy of Thomas Edison National Historical Park, January 5, 1881. Central Grid System data from Hughes (1993) p. 39, figure II.6.

3. One of the first examples of a multi-utility grid was the 1927 connection of the electric power systems of Philadelphia, Pennsylvania, and Newark, New Jersey, several large steam plants, and a huge new hydroelectric station in Maryland, the Conowingo plant. This was the first multi-state grid in the United States, formed to take advantage of the size advantages of Conowingo and greater load diversity between the two utility service areas. This "Pennsylvania–New Jersey–Maryland" interconnection was the ancestor of today's PJM power pool, the largest electricity market in the world. Soon after, governments and utilities all over the United States and Europe began to plan and later implement regional grids, such as the Bayernwerk grid in Bavaria (Hughes, 1993, p. 326).

4. Heidel and Miller (2017).

5. Bronski et al. (2014).

6. Kantamneni et al. (2016).

7. See Rocky Mountain Institute (2015) and a subsequent 2017 RMI blog post by James Mandel and Mark Dyson, "Why We Still Need to Discuss Grid Defection," March 2, 2017, available at https://rmi.org/still-need-discuss-grid-defection/.

8. Borenstein (2015).

9. A traditional blackout occurs when a large power source experiences a sudden outage, creating an immediate mismatch between supply and demand that is resolved by suddenly disconnecting enough customers to rebalance the remaining supply and load. Chapter 5 examines the causes of blackouts and the reliability dimensions of the grid in more detail.

10. U.S. Department of Energy (2016) p. 50, figure 42. Small projects are less than 5 MW—the size that would tie into distribution systems; large projects are more than 200 MW.

11. Lazard (2017) pp. 9, 12. In a faint echo of rooftop solar, the report also notes (on p. 17) that both residential and commercial (i.e., smaller) installations were not declining in cost as quickly as utility-scale units, and in some cases were actually increasing.

12. If the generation is all emissions free, there are no emissions externality differences.

13. For a new and thorough discussion of the benefits of large-scale grids, see American Wind Energy Association (2019). See Chapter 7 for a more complete discussion.

14. The benefits of aggregation via trading are by no means limited to short-term hourly trades, known as *spot* or *imbalance* markets in the power industry. The benefits of wider aggregation and trading of loads also apply to longer, bilateral trades and contracts, which constitute the majority of power trading by volume.

15. It is worth noting that some of the studies of trading benefits come from systems that still operate under the original system balancing paradigm. This approach achieves moment-to-moment balance by adjusting power plants up and down to match aggregated load, which is almost entirely outside the control of the system operator, except in emergencies. In other words, the design paradigm is to treat demand as uncontrollable and adjust supply to match it. This is still by far the dominant paradigm at this moment, but it is gradually changing. Fully developed smart grids will allow customers to adjust their loads up and down in response to price changes or other signals. System operators will then have two degrees of freedom to balance momentary load: turn generators up or turn loads (who choose to allow it) down. This is a now-familiar tool for utilities in the developed world called DR; several grid operators match supply and demand using both supply and demand resources today. However, the idea is still relatively undeveloped, and only a small fraction of customers and electrical applications allow their loads to be controlled. In contrast, the smart grid and Internet of Things will enable nearly every device

to be controlled using artificial intelligence–enabled systems, and onsite storage will enable customers to vary their loads much more willingly, often without sacrificing immediate use. Many studies show that electricity trading benefits go up substantially when both demand and supply contribute, and they also suggest that the benefits continue until grids become regional in size. To cite one example, DNV GL Energy (2014, figure 8) finds that DR reduces the need for backup power on large grids by a full 25 percent.

16. In addition, the study found that California would be able to export more of its own renewables surplus, use less endangered habitat, increase reliability, and gain several other benefits. See Brattle Group (2016). For a short summary, see Aydin, Pfeifenberger, and Chang (2017).

17. The full shape is due to the following sequence of daily events: Load remains low throughout the late-night hours, when there is no solar or much in-state wind generation, so balancing the load is left to base-load nuclear and gas plants. Then both load and solar generation increase in the morning as everyone awakens and heads to work. Initially, load grows more rapidly than solar production, but then in the middle of the day solar production begins to outstrip load growth. Since most solar production in California is distributed, this solar production met the demands of its local customers and the grid operator saw this as a net decrease in load. In the late afternoon into the evening, solar production dwindles but load increases. As load goes up and solar production goes down, net demand on the system goes up steeply until the middle of the evening, when demand declines into nighttime levels and the base-load plants again balance load.

18. MacDonald et al. (2016).

19. Summaries can be found in Mills and Wiser (2010), Bird, Milligan, and Lew (2013), Chang et al. (2014), and Leisch and Cochran (2015) from the National Renewable Energy Laboratory.

20. Put differently, if the addition of DG avoids the need to build a new transmission line, this saving is reflected in the tally of total costs.

21. In most studies to date, “backup power” is in the form of natural gas–fired generation rather than oversized storage. However, unless small–scale carbon-free generation (such as gas with CCS) achieves cost breakthroughs, local backup will eventually have to become some sort of storage. Because the costs of the gas backup generation assumed in these studies is very low compared to storage, the result that smaller grids with higher levels of DG are more costly is (for now) even stronger were we to assume that local storage is the main form of backup power.

22. For the record, they also find that energy efficiency and DR are both very important to reducing costs. As an example, the ICL-NERA-DNV study of the European grid finds that DR could reduce the need for backup generation or storage by 25 percent—a gigantic savings in cost (DNV GL Energy, 2014, figure 8).

23. DNV GL Energy (2014).

24. DNV GL Energy (2014) figure 8.

25. DNV GL Energy (2014) p. 90.

26. DNV GL Energy (2014) p. 90. One reviewer of this book has pointed out correctly that this loss of value from local renewables will be offset by local DER tariffs they pay DERs for other grid services, thus increasing the amount of local DERs and the surplus energy they will have available for trading. But this does not change the fundamental scale economics of trading.

27. DNV GL Energy (2014) p. 187.

28. The fact that larger markets and grids have an overall lower level of cost does not mean that individual customers will experience lower prices under this arrangement or that the industry will choose this least-cost route. The choice between large and small power sources generally, and which combination of generation and transmission to combine, isn't (and shouldn't be) made based solely on what is lowest in measurable dollar cost.

Although measured total cost is extremely important, and should always be understood, other "unpriced externalities" enter into electricity and energy system designs and policies. This sometimes warrants pursuing an option that has higher measured costs than the alternative. For example, there are economic opportunity and development benefits from local energy production that are important to consider.

Additionally, when studies such as this show that one option has lower costs, they often embed assumptions that the industry, working within its overall economy, will be forced by policies and market forces to create an operating industry that will actually function at this efficient level of costs, and presumably pass these costs on to customers in the form of prices consistent with these cost levels, rather than prices that are inflated by inefficiency and / or excess profits. For this reason, any conclusion that costs are lower should consider the extent to which those projected costs will actually be realized and passed on to consumers.

29. See, for example, Bloom et al. (2016, table 53, p. 162), Tsuchida et al. (2015), and Hledick, Tsuchida, and Palfreyman (2018).

30. Fares and King (2017).

31. Black and Veatch (2014).

32. Osborn and Waight (2014) slide 21.

33. MacDonald et al. (2016). This study uses an approach that is unique in many ways and is subject to many caveats. However, it is probably the most sophisticated treatment of geographic renewables resource diversity per se completed to date.

34. The Climate Institute's work on the North American Supergrid can be found at http://www.cleanandsecuregrid.org; EU studies are collected by the European Network of Transmission System Operators at http://www.entsoe.eu/outlooks/ehighways-2050/.

35. For the sake of full disclosure, I lead a Boston University research team working with GEIDCO to examine these ideas.

36. It is fascinating to realize that the economic case for a large grid relies so much less on its traditional foundation, which was preventing outages caused by large fossil plants suddenly going out of service. We may not have many such plants online in the future, and we'll have much more storage on this system to buffer them in any case. Paradoxically, the benefits of the future large grid come from scale economies in producing and integrating and diversifying variable supply sources—things power engineers of the 1940s never worried about.

5. The Fragmented Future

1. So-called major events (storms) cause about 58 percent of all outages and 87 percent of all large-scale blackouts; most of the rest are triggered by trees falling on local lines, animals, drivers hitting utility poles, and other local causes (Executive Office of the President, 2013, p. 8, citing DOE form OE-17). Rouse and Kelly (2011, table 7) report that storms cause 80 percent of all outages, which is roughly consistent.

2. For the canonical explanation of why plants can never be made failproof, see Lovins and Lovins (1982) chapter 3.

3. See Webber (2016) for a very good overview of this subject.

4. Abi-Samra (2017) pp. 7–15.

5. For the best book-length treatment of this topic, see Webber (2016).

6. Abi-Samra (2017) p. 175.

7. Abi-Samra (2017) p. 185.

8. Centra Technology (2011) p. 25. Also see Kapperman (2010). There is also a good discussion of transformer vulnerability in Mission Support Center (2016) p. 10.

9. Mission Support Center (2016) p. 19.

10. NERC Reliability Standard TPL-007-1, adopted September 2016, sets forth the requirements. For a useful overview, see Koza (2017).

11. Florida Power and Light Company's Amended Response to Staff's Second Data Request No. 29, Docket 20170215-EU (2018) p. 68.

12. Abi-Samra (2017, p. 257) contains a good concise summary of distribution system undergrounding. Larsen (2016) contains a useful framework for comparing the costs and benefits of undergrounding. For a nice, short non-technical treatment, see Xcel (2014).

13. Kossin et al. (2017) p. 257.

14. Wikipedia (n.d.-a).

15. Kron (2016) pp. 35–39.

16. Geophysical Fluid Dynamics Laboratory (2019).

17. Kossin et al. (2017) p. 237.

18. Kossin et al. (2017) p. 249.

19. Kossin et al. (2017) p. 240.

20. Climate Ready Boston (2016).

21. U.S. Global Change Research Program (2018) p. 30.

22. NOAA Office for Coastal Management (n.d.) and NOAA National Centers for Environmental Information (2019).

23. Lacey (2014).

24. See Pacific Gas and Electric Company (n.d.) for clear definitions of these terms.

25. See, for example, New York State Department of Public Service (2018).

26. Retière et al. (2017). Other very similar definitions are in Cuadra et al. (2015) and Miller et al. (2014), among others. The concept comes originally from ecology and systems theorists (see, e.g., Gao, Barzel, and Barabási, 2016), and is now extended to human communities (see, e.g., All Aces, Inc., n.d.).

27. Tierney (2017) and Abi-Samra (2017) p. 271. Definition includes the same four elements.

28. Lovins and Lovins (1982).

29. Lovins and Lovins (1982).

30. Abi-Samra (2017) pp. 271–272. Also see Castillo (2014), Mukhopadhyay and Hastak (2016), Panteli and Mancarella (2015), and Ouyang and Dueñas-Osorio (2014).

31. National Electrical Safety Code (2017). Starting in 1977, the code added stronger standards for poles and lines in high wind areas and has now added standards for areas with wind and ice. Abi-Samra (2017, p. 193) reports that distribution system parts are typically rated at levels where there is a 3–5 percent chance that their ratings will be exceeded during their anticipated service life.

32. The first phase of PSEG's hardening program is described in Fox-Penner and Zarakas (2013). PSEG has recently proposed adding $2.5 billion in additional hardening investments (PSEG, 2018).

33. A recent FERC order describes these arrangements. See Order Authorizing Acquisition and Disposition of Jurisdictional Facilities, 163 FERC 61,005 (2018).

34. The investor-owned mutual assistance groups are regional, operating through guidelines set by the Edison Electric Institute (n.d.-b). For the public power industry and rural electric cooperatives, the American Public Power Association maintains a mutual aid network (American Public Power Association, n.d.-a).

35. Sandalow (2012).

36. @inside FPL, tweeted September 13, 2017, quoted in Iannelli (2017).

37. Iannelli (2017).

38. Salisbury (2010).

39. Kaye (2015) and Pounds and Fleshler (2017).

40. Florida Power and Light Company's Amended Response to Staff's Second Data Request No. 29, Docket 20170215-EU (2018) p. 69.

41. Committee on Analytical Research Foundations for the Next-Generation Electric Grid (2016) p. 88.

42. Abi-Samra (2013b).

43. Superstorm Sandy affected 8.5 million customers, less than one third of the thirty million affected by the 1965 blackout or the fifty-five million who lost service in the United States in 2003. A list of the top ten largest blackouts in the world, all of which are cascading failures of one form or another, shows that these two U.S. events are among the smallest of this group (Wikipedia, n.d.-b). See also Fairley (2004).

44. For further background, the Microgrid Institute (n.d.) is a useful resource. Hirsch, Parag, and Guerrero (2018) is also helpful.

45. Miller et al. (2014, p. 29) lays out the technical principles for true fractal design and operation. Jeffrey Taft, Chief Grid Architect at PNNL, notes that practical grids will rarely act as true fractal structures because "there is not really much that is . . . 'fractal-dimensional' about the grid being organized into cells that can be redefined on the fly." In other words, the real power grid won't be able to reproduce itself into identical copies of itself at different scales, which is the core feature of fractal structures (e-mail from J. Taft to the author, May 11, 2018). Nonetheless, research on how to make the grid behave like a fractal structure continues, such as the French National Research Agency's Fractal Grid project (http://fractal-grid.eu/#menu; see Retière et al. 2017).

46. Committee on Analytical Research Foundations for the Next-Generation Electric Grid (2016) p. 86.

47. See Marnay (2016).

48. Gholami, Aminifar, and Shahidehpour (2016). These authors provide an excellent, understandable review of microgrid benefits, as well as their implementation challenges.

49. Abi-Samra (2013b).

50. Kelly (2014).

51. Wood (2018).

52. See Fairley (2004).

53. See Miller et al. (2014) p. 19.

54. See Marnay (2016).

55. Wood (2019) and St. John (2018b).

56. Personal communication with the author, John Bruns and Riney Cash, Chief Engineer, White River Electric Co-op, January 3, 2019.

57. 2018 State of the Electric Utility Survey (2018) p. 60.

58. Babloyan (2014).

59. For example, the California Public Utilities Commission recently concluded that "[T]o date, impact on grid reliability attributable to physical security incidents has been rare and limited" (Battis, Kurtovich, and O'Donnell, 2018, p. 44).

60. Sanger (2018).

61. PwC (2017) p. 3.

62. Assante, Roxey, and Bochman (2015).

63. Federal security experts noted in 2016 that "no lasting damage—physical, cyber-physical, or otherwise—to U.S. utilities as a result of a cyber-attack has yet been reported publicly" (Mission Support Center, 2016, p. 4).

64. For a good technical overview, see Sun, Hahn, and Liu (2018). Battis, Kurtovich, and O'Donnell (2018, pp. 12 and 38) summarize the case for distribution substations being low-value targets by noting that these installations are: too numerous to attack at once (there are more than 300 in California alone), "moderately to highly redundant," located mainly in dense urban areas, compact in footprint, and able to be repaired quickly, typically within a week or less.

65. Assante, Lee, and Conway (2017) p. 3. However, just as these substations are ideal attack targets, they may also be ideal lines of defense. Erfan Ibrahim (see note 64 above) argues that each distribution system, which is typically served by at most a handful of substations, can be cyber engineered so that the substations serve as firewalls between all microgrids downstream of the sub and also between the distribution system and the bulk power system. Properly engineered and maintained, this architecture would prevent cyberattacks on any one microgrid from affecting another or moving upstream. The key feature of this view is that Ibrahim views the scale of a typical substation coverage—about 150 MW of load, and 25,000–50,000 customers—as the efficient scale for local cyber defense.

66. Assante, Roxey, and Bochman (2015) p. 6.

67. Assante, Roxey, and Bochman (2015) p. 6 and Dunietz (2017) p. 4.

68. For a recent, complete overview of digitization impacts on utilities, see Sivaram (2018a).

69. Mission Support Center (2016) p. 14. Securing the supply chain and the many networks that form the communications part of the smart grid are two of the main cybersecurity challenges at the forefront of current standard-setting.

70. Erfan Ibrahim (www.tbbllc.com). Interview with the author, May 17, 2018.

71. See, for example, S&C Electric (2017).

72. The cybersecurity of distribution systems in the absence of enforceable standards and microgrid security are frequent subjects of discussion in the industry; see Weigert (2017) p. 6, Accenture (2016), Ellis (2016), S&C Electric (2017), and author communications with industry cybersecurity consultant James P. Fama, June 2, 2018.

73. Accenture (2016).

74. Greenberg (2018).

75. "Several cyber-attacks targeting the AMI (automated metering infrastructure, or smart meter systems) have been identified, including energy theft, false

data injection, and leakage of customer information" (Sun, Hahn, and Liu, 2018, p. 49).

76. Weigert (2017) p. 6.

77. At present, it is also the case that the absence of substantial distributed generation (DG) allows the transmission and distribution systems to be planned and operated with moderate overlap and little real-time feedback. One common element in all new grid architectures is the need to design and operate with real-time feedback between the T and D layers.

78. See De Martini and Kristov (2015), in which this concept is called layered optimization. E-mail from L. Kristov to the author, October 24, 2018.

79. Taft (2016) provides an excellent, relatively technical overview of similarities and differences between seven different control architecture proposals. De Martini, Kristov, and Taft (2019) have produced a somewhat less technical review of architecture proposals from system operators all over the world. These two documents, along with De Martini and Kristov (2015) in note 74, are absolute "must-reads" for serious students of grid architecture. Also see Hirsch, Parag, and Guerrero (2018) section 6.1.

80. Martin (2016).

81. Miller et al. (2014).

82. Committee on Analytical Research Foundations for the Next-Generation Electric Grid (2016).

83. Battis, Kurtovich, and O'Donnell (2018; emphasis added). An early EU document, Strbac, Mancarella, and Pudjianto (2009), describes the stages of the grid evolution from the current grid to a "microgrid-based integrated energy system" (their term for a full sectionalized grid), concluding that the earliest the transition could occur would be 2030. Moreover, the authors confessed that they could not estimate infrastructure costs, so their analysis did not take transition costs into consideration.

6. Decarbonizing the Big Grid

1. International Energy Agency (n.d.-b).

2. International Energy Agency (n.d.-b).

3. The continued use of existing fossil plants will hinge on whether carbon capture can be added to them more economically than other carbon-free alternatives. It is widely agreed that adding capture to an existing plant will be substantially more expensive than building it into a new plant from the start.

For this reason, I discount the likelihood of widespread retrofitting. However, if carbon capture and storage (CCS) retrofits become economical, power decarbonization becomes that much easier. For a recent study arguing that CCS retrofits will be economical, see Nagabhushan and Thompson (2019).

4. Note that on February 16, the sum of wind and baseload generation was greater than demand. In instances when generation is greater than demand, NorthWestern must sell the surplus power off-system, pay a third party to take the surplus power (if the region is energy rich at the time), or request owners curtail (turn off) their generation and pay them for lost production.

5. Cleveland et al. (2019b).

6. Auffhammer, Baylis, and Hausman (2017).

7. Véliz et al. (2017) p. 8, sec. 3.1. This problem is truly global. University of Chicago professor Michael Greenstone has written about the conflicting benefits and costs of India's increased use of air conditioning. Whereas air-conditioning use helps reduce heat-related deaths, it also adds to electricity use and emissions, creating a large-scale catastrophe. "So we are in a difficult position," Greenstone wrote in the *New York Times* in 2016. "The very technology that can help to protect people from climate change also accelerates the rate of climate change" (Greenstone, 2016).

8. The benefits of energy efficiency (EE), properly done, go beyond system-wide cost reductions to include improvements in the quality of housing—a public need as pressing as decarbonization in many cities—as well as local jobs and economic development. At the same time, traditional EE programs must also now incorporate the temporal value of savings for both cost and carbon—that is, savings during certain periods are more valuable than savings in others. For a discussion of this topic, see Baatz, Relf, and Nowak (2018).

9. Solar District Heating (n.d.).

10. For a fascinating, far-reaching, highly technical discussion of all-renewable systems, see Taylor, Dhople, and Callaway (2016). Among other things, the authors observe that future all-renewable systems may switch to entirely direct current transmission systems and abandon locational marginal pricing.

11. The language describing the new design paradigm is evolving and unstandardized, in part reflecting an emphasis on different combinations of future resource options. This is illustrated in Table D-1 of Appendix D. In this table, column 1 shows the three types of energy needed to power grids. Column 2 shows the type of plants providing each type of power in traditional systems.

Column 3 shows my simplified description of how this energy will be made in systems that consist primarily of wind, solar, and storage. I acknowledge in the main text that fossil-fueled generators with economical, widely available CCS may also be significant sources of clean baseload power. The latter plants, labeled firm low-carbon resources, are the main technology for baseload energy emphasized by Sepulveda, Jenkins, and Sisternes. (2018; column 4). There is greater agreement in terminology between my new paradigm and Sepulveda et al. in the intermediate energy category: we both see these coming mainly from variable renewables and storage. We also agree on the new technologies that will supply peaking power, which I generally label storage and they call fast burst variable resources.

12. This is based on the ISO New England (2018b) tariff Section III.13.1.4 as of March 2019.

13. Goldenberg, Dyson, and Masters (2018).

14. Lubershane (2019) and Solar Energy Industries Association (2019).

15. Lubershane (2019).

16. International Energy Agency (2019) and U.S. Department of Energy (2018) p. 15.

17. Traditional PSH plants could not always respond on a second-by-second basis, but the most modern PSH plants use a new variable-speed turbine that is able to respond on a sub-second basis.

18. U.S. Department of Energy (2018) p. 2 and Roach (2015) p. 3.

19. U.S. Department of Energy (2018) p. 2 and Roach (2015) p. 3.

20. International Hydropower Association (2019) p. 12.

21. New technologies are attempting to solve PSH's geographic limits. One German company has tested underwater tanks in place of a second above-ground reservoir, and other similar ideas are under exploration (Slashdot, 2017).

22. There is very interesting technology development going on in compressed air storage, including storage of pressurized air in flexible underwater tanks and a company running railroad trains up mountains as a form of short-term storage. See Frost & Sullivan (2017) slides 19 and 21.

23. Traditional lead-acid auto batteries and rechargeable nickel-cadmium batteries are also in this category, but they are not expected to play significant roles in power systems.

24. Weston (2018) and Spector (2018).

25. Chew (2015).

26. For a thoughtful, well-written discussion of the limitations of lithium-ion technologies versus other forms of electrochemical and mechanical short-term storage, see Hart, Bonvillian, and Austin (2019).

27. Shaner et al. (2018) discussion section.

28. Hart, Bonvillian, and Austin (2019) figure 5. This cost level is for batteries within EVs; according to Hart et al., utility lithium-ion batteries, which have different scale effects, durability requirements, and balance-of-system costs, will still cost several hundred dollars per kilowatt-hour by 2040.

29. Frost & Sullivan (2017) p. 10 and St. John (2018b and 2018c).

30. Frost & Sullivan (2017) p. 10 and St. John (2018b and 2018c).

31. For a good (though slightly dated) graphical description of emerging storage technologies, see Slaughter (2015) p. 6.

32. Retrieved June 29, 2019, from: www.formenergy.com.

33. See Penn (2018) and EOS Energy Storage (n.d.) for examples of zinc-air technologies.

34. Frost & Sullivan (2017) p. 35. This may be a little exaggerated, as Frost & Sullivan's 2017 forecast for flow batteries foresees only a slight increase in market share for flow batteries, from 3 to 8 percent of the market by 2023. Also see Hart, Bonvillian, and Austin (2019).

35. IRENA (2017) p. 3.

36. Frost & Sullivan (2017) p. 20.

37. Harvey (2017).

38. Siemens Gamesa (n.d.).

39. Deign (2019).

40. Crees (n.d.).

41. St. John (2018c).

42. Wesoff (2015).

43. The situation is vastly different in Southeast Asia, Latin America, and Africa. See, for example, International Rivers (2019) and Ledec and Quintero (2003).

44. U.S. Department of Energy (2018) p. 1 and International Hydropower Association (2019) p. 13.

45. Global CCS Institute (2018).

46. For a good but slightly dated outlook, see U.S. Department of Energy (2017).

47. 8 Rivers (n.d.).

48. Inventys (n.d.).

49. See the discussion in Majumdar and Deutch (2018).

50. Majumdar and Deutch (2018).
51. S. Anderson (2017) p. 108.
52. Aminu et al. (2017) p. 1404.
53. Aminu et al. (2017) and Zoback and Gorelick (2012).
54. Zoback and Gorelick (2012) table 2.
55. Breakthrough Energy Coalition, Vancouver 2019 meeting materials; cited by permission.
56. Townsend and Havercroft (2019) p. 4.
57. Townsend and Havercroft (2019) p. 4.
58. Pawar et al. (2015, pp. 304–305) contains a nice concise picture of the commercial storage development business challenges.
59. Nuclear Energy Institute (2019).
60. Kempner and Ondieki (2018).
61. Financial Tribune (2019).
62. The case for public support of nuclear power for grid decarbonization is made in Shellenberger (2017).
63. NuScale (2019).
64. U.S. Department of the Interior Bureau of Reclamation (2019) and Bonneville Power Administration (2017).
65. Michael Goggin, personal communication, March 20, 2019.
66. See Shaner et al. (2018) esp. figure 3.
67. Shaner et al. (2018) discussion section.
68. American Wind Energy Association (2019) contains a good explanation of the relationship between wind and solar expansion and transmission.
69. Roberts (2018c). The Department of Energy's technical targets for the cost of distributed electrolysis are a 45 percent decline between 2011 and 2020 and more than 50 percent for large-scale plants (U.S. Department of Energy's (DOE) Office of Energy Efficiency and Renewable Energy, n.d.-a).
70. Kraftwerk Forschung (n.d.) and GE Power (n.d.).
71. A second, equally important conversion technology is fuel cells, which convert hydrogen and oxygen into water vapor and electricity. There are a variety of different types of cells, and several are fully commercial today, operating on hydrogen that is made from splitting natural gas (and thus, with no CCUS available, creating carbon pollution as they operate; U.S. Department of Energy's (DOE) Office of Energy Efficiency and Renewable Energy, n.d.-a). See Bloomberg (2019) for one company. For views doubting the economic viability of hydrogen, see Romm (2004).

72. van Leeuwen and Mulder (2018) p. 258.

73. Glenk and Reichelstein (2019) p. 220 (emphasis supplied).

74. Editorial (2019).

75. Caughill (2017).

76. See, for example, Hezir et al. (2019) and Rissman and Marcacci (2019).

77. ARPA-E (n.d.).

78. Breakthrough Energy (n.d.).

79. Smith (2015).

80. Mission Innovation (n.d.).

81. International Energy Agency (n.d.-a).

82. Nelsen (2016) and Phillips (2015). Costa Rica, a tiny system blessed with good hydro, strong winds, and great solar, has run for seventy-five days straight on all renewables and 266 days total, as of March 2019 (Rote, 2019).

83. Diesendorf and Elliston (2018).

84. Carbon Neutrality Coalition (n.d.).

85. Diesendorf and Elliston (2018) and Brown et al. (2018) describe many of these studies. See also Jacobson et al. (2018a and 2018b) and MacDonald et al. (2016) for studies that have attracted particularly wide notice.

86. Two thoughtful criticisms can be found in Loftus et al. (2014) and Heard et al. (2017).

87. Many studies are also criticized for: (1) using electricity demand forecasts that are too low, (2) failing to recognize the large increases in electricity demand that will come from climate policies (as per Chapter 2), (3) failing to simulate the intra-hour fluctuations in wind and solar that destabilize a system without more storage and operating control, (4) failing to measure and recognize the difficulties of building transmission and distribution, and (5) failing to incorporate other technical aspects of grid operations (ancillary services). Brown et al. (2018) provide an extensive, point-by-point rebuttal to these criticisms, and Diesendorf and Elliston (2018) follow suit. Chapter 7 of this book discusses the transmission aspects of these simulations, effectively agreeing with the point that expanding transmission is particularly difficult and time-consuming.

88. Shaner et al. (2018). This point is also made in Hausker (2018) and in Safaei and Keith (2015).

89. For example, Loftus et al. (2014, p. 1), referring to carbon-free grid simulations as scenarios, write that "To be reliable guides for policymaking, scenarios such as these need to be supplemented by more detailed analyses realistically addressing the key constraints on energy system transformation."

7. Not in My Backyard-State-Region

1. Two Brattle Group colleagues, Judy Chang and Johannes Pfeifenberger, studied the use of competitive bidding for new transmission lines to reduce the cost of line construction. They concluded that bidding reduced the capital costs of new lines by 20–30 percent in the sample they examined (Pfeifenberger et al., 2019). The topic was covered in Wilson (2019, p. B1). Data for the United States were used from the U.S. Energy Information Administration (2019a), for the EU from Eurostat (n.d.), and Australia from the Energy Networks Association (2014).

2. Fox-Penner (2014) p. 80.

3. U.S. Department of Energy (2018) pp. 8, 11.

4. See Chang and Pfeifenberger (2014) for an early discussion.

5. MacDonald et al. (2016) cited in Shaner et al. (2018) fn. 22.

6. Trabish (2017).

7. ENTSO-E (2019) chapter 5 and ENTSO-E (2018b) chapter 4. Also see Fürsch et al. (2013).

8. Australia's 2017 Electricity Network Transformation Roadmap (Crawford et al., 2017) reflects the state of the art in a long-term decarbonization planning exercise that begins to integrate traditional transmission planning with climate policies, generation planning, and a strong emphasis on distributed generation, downstream markets, and new distribution business models. Even here, however, individual grid regions were not modeled separately, and several of the main recommendations coming from the study involve improving current transmission planning procedures, such as "develop new tools . . . to assess, test, and expand power system planning approaches" (p. 62).

9. Scott and Bernell (2015) provide a good overview of various agency and constituency positions on transmission plans.

10. Sierra Club (n.d.).

11. A good discussion of the process issues involved is found in Wellinghoff and Cusick (2017).

12. See Pfeifenberger, Chang, and Sheilendranath (2015) table 1.

13. One encouraging bit of evidence comes from recent work by Carley, Konisky, and Ansolabehere (2018), who find that citizen support for transmission lines increases significantly if they understand that the line will be used to carry only solar or wind power.

14. ENTSO-E (2016).

15. Federal Energy Regulatory Commission (2019a). For a description of PJM's most recent planning process, see PJM (2017). Europe's planning process is described in the EU's 2018 Ten Year Network Development Plan. See Scheibe (2018).

16. A one-time multi-regional transmission planning group known as the Eastern Interconnection Planning Collaborative was funded by the Obama administration in 2009. This group was a first move toward interregional planning, but it had no authority to produce an enforceable plan and did not yield large lasting changes in either U.S. processes or the national system (EIPC, n.d.).

17. ENTSO-E (2018b) p. 2. As explained by Scheibe (2018, p. 25), "The plan now describes two *Best Estimate* scenarios for the short (2020) and medium term (2025) respectively. These two scenarios follow a bottom-up logic and are based on the TSOs' estimates, and consider current national and European regulations. Their purpose is to map the development of the European energy system up to the point where uncertainties over possible future pathways become too great after 2025. For the year 2030, . . . ENTSO-E developed two scenarios, *Sustainable Transition and Distributed Generation* that differ from one another with regard to a number of criteria such as: general economic conditions, growth of electric vehicle use, wind and solar generation capacity, and demand flexibility, among others. In addition, the EC's *EUCO Scenario* was included as a third pathway toward 2030 . . . The aim of the Reference Scenario was to take into consideration current policies and market trends, and to project their impact on the development of the European energy system up to 2050. The scenario assumes the achievement of both the 2020 and 2030 EU energy policy goal."

18. Scheibe (2018) section 3.

19. Scheibe (2018).

20. Scheibe (2018) pp. 8, 22.

21. Scheibe (2018) p. 21.

22. Scheibe (2018) p. 27.

23. ENTSO-E (2018a) section 3.6.

24. Oregon Department of Energy (2017).

25. Utah Office of Energy Development (2013) and California Public Utilities Commission (2009).

26. TransWest Express LLC (n.d.).

27. Guidelines for the trans-European energy infrastructure (EUR-Lex, 2013).

28. Glachant, Rossetto, and Vasconcelos (2017) section 2.1.3.

29. ENTSO-E (n.d.-b).

30. In the Ten-Year Network Development Plan 2018, Scheibe writes that "project promoters of PCIs can apply for the European Investment Bank's (EIB) European Fund for Strategic Investments (EFSI) scheme" (Scheibe, 2018).

31. Van Nuffel et al. (2017).

32. For distances of less than about 200 miles for 3,000 MW, AC lines are still cheaper, unless the line is running underwater in which case DC is also cheaper (Siemens, 2014).

33. Black & Veatch (2014) table 2.1. Costs are 2014 dollars for baseline only, excluding multipliers for terrain, conductors, and towers. Line capacities from Figure 5-2 in this reference. AC substations cost $2 million while DC converter stations cost $500 million.

34. For useful overviews on this topic, see Gonzalez-Longatt (2015) and Wolf (2018).

35. ENTSO-E (n.d.-a).

36. ASEA Brown Boveri (2018).

37. Fairley (2019).

38. Simon (2018).

39. Wikiquote (n.d.).

40. Using a relaxed definition of supergrids, Bernard (2015) reviews many partial efforts and proposals as of 2015, including the proposed Atlantic Wind Connection off the East Coast of the United States and lines on the Arabian Peninsula and Central America. A supergrid for Europe is discussed by Hadarau (2016) and Van Hertem, Gomis-Bellmunt, and Liang (2016).

41. North American Supergrid (2017).

42. North American Supergrid (2017) p. 1.

43. North American Supergrid (2017) p. 1.

44. One key element of the North American Supergrid idea is to place as much of the system underground along existing infrastructure right-of-ways as is feasible—about 75 percent, according to proponents' studies. Although underground lines cost three times as much to build or more, they provide some potential benefits as well. Underground DC lines are obviously not visible and have far lower electromagnetic fields and other features that disturb neighbors and lower property values. They are much less vulnerable to

electromagnetic pulses from natural causes or man-made nuclear weapons that can cripple grids. Most of all, its proponents believe that siting permissions will be able to be greatly streamlined.

45. Of course, these implementation challenges grow even larger when considering transcontinental and transoceanic supergrids such as those proposed by GEIDCO. When electricity is supplied by another country thousands of miles away, the geopolitical risks and security-of-supply concerns associated with electricity begin to resemble the long-standing issues faced by oil- and gas-dependent countries such as China (for oil) and Germany (for gas). In addition to their land use impacts, supergrids also often rely on large hydroelectric resources, which have their own major environmental issues, and can willfully or inadvertently promote fossil generation. For a brief glimpse of these issues, see Simonov (2018). For a more thorough discussion and case study, see Downie (2019a and 2019b).

46. Fox-Penner (2001) and Scott and Bernell (2015).

47. Pfeifenberger et al. (2016).

8. The Big Grid Bucks Stop Here

1. According to Meyer and Pac (2013), state-owned utilities in Eastern Europe owned more than 251 power plants as of 2010.

2. S&P Global Platts (2019).

3. Patel (2019).

4. U.S. Energy Information Administration (2017) and S&P Global Platts (2018).

5. U.S. Energy Information Administration (2017) and S&P Global Platts (2018).

6. U.S. Energy Information Administration (2017) and S&P Global Platts (2018).

7. Benn et al. (2018).

8. In addition to Benn et al. (2018), Sartor (2018) contains a good general discussion.

9. Sartor (2018) and Varadarajan, Posner, and Fisher (2018). Both of these reports are highly readable and are good summaries of financial approaches to early closures. Other financial measures include applying tax over-collection, issuing Green Bonds with capital recycling, and adopting green tariffs. Personal communication, Robert Mudge, the Brattle Group, December 7, 2018.

10. As of this writing, a bill pending before the Colorado legislature, HB17-339, would divert 15 percent of the proceeds from securitized plant-closure bonds to fund the Colorado Energy Impact Assistance Authority, which would "aid displaced workers and backfill a portion of the affected communities' lost property tax through extra funds generated by the refinancing" (Gimon, 2017a).

11. Benn et al. (2018) has four global case studies; Boettner et al. (2000) contains ten within the United States.

12. In some cases, policy makers have mandated that conversion of coal plants to gas be given "primary consideration," as in the State of Colorado (2010).

13. Personal communication with Robert Hemphill, April 11, 2019, and Brayton Point Commerce Center (n.d.).

14. Appalachian Regional Commission (n.d.). There is a burgeoning literature on just and sustainable energy transitions for workers, an example of which is Johnstone and Hielscher (2017).

15. Jakubowski (2018).

16. CBC (2017).

17. Gimon (2017a) and Thorp (2019).

18. Pyper (2018b) and CPR News (2018).

19. Hood (2018).

20. Hood (2018).

21. Weiser (2018).

22. Weiser (2018).

23. Simeone (2017b).

24. Pyper (2018b).

25. Europe beyond Coal (2019).

26. Bloomberg (2019). See also Plumer (2019).

27. Formally, utility regulation offers the regulated company only a reasonable opportunity to earn a fair return on invested capital, not a guarantee. See, for example, Kahn (1988) v. 1, p. 42ff. There is always a significant chance that regulators will reduce returns due to dissatisfaction with a utility's performance, or that they'll give returns higher than are necessary.

28. This is the capital attraction role of regulation; see, among others, Bonbright, Danielsen, and Kamerschen (1988) p. 111.

29. For the official overview of California's resource adequacy requirements, as one example, see California Public Utilities Commission (n.d.).

30. Another approach to making specific types of power generators financeable is to offer them automatic contracts with guaranteed prices, or even to simplify contracting to the point of treating the prices a utility pays for buying all power from a generator under a tariff. In Europe, these are called *feed-in tariffs*. Because they are especially inflexible, technology-specific, and do not adjust to market circumstances, they have fallen out of favor and are now rarely used. See the discussion of PURPA contracts a little later in this chapter.

31. Hogan's extensive library of publications is available at Harvard Electricity Policy Group (n.d.). Many of these are quite technical; one treatment of scarcity pricing comes from a 2016 slide deck (Hogan, 2016).

32. The question of whether high energy prices were appropriate signals to build more plant or windfall profits that were the result of undue levels of market power was the central economic issue litigated in the wake of the California energy crisis of 2000–2001. Although U.S. federal regulators and courts never clearly resolved the issue, both the state and federal regulators continued to rely on price caps and added a resource adequacy requirement on utilities described in Tierney (2018) and Joskow (2019). For contrasting views on the California crisis, see Sweeney (2002), Harvey and Hogan (2001), and Taylor et al. (2015).

33. Milligan et al. (2016) table 5. Good examples of popular objections to price spikes can be found in Orvis and O'Boyle (2018).

34. Bureau of Economic Geology. Center for Energy Economics (2013). See ERCOT (n.d.-a) for recent capacity changes by fuel type.

35. Personal communication with Chris Hunt, Managing Director, Riverstone LLC, London, April 16, 2019.

36. Milligan et al. (2016) p. 34.

37. Joskow (2019) pp. 53–54.

38. Modern buyers of power on the Big Grid routinely use competitive tender processes to harness market forces and achieve efficient prices. There are also many consultants and price reporting services in each market that help buyers and sellers benchmark their offers.

39. ERCOT (n.d.-b) slide 8, Monitoring Analytics LLC (2019), and Trabish (2012).

40. About 45 percent of all renewable generation added in the United States since 2000 can be attributed to RPS rules. While some of this would have probably occurred anyway, much of the balance was anchored by contracts outside the RPS process, such as by voluntary utility or corporate purchases (Barbose, 2018, p. 13).

41. U.S. Energy Information Administration (2019b) and Renewable Energy Buyers Alliance (n.d.).

42. There is an extensive literature on the benefits and costs of contracts in the power industry, with many studies concluding that contracts do provide economic benefits that other mechanisms can't match. See, for example, Parsons (2008).

43. See, for example, Mufson (2011).

44. Gabaldón-Estevan, Peñalvo-López, and Solar (2018).

45. See, for example, Hirsch (1999).

46. The three centralized markets are operated by ISO New England, the New York independent system operator, and PJM. Other mechanisms that function like capacity markets are examined in a moment. See Federal Energy Regulatory Commission (2019b). EU statistics come from personal communication with Kathleen Spees, July 2, 2019. See also European Commission (2018).

47. PJM (n.d.-b).

48. PJM (n.d.-b).

49. Re-adding storage, distributed energy resources.

50. The more general question of whether restructured or integrated markets are more able to decarbonize can't be answered, but there is certainly evidence on both sides. Some traditional utilities may be among the first to reach 100 percent carbon-free systems, but restructured markets often provide greater liquidity and access. See, for example, Pfeifenberger et al. (2016).

51. See Newell et al. (2018) for the United States and Aurora Energy Research (2018) pp. 11–16. You can also integrate wind or solar with storage and bid the combination as a capacity resource. This is happening more and more, but remember that there is little storage that provides year-around performance, so this is far from a perfect solution.

52. For a concise but thorough recent overview, see Goggin et al. (2018).

53. PJM (2019) p. 29.

54. PJM (2019) p. 29.

55. Vaughan (2016).

56. Goggin et al. (2018) p. 26.

57. Newell et al. (2018).

58. These names come from Hogan (2018), but see note 59 for a number of market design change proposals.

59. Calls for greater capacity market flexibility emanate from a wide range of experts, including Steven Corneli, Bruce Ho, Michael Hogan, Kathleen Spees,

Sam Newell, Bloomberg New Energy Finance, Michael Goggin, Rob Gramlich, Robbie Orvis, and Sonya Aggarwal. See, among others, Goggin et al. (2018), Orvis and Aggarwal (2018), Gimon (2017b), and Cheung (2017). "Long Term Visions for Wholesale Electricity Markets in a Future Dominated by Zero Marginal Cost Resources," unpublished PowerPoint by Sonia Aggarwal and Rob Gramlich, Golden, CO, February 26–28, 2018.

60. California ISO (2016b).

61. World Bank Group (2018) pp. 9–11.

62. See Newell et al. (2017) and Spees et al. (2018) for two examples. Palmer, Burtraw, and Keyes (2017) also discuss "carbon market adders," as they are referred to among power market designers.

63. Last June, the Federal Energy Regulatory Commission released its most extensive proposal to allow sellers and buyers to opt out of PJM's capacity market—a significant, long-sought change in U.S. policy. Order, Docket No. EL-16-49-000, Federal Energy Regulatory Commission, 163 FERC 61,236. For useful discussion of this order, see Chen (2018).

64. Unfortunately, structural market power concerns often do crop up in capacity auctions due to limits on transmission capacity that tend to define constrained areas that may only have a few sellers or buyers. PJM, which operates the largest auction in the world, recently concluded that its entire 2019–2020 auction could not be certified as fully competitive, and therefore applied special market power mitigation policies to *all* capacity offers (PJM, n.d.-a).

65. Farmer (2018).

66. Appunn (2018).

67. See Aggarwal et al. (2019). All of these proposals include and strongly favor centralized energy spot markets, with some improvements.

68. Gramlich and Hogan (2019).

69. Corneli, Gimon, and Pierpont (2019).

70. Many of these requirements are set by the European Commission; see note 47.

9. The Utility Business in Three Dimensions

1. National Academies of Sciences, Engineering, and Medicine (2017) p. 19. See table 2.1, citing EIA Form 861 data. Cooperative numbers from the National Rural Electric Cooperative Association (n.d.). Public power systems numbers from American Public Power Association (n.d.-b).

2. Confusingly, the Europeans use the acronym DSO to refer to what we in the United States refer to simply as electric distribution utilities or distributors. In the United States, DSO stands for an independent nonprofit system operator that is distinct from the utilities whose system it operates. The latter concept is discussed later in this chapter. Council of European Energy Regulators (2016) p. 5.

3. Groebel (2013) slide 29, and Ofgem (2017) p. 1.

4. Data compiled by the author from multiple sources in Asia Australia Distribution.xls available from the author. China has about 3,000 distinct distribution companies, but it appears that about 2,700 are owned and controlled by the two large state-owned utilities, State Grid and China Southern Power Grid. The remainder appear to be regionally owned. However, utilities in China and other emerging economies do not have the challenges outlined in this chapter (CCCL, 2010, and referred by Caixin, 2008).

5. PwC (2015) p. 4.

6. Edison Electric Institute (2018).

7. Retrieved September 27, 2018, from: https://wolferesearch.com/utilities-peg-batting-clean-njs-energy-plan-management-visit.

8. Edison Electric Institute (n.d.-a).

9. The language and literature surrounding utility business models has exploded in recent years. Many utilities and consultants have proposed other ways to visualize utilities' and regulators' strategic choices.

10. Smart Energy Consumer Collaborative (2018) p. 4.

11. Trabish (2014).

12. See Tong and Wellinghoff (2014) for the seminal article; also see Trabish (2014) and Bade (2015) for subsequent blog posts.

13. This and other related architectural and operation concerns are discussed at length in De Martini and Kristov (2015).

14. For example, ISO New England's latest annual budget is about $170 million, or 0.2 cents per kilowatt-hour transmitted (ISO New England, 2018a). They typically service a huge area as large as one country or one region of a large country. So, their costs are spread across a large number of transmission transactions. At this scale, U.S. TSOs add only a fraction of a percent to the cost of power (about 0.03–0.2 cents per kilowatt-hour); Federal Energy Regulatory Commission (2016) figure 37.

15. The commission wrote an unusually lengthy and detailed explanation for their rejection of a DSO; see pp. 45–53 of Order Adopting Regulatory Policy

Framework and Implementation Plan, New York Public Service Commission, 14M-0101 (2015).

16. AEMO and Energy Networks Australia (2018). See also Burger et al. (2019) on this point.

17. Chapter 13 discusses these movements in more detail. As noted, however, they are not substitutes for a distribution business model. They are a new type of supply arrangement applicable only where the industry has unbundled and adopted retail choice.

18. Cross-Call et al. (2018) pp. 20–21.

19. City of Fort Collins (2015).

20. See Zummo (2018) for an excellent overview of Public Power's view of its future business model.

21. National Rural Electric Cooperative Association (2015).

22. This isn't an unheard of outcome in the utility sector, though it is pretty rare. Public Service New Hampshire went bankrupt in 1988 due to large cost overruns that regulators would not allow to be recovered in rates; it sold itself to another larger New England utility. In 1996, the Long Island Lighting Company (Lilco), also in dire financial circumstances over high costs, was essentially forced to transfer itself to the Long Island Power Authority, a new municipal utility created from scratch. With the onset of retail choice, the Australian utility AGL sold its distribution system and became exclusively a power generator and competitive retailer, as did the UK utility Centrica. In Chapter 11, we examine past and ongoing attempts to covert IOU grids to public ownership.

23. Chen (2017). Similar conclusions emerge from Bloomberg New Energy Finance's proprietary data on utility venture capital investments between 2008 and 2017 (Annex, 2018).

24. Accenture (2017) p. 18.

25. See Energy Impact Partners (n.d.). ClearSky (n.d.) is another coalition of utilities investing in new, unregulated businesses. One U.S. utility, Avista, has an early history of incubating and spinning off successful startups and also runs a modern incubator.

26. Enel (2017) pp. 431–474.

27. S. Byrd, Morgan Stanley. Personal communication with the author, December 20, 2018.

28. Burger and Weinmann (2016) pp. 304–308.

29. Enel (2017) p. 8.

30. Heath (2018).

31. Smart Energy Consumer Collaborative (2018).
32. Accenture (2017) p. 33.
33. Smart Energy Consumer Collaborative (2018).
34. Smith and MacGill (2016) provide an excellent view of electricity customer evolution.

10. The *Really* Smart Grid

1. Industry insiders may find the term ESCO confusing, since ESCOs have traditionally referred to companies that rehabilitate energy systems in large buildings to make them energy efficient, especially buildings in the public sector. I intend the term to refer to a much broader set of companies that sell all types of energy services to all types of retail customers. The term *aggregators* is often used to refer to the companies I call ESCOs because one of their future roles will be to aggregate their customers' surplus self-generation and remarket it on their behalf. This is only one of ESCOs' many future functions, and so I use the term ESCO instead.
2. Accenture (2013) p. 21.
3. Public power and cooperatives usually do not have such extended approval processes and have demonstrated greater agility in many cases. However, they typically also cannot move as fast as modern private ESCOs.
4. California 2014 Smart Inverter Working Group Recommendations, quoted in Corneli and Kihm (2015) p. 29.
5. This figure is nearly identical to illustrations of the industry under retail electricity deregulation (so-called retail choice) which has operated since the 1990s. Fox-Penner (1997). See, for example, figure 9-4 in *Electric Utility Restructuring,* my 1997 book on retail choice. The difference now is that the deregulated retailers no longer sell only electricity, and are therefore no longer referred to as retail electricity sellers. The other related services they sell will hopefully become critical to commercial success.
6. Cooper (2016) p. 99.
7. ElectricityPlans (n.d.).
8. See Graves et al. (2018, slide 9) for a very brief overview, and Rai and Zarnikau (2016) for a thorough discussion as of 2016.
9. A good discussion on subscriptions as a means of enabling management / opt of network can be found in Trabish (2018).
10. This concept is not new. In *Smart Power,* I describe how the original gas and electric utilities sold lighting, not kilowatt-hours. Chicago's first lighting

salesman sold the light from fifty arc lights for fifteen cents an hour of operation. Roger Sant and Amory Lovins are credited with reviving the idea in the 1970s (Fox-Penner, 2014, pp. 200–201).

11. Navigant's excellent work on EAAS, published in 2017, is behind a pay wall (Navigant, 2017), but is cited in Maloney (2017).

12. Bonbright (1961).

13. Order, Case No. 15-M-0127, Order Resetting Retail Energy Markets and Establishing Further Process, New York Public Service Commission (2016).

14. Littlechild (2018) p. 3.

15. Littlechild (2018) p. 3.

16. National Conference of State Legislatures (2019).

17. See Tabors et al. (2017) for a good simple overview of DLMPs. See also Edmunds, Galloway, and Gill (2017) for a good academic review. Tierney (2016) provides a good broad discussion on how to price distribution.

18. See Chapter 11 for further discussion of DLMPs' regulatory challenges.

19. Knieps (2016) argues for the necessity of model pricing in efficient distribution systems.

20. Regulators could probably require that delivery rates be passed through exactly as they are charged, but it would be extremely difficult to prevent ESCOs from offering pricing for the balance of each customer's charges that offset or hedged delivery rate changes. In fact, customers would probably find such hedging quite attractive and would not want regulators to interfere with ESCOs' ability to offer it. Moreover, if regulators start putting constraints of their nature on ESCOs, the latter cease to be unregulated entities—a distinct possibility, but then an industry model closer to the other more regulated end of the business model spectrum.

21. There is a reason I used the term market-determined rather than competitive. Not all bulk power markets have sufficient liquidity to be classified as workably competitive. In markets where liquidity is too low, regulators have devised ways to price power in ways that are influenced by market circumstances, but are still designed to protect buyers and sellers from undue levels of market power.

22. A lively debate has broken out among grid design experts and stakeholders regarding the best approach to integrating ESCO-distributor, ESO-bulk market, and distributor-bulk market transactions. (In the industry, these are referred to as questions of grid architecture and the hierarchy of control.) One school of thought suggests merging the control functions of the bulk power

market operator and distribution utilities, so that there is only one entity responsible for balancing controls. Another school of thought argues for a layered control approach, in effect solving for a local grid optimum, sending this solution up to the Big Grid, letting it solve, and then repeating the process. See the work of CAISO, PG&E, SCE, SDG&E with support from More than Smart (2017) and AEMO and Energy Networks Australia (2018) among others.

23. Littlechild (2018) p. 3. For excellent U.S. overviews, see Graves et al. (2018) and Graves, Carroll, and Haderlein (2018).

24. See Silverstein, Gramlich, and Goggin (2018) p. 15.

25. See Graves et al. (2018) slides 12 and 13.

26. Forbes (2018).

27. Kellner (2016).

11. Governing a Really Smart Grid

1. It is also important to note that these are just the high-level objectives. There are many challenging implementation tasks. One expert, Steve Corneli, lists the following key detailed implantation tasks for regulators in an SI industry: regular scrutiny of the underlying cost structure of the utilities they regulate for insights into the extent of natural monopoly, whether it is strong or weak; evaluation of the potential for utilities to put capital at excessive risk by increasing costs above levels that result in sustainable prices, creating transparency and value-producing opportunities for competitive DERs; leading in the implementation and continual improvement of low-cost approaches to DER optimization and control; developing fair rules for interconnection, operation, and optimization; establishing clear affiliate interest rules; making high priorities in rate-making criteria for both sustainable pricing (rates should not exceed either the utility's cost or the price of competitive alternatives) and economic efficiency (rates should be based on long-run marginal cost, not short-run); and considering potential benefits of distribution utility consolidation (Corneli and Kihm, 2015, pp. 35–36).

2. Hawaii State Legislature (2018).

3. The UK's RIIO regulatory system, which reportedly sets the lower end of the band at a 2 percent return on capital (far below the market requirement), is the most extreme penalty scheme extant, but so far, UK regulators have not needed to penalize utility performance at anywhere near this lower level of returns (Stone, 2016).

4. The acronym stands for Revenues equal Incentives plus Innovation plus Outputs. A condensed summary of the approach is that revenues are determined by a combination of measuring the outputs of the utility, applying incentives to those outputs, and rewarding the utility for innovation. See Ofgem (2010), and Fox-Penner, Harris, and Hesmondhalgh (2013). For a good recent discussion, see Meeus and Glachant (2018).

5. Ofgem (2017 and 2018b).

6. Ofgem (2017) pp. 3–9.

7. These points are discussed by Littell et al. (2017) section 3. Recognizing these challenges, an expert group commissioned by Ofgem to evaluate the entire RIIO scheme concluded in 2018 that "we do not think that the truly high risk-reward profile envisaged for RIIO can be realistically achieved under the current framework." This should not be read as a rejection of the approach, but rather an admission that any PBR scheme has inevitable inaccuracies, so that it can't be assured of high achievement of all its many goals (Ofgem, 2018a, p. 8).

8. Regulators in Germany and some other jurisdictions use another element of the RIIO scheme: a cap on utility revenues for several years. This gives utilities an incentive to find cost savings, but it is not really a scheme to reorient utilities' objectives (Groebel, 2017). As of 2019, twenty-four states in the United States used a form of revenue-cap incentive more commonly called *decoupling* to counter traditional regulation's tendency to encourage utilities to sell more power rather than help customers conserve energy (Center for Climate and Energy Solutions, 2019).

9. Littell et al. (2017) p. 63.

10. Trabish (2017).

11. For recent discussions, see Nelson and MacNeill (2016) and Biggar and Reeves (2016).

12. As noted, DLMPs are something like a miniature auction for new power supplies at each location on the distribution system. As we all know, auction outcomes can sometimes be manipulated by sellers or buyers who hold large market shares or otherwise game the auction rules. When DLMPs are used for setting prices, semi-automated rules and procedures for overseeing hundreds of mini-auctions on each distribution system will have to be in place. In addition, some experts suspect (myself included) that DLMP markets will generally be less liquid than the bulk power market, making processes to monitor trading and substitute regulated prices much more important.

13. Faruqui and Palmer (2011) and Faruqui and Aydin (2017).

14. See Faruqui, Grausz, and Bourbonnais (2019) and Faruqui and Aydin (2017) p. 47.

15. Also see Trabish (2019) for a recent update.

16. See, for example, Burger et al. (2019).

17. Wood et al. (2016) p. 20. One solar company writes, "All members of society will be better served if solar households remain connected to our broader energy system. It is more efficient for the household and the system" (Sunrun, 2018, p. 12).

18. John Howat, Senior Energy Analyst, National Consumer Law Center, writing in Wood et al. (2016) p. 26.

19. For a supurb, fulsome debate on fixed cost recovery approaches, see Wood et al. (2016). For a more formal economic discussion, see Burger et al. (2019).

20. For a good short overview, see Hanser and Van Horn (2016).

21. This may be an example of the "governance paradox," a term coined by economic sociologist Vili Lehdonvirta. As summarized by *New York Times* columnist James Ryerson, "the very mechanisms needed to make a decentralized system desirable—namely, apparatuses of oversight and regulation—appear to make the system no longer decentralized" (Ryerson, 2019, p. 15).

12. The Business and Regulation of Energy Service Utilities

1. See, for example, Viscusi, Vernon, and Harrington Jr. (1998, p. 322) or Kahn (1988, p. 3), as well as further discussion later in this chapter.

2. More completely, the economic regulation unique to utilities includes an exclusive franchise (barriers to entry, as some put it), control over prices, control over the terms of service, and an obligation to serve all customers. See Kahn (1988) p. 3.

3. Hawaii State Legislature (2018).

4. See National Rural Electric Cooperative Association (2017) and Zummo (2018).

5. This discussion is no substitute for a concrete analysis of the potential for competitive energy services in any one city, town, or rural area—an unsettled question. Each energy service has its own set of potential customers, its own cost structure, and its own profit potential as a function of the size and density of a geographic market. See the discussion later in this chapter on the tension between regulatory capture and the optimization of the system.

6. Wimberly and Treadway (2018) p. 5.

7. CPUC (2018).

8. Coase (1937) and Williamson (1975 and 1979). The value of vertical integration in the modern vehicle electricity service industry is discussed in the pioneering work of Joskow and Schmalensee (1983). See note following in a brief survey in Fox-Penner (2014) pp. 160–164.

9. In *Markets for Power,* written before retail choice was common, two of the leading electric power economists of their era, Paul Joskow and Richard Schmalensee, analyzed electricity deregulation from the transactions cost standpoint. They presciently predicted that many customers might not perceive enough gains from shopping for power and would rather reduce transaction costs by continuing to buy from their known utility. See Joskow and Schmalensee (1983).

10. Corneli and Kihm (2015) p. 45.

11. See Pechman (2016), especially section 4.1.

12. The proposal by Brendan Pierpont in Aggarwal et al. (2019), mentioned briefly in Chapter 8, is an aggregated-demand central market for multiple utilities, much as is described here.

13. Pyper (2015).

14. Pyper (2015) p. 2.

15. "From a Regulation Mindset to an Entrepreneurial Orientation? Can (Should) Utilities Make the Switch?" presented September 18, 2018, at Wisconsin Public Utility Institute (Kihm, 2018). Another interesting experiment was conducted by a team of EU management science academics, who surveyed 129 EU distribution utilities from nearly all EU countries. These researchers assessed the relative ability of the managements of each utility to adopt new business models and practices via their survey instrument. They found a wide range of adaption capabilities, and also found a positive correlation between management agility and operating performance. See Pereira, Pereira da Silva, and Soule (2019).

16. For the sake of transparency, I note that Innowatts is one of Energy Impact Partners' portfolio investments.

17. Fast Company (n.d.).

18. During Mary's tenure, GMP also became the first utility B Corps corporation that consciously adopts social and environmental goals alongside profit maximization (B Corps, n.d.).

19. Furthermore, cultural change does not belong solely to ESUs; some highly evolved, customer-centric utilities may opt to become SIs or be forced

into it by policy makers. Public power utilities and cooperatives also must grapple with similar cultural change to accompany business model changes; to see one such story for a public power utility (see Williams, 2018).

20. See Green Mountain Power (2017).

21. See Green Mountain Power (2017) Att. 1.

22. Corneli and Kihm (2015) and former FERC chair Jon Wellinghoff (Wellinghoff and Tong, 2015) both warn of cross-subsidization in their discussion of ESUs. Professor Lynn Kiesling, a leading electricity economist, describes the risk as one of vertical foreclosure (Kiesling, Munger, and Theisen, 2018).

23. Recognizing that utilities might have a useful role to play, it quickly lifted this ban, and utilities have been involved in regulated EE activities ever since. See Pechman (2016) p. 9.

24. Council of European Energy Regulators (2016) p. 27.

25. Council of European Energy Regulators (2016) p. 27.

26. The erosion of the utility natural monopoly for traditional and new services is illustrated and discussed extensively in Corneli and Kihm (2015) pp. 13–25.

27. It is also worth noting that this problem is less stringent for public power utilities and rural cooperatives. The latter two types of utilities are nonprofit. So, they can't gain any profits by cross-subsidizing competitive products at the expense of captive regulated customers. They can still cross-subsidize, as all companies of all types do to a degree, but they have no incentive other than to serve their customers in the best overall manner. This may be another reason why the ESU model has so far seen more traction among these types of utilities.

28. Griswold and Karaian (2018) and Kopnecki (2018).

13. Forces and Fault Lines beyond the Industry

1. Galloway (2018) p. 6.

2. Pepall and Richards (2019).

3. Wu (2018) p. 18. For additional arguments, see Jonathan Tepper (n.d.) and the writings of *Washington Post* economic columnist Steve Pearlstein (e.g., Pearlstein, 2019).

4. Khan (2018).

5. Lanier (2011).

6. Silverman (2019) p. 10.

7. EU GDPR.ORG (n.d.).

8. See Wakabayash (2018).

9. Ghosh (2018).

10. Cameron (2018). Also Phelan (2015) slide 5. California's rules go back to 2011, long before more general privacy regulations were enacted, namely the COM / MP1 / tcg / jt2 issued on July 29, 2011 [Rulemaking 08-12-009 (filed December 8, 2008)].

11. See ACEEEE (2016) and Decision Adopting Rules to Protect the Privacy and Security of the Electricity Usage Data of the Customers of Pacific Gas and Electric Company, Southern California Edison Company, and San Diego Gas & Electric Company, Public Utilities Commission of the State of California, Decision 11-07-056, Public Utilities Commission of the State of California (2011).

12. Levenda, Mahmoudi, and Sussman (2015) p. 615.

13. Levenda, Mahmoudi, and Sussman (2015) p. 629.

14. Personal communication, Stephen Comello, Stanford University, April 1, 2019.

15. Public Power for Your Community (n.d.).

16. See Burke and Stephens (2017), Szulecki (2018), and Institute for Local Self-Reliance (n.d.) for good definitions and discussions of the overall movement.

17. Although the trends supporting ED stretch back to the 1970s, when E. F. Schumacher wrote *Small Is Beautiful* and Amory Lovins began explaining the benefits of small-scale systems, some scholars trace the recent emergence of the ED movement to Germany's Energiewende, which ignited strong interest in small electricity cooperatives that could take advantage of the solar feed-in tariff, and to the climate justice movement, which forced climate policy thinkers to consider just approaches to expanding clean energy around the world.

18. Szulecki (2018).

19. Burke and Stephens (2017) p. 6, citing www.energydemocracyinitiative .org.

20. Knight (2011).

21. See Walton (2018) and Energy Transition (2014).

22. N. Johnson (2018) pp. 6–7.

23. It is also called municipal aggregation, community aggregation, community choice energy, and government aggregation.

24. U.S. Environmental Protection Agency (n.d.).

25. Lean Energy U.S. (n.d.).

26. Blaylock (2019).

27. McKibben (2015).
28. Curren (2017).
29. Manjoo (2017), Ferenstein (2017), and S. Johnson (2018).
30. Newsham (2015).
31. Palast, Oppenheim, and MacGregor (2003).

14. Money Talks

1. Simmons (2018).
2. Strauss (2018).
3. Kenny (2019).
4. Harris et al. (2017).

15. Power without Carbon

1. Vrins et al. (2015).
2. McKibben (2015).

Appendix A

1. As explained in chapter 10 of *Smart Power* (Fox-Penner, 2014), utilities have some unique advantages for providing comprehensive energy services, provided their regulatory model gives them proper incentives and oversight. Other entities can be given the responsibility for energy efficiency as part of a new regulatory and market scheme, including government, nonprofit, and purely commercial agencies. The critical factors are that the responsible party be able to harness proper price signals and incentives to the ultimate customer, have access to ample low-cost (patient) capital, and have strong technical expertise and oversight.

Appendix B

1. Joskow (2019) p. 13.

2. Wind and solar plants typically sign contracts for most of their output, but often they have surplus generation. In addition, the mechanics of many contractual sales involve these plants bidding their output into spot markets, even though a buyer has purchased this output in advance.

3. Bushnell and Novan (2018).
4. Seel et al. (2018).
5. Starn (2018).

Appendix C

1. Mai et al. (2018) p. xiv, figure ES-3.
2. The White House Washington (2016) p. 48, figure 4.6.
3. Electric Power Research Institute (2018) p. 7.
4. Statistics Canada (2018), table 25-10-0021-01.
5. Bataille, Sawyer, and Melton. (2015) p. 43, figure: Energy Supply Pathways by Resource.
6. Hillebrandt et al. (2015) p. 23.
7. Pye et al. (2015).
8. Ekins et al. (2013) p. 10.
9. Ekins et al. (2013) p. 11.
10. HM Government (2010) pp. 17–21.
11. National Statistics Digest of UK Energy Statistics (2012).
12. SEMARNAT-INECC (2016), p. 81, figure 29.
13. Altieri et al. (2015) p. 33, figure 23.
14. Hopkins et al. (2017) p. 52.
15. Kwok and Haley (2018) p. 42, figure 28.
16. Mahone et al. (2018) p. 39.
17. Amorim et al. (2014) figure 3.
18. Ming et al. (2019).

References

8 Rivers. (n.d.). The Allam Cycle and NET Power. Retrieved April 2, 2019, from: https://8rivers.com/portfolio/allam-cycle/

2018 State of the Electric Utility Survey. (2018). Retrieved July 8, 2019, from: https://www.utilitydive.com/library/2018-state-of-the-electric-utility-survey-report/

Abi-Samra, N. (2013a). Extreme Weather Effects on the Energy Infrastructure, EIC Climate Change Technology Conference, May 2013. Retrieved July 8, 2019, from: http://www.cctc2013.ca/Papers/CCTC2013%20EXT1-3%20Abi-Samra.pdf

Abi-Samra, N. (2013b). One Year Later: Superstorm Sandy Underscores the Need for a Resilient Grid. Retrieved July 8, 2019, from: https://spectrum.ieee.org/energy/the-smarter-grid/one-year-later-superstorm-sandy-underscores-need-for-a-resilient-grid

Abi-Samra, N. (2017). *Power Grid Resiliency for Adverse Conditions.* Norwood, MA: Artech House.

Accenture. (2013). The New Energy Consumer Handbook. Retrieved May 15, 2019, from: https://www.accenture.com/_acnmedia/Accenture/next-gen/insight-unlocking-value-of-digital-consumer/PDF/Accenture-New-Energy-Consumer-Handbook-2013.pdf

Accenture. (2016). Cyber-Physical Security for the Microgrid: New Perspectives to Protect Critical Power Infrastructure. Retrieved July 8, 2019, from: https://www.accenture.com/_acnmedia/PDF-27/Accenture_DEG-Microgrid-Security_POV_FINAL.pdf

Accenture. (2017). New Energy Consumer: New Paths to Operating Agility. Retrieved May 24, 2019, from: https://www.accenture.com/t20171113T063921Z__w__/us-en/_acnmedia/Accenture/next-gen-5/insight-new-energy-consumer-2017/Accenture-NEC2017-Main-Insights-POV.pdf

ACEEEE (2016). State and Local Policy Database. Retrieved October 27, 2019, from: https://database.aceee.org/state/data-access

AEMO and Energy Networks Australia. (2018). Open Energy Networks: Consultation on How Best to Transition to a Two-Way Grid That Allows Better Integration of Distributed Energy Resources for the Benefit of All Customers. Consultation Paper. Retrieved June 14, 2019, from: https://www.energynetworks.com.au/sites/default/files/open_energy_networks_consultation_paper.pdf

Aggarwal, S., Corneli, S., Gimon, E., Gramlich, R., Hogan, M., Orvis, R., and Pierpont, B. (2019). *Wholesale Electricity Market Design for Rapid Decarbonization.* San Francisco, CA: Energy Innovation Policy & Technology LLC.

Alexander, M. (n.d.). Air Conditioners Really Are Getting Better. Retrieved June 4, 2019, from: https://www.thisoldhouse.com/ideas/air-conditioners-really-are-getting-better

All Aces, Inc. (n.d.). Retrieved June 14, 2019, from: https://www.allacesinc.com

Allcott, H., and Greenstone, M. (2017). Measuring the Welfare Effects of Residential Energy Efficiency Programs. Working Paper 23386. Retrieved June 14, 2019, from: https://www.nber.org/papers/w23386

Altieri, K., Trollip, H., Caetano, T., Hughes, A., Merven, B., and Winkler, H. (2015). Pathways to Deep Decarbonization in South Africa. ZA 2015 Report. Retrieved July 8, 2019, from: http://deepdecarbonization.org/wp-content/uploads/2015/09/DDPP_ZAF.pdf

American Planning Association. (n.d.). Knowledgebase Collection: Autonomous Vehicles. Retrieved July 18, 2019, from: https://www.planning.org/knowledgebase/autonomousvehicles/

American Public Power Association. (n.d.-a). Disaster Planning and Response: Join the Mutual Aid Network. Retrieved November 8, 2018, from: https://www.publicpower.org/disaster-planning-and-response

American Public Power Association (n.d.-b). Our Members. Retrieved November 15, 2019, from: https://www.publicpower.org/our-members

American Wind Energy Association. (2019). Grid Vision: The Electric Highway to a 21st Century Economy. Retrieved July 1, 2019, from: https://www.awea.org/Awea/media/Resources/Publications%20and%20

Reports/White%20Papers/Grid-Vision-The-Electric-Highway-to-a-21st-Century-Economy.pdf

Aminu, M., Ali Nabavi, S., Rochelle, C., and Manovic, V. (2017). A Review of Developments in Carbon Dioxide Storage. *Applied Energy,* 208, 1389–1419. DOI: 10.1016/j.apenergy.2017.09.015.

Amorim, F., Pina, A., Gerbelova, H., Da Silva, P., Vasconcelos, J., and Martins, V. (2014). Electricity Decarbonisation Pathways for 2050 in Portugal: A TIMES (The Integrated MARKAL-EFOM System) Based Approach in Closed Versus Open Systems Modelling. *Energy,* 69, 104–112. DOI: 10.1016/j.energy.2014.01.052.

Anandarajah, G., Strachan, N., Ekins, P., Ramachandran, K., and Hughes, N. (2009). Pathways to a Low Carbon Economy: Energy Systems Modelling: UKERC Energy 20150 Research Report 1. Retrieved July 8, 2019, from: http://www.ukerc.ac.uk/publications/pathways-to-a-low-carbon-economy-energy-systems-modelling-ukerc-energy-2050-research-report-1-.html

Anderson, C. (2017). 2016 Solar Penetration by State. Retrieved June 14, 2019, from: https://www.ohmhomenow.com/2016-solar-penetration-state/

Anderson, S. (2017). Risk, Liability, and Economic Issues with Long-Term CO2 Storage—A Review. *Natural Resources Research,* 26(1), 89–112. DOI: 10.1007/s11053-016-9303-6.

Andrae, A. (2017). Total Consumer Power Consumption Forecast, Conference: Nordic Digital Business Summit. Project: Global Forecasting of ICT Footprints. Retrieved June 14, 2019, from: https://www.researchgate.net/publication/320225452_Total_Consumer_Power_Consumption_Forecast

Andrae, A. (2019a). *Drawing the Fresco of the Electricity Use of Information Technology—Part I.* Kista, Sweden: Huawei Technologies. DOI: 10.13140/RG.2.2.18392.14080.

Andrae, A. (2019b). Prediction Studies of Electricity Use of Global Computing in 2030. *International Journal of Science and Engineering Investigations,* 8(86), 64–79.

Angel, S., Parent, J., Civco, D., Blei, A., and Potere, D. (2011). The Dimensions of Global Urban Expansion: Estimates and Projections for All Countries, 2000–2050. *Progress in Planning,* 75(2), 53–107. DOI: 10.1016/j.progress.2011.04.001.

Annex, M. (2018). *Utility M&A and Venture Capital: 2008–17 Digital Ventures Take the Lead.* Bloomberg New Energy Finance.

Appalachian Regional Commission. (n.d.). POWER Initiative. Retrieved April 20, 2019, from: https://www.arc.gov/funding/power.asp

Appunn, K. (2018). EU Deal on Power Market Rules and Capacity Mechanisms Criticised for Coal Subsidies. Retrieved June 14, 2019, from: https://www.cleanenergywire.org/news/eu-deal-power-market-rules-and-capacity-mechanisms-criticised-coal-subsidies

ARPA-E. (n.d.). Retrieved June 12, 2019, from: https://arpa-e.energy.gov/?q=arpa-e-site-page/about

ASEA Brown Boveri. (2018). Historical Power Interconnector in Canada Achieves Key Milestone. Retrieved June 14, 2019, from: https://new.abb.com/news/detail/3462/historical-power-interconnector-in-canada-achieves-key-milestone

Assante, M., Lee, R., and Conway T. (2017). Modular ICS Malware. White Paper. Retrieved July 8, 2019, from: https://ics.sans.org/media/E-ISAC_SANS_Ukraine_DUC_6.pdf

Assante, M., Roxey, T., and Bochman, A. (2015). The Case for Simplicity in Energy Infrastructure: For Economic and National Security. Retrieved July 8, 2019, from: https://csis-prod.s3.amazonaws.com/s3fs-public/legacy_files/files/publication/151030_Assante_SimplicityEnergyInfrastructure_Web.pdf

Auffhammer, M., Baylis, P., and Hausman, C. (2017). Climate Change Is Projected to Have Severe Impacts on the Frequency and Intensity of Peak Electricity Demand across the United States. *Proceedings of the National Academy of Sciences of the United States of America,* 114(8), 1886–1891. DOI: 10.1073/pnas.1613193114.

Aurora Energy Research. (2018). *Capacity Market 2018: Results, Implications, and Potential Policy Reforms. Public Report.* Berlin, Germany: Aurora Energy Research.

Aydin, O., Pfeifenberger, J., and Chang J. (2017). *Western Regional Market Developments: Impact on Renewable Generation Investments and Balancing Costs.* San Diego, CA: Brattle Group.

B Corps. (n.d.). About B Corps. Retrieved on May 29, 2019, from: https://bcorporation.net/about-b-corps

Baatz, B., Relf, G., and Nowak, S. (2018). The Role of Energy Efficiency in a Distributed Energy Future. Research Report U1802. Retrieved July 17, 2019, from: https://aceee.org/research-report/u1802

Babloyan, A. (2014). *Keeping the Lights On: Why Utilities Need to Integrate Physical and Cyber Security.* Armonk, NY: IBM Corp.

Bade, G. (2015). Who Should Operate the Distribution Grid? Retrieved May 24, 2019, from: https://www.utilitydive.com/news/who-should-operate-the-distribution-grid/376950/

Ball, J. (2018). Why Carbon Pricing Isn't Working: Good Idea in Theory, Failing in Practice. *Foreign Affairs,* July / August, 134–146.

Barbose, G. (2018). U.S. Renewable Portfolio Standards 2018 Annual Status Report. Retrieved July 8, 2019, from: http://eta-publications.lbl.gov/sites/default/files/2018_annual_rps_summary_report.pdf

Barrón, K. (2018). *Environmental Policy and Competitive Markets in Harmony.* Chicago: Exelon.

Bartlett, J. (2018). Reducing Risk in Merchant Wind and Solar Projects through Financial Hedges. Working Paper (19–06). Retrieved July 23, 2019, from: https://media.rff.org/documents/WP_19-06_Bartlett.pdf

Bataille, C., Sawyer, D., and Melton, N. (2015). Pathways to Deep Decarbonization in Canada. CA 2015 Report. Retrieved July 8, 2019, from: http://deepdecarbonization.org/wp-content/uploads/2015/09/DDPP_CAN.pdf

Battis, J., Kurtovich, M., and O'Donnell, A. (2018). Security and Resilience for California Electric Distribution Infrastructure: Regulatory and Industry Response to SB 699. A California Public Utilities Commission (CPUC) Staff White Paper. Retrieved July 8, 2019, from: https://www.eisac.com/cartella/Asset/00006887/CPUC_Physical_Security_White_Paper_January_2018.pdf?parent=114006

Ben-Joseph, E. (2012). *Rethinking a Lot: The Design and Culture of Parking.* Cambridge, MA: The MIT Press.

Benn, A., Bodnar, P., Mitchell, J., and Jeff, W. (2018). *Managing the Coal Capital Transition: Collaborative Opportunities for Asset Owners, Policymakers, and Environmental Advocates.* Boulder, CO: Rocky Mountain Institute.

Bernard, P. (2015). Supergrids of the World GCCIA, SIEPAC, & the AWC Projects: An Overview. In Aguado-Cornago, A., ed. *The European Supergrid.* European Energy Studies. Deventer, Netherlands: Claeys & Casteels Law Publishing, pp. 27–36.

Biggar, D., and Reeves, A. (2016). Network Pricing for the Prosumer Future: Demand-Based Tariffs or Locational Marginal Pricing? In Sioshansi, F., ed. *Future of Utilities—Utilities of the Future: How Technological Innovations in Distributed Energy Resources Will Reshape the Electrical Power Sector.* Cambridge, MA: Academic Press, pp. 247–266.

Bird, L., Milligan, M., and Lew, D. (2013). *Integrating Variable Renewable Energy: Challenges and Solutions.* Golden, CO: National Renewable Energy Laboratory.

Black & Veatch. (2014). Capital Costs for Transmission and Substations: Updated Recommendations for WECC Transmission Expansion Planning.

Retrieved July 1, 2019, from: https://www.wecc.org/Reliability/2014_TEPPC_Transmission_CapCost_Report_B+V.pdf

Blaylock, D. (2019). The Power of Local Solutions. Retrieved June 25, 2019, from: https://www.publicpower.org/periodical/article/power-local-solutions

Bloom, A., Townsend, A., Palchak, D., Novacheck, J., King, J., Barrows, C., . . . Gruchalla, K. (2016). Eastern Renewable Generation Integration Study, NREL / TP-6A2064472-ES. Retrieved July 8, 2019, from: http://www.nrel.gov/docs/fy16osti/64472-ES.pdf

Bloomberg, M. (2019). Our Next Moonshot: Saving Earth's Climate. Retrieved July 8, 2019, from: https://www.bloomberg.com/opinion/articles/2019-06-07/michael-bloomberg-at-mit-how-to-save-planet-from-climate-change

Boettner, F., Fedorko, E., Hansen, E., Goetz, S., Han, Y., . . . Zimmerman, B. (2000). *Strengthening Economic Resilience in Appalachia.* Washington, DC: Appalachian Regional Commission.

Bonbright, J. (1961). *Principles of Public Utility Rates.* New York: Columbia University Press. Reproduced online. Retrieved May 21, 2019, from: http://media.terry.uga.edu/documents/exec_ed/bonbright/principles_of_public_utility_rates.pdf

Bonbright, J., Danielsen, A., and Kamerschen, D. (1988). *Principles of Public Utility Rates.* 2nd ed. Reston, VA: Public Utilities Reports, Inc.

Bonneville Power Administration. (2017). Fact Sheet: BPA Sets Wholesale Rates for Fiscal Years 2018–2019. Retrieved March 25, 2019, from: https://www.bpa.gov/news/pubs/FactSheets/fs-20170726-BPA-sets-wholesale-rates-for-fiscal-years-2018-2019.pdf

Borenstein, S. (2015). Is the Future of Electricity Generation Really Distributed? Retrieved June 19, 2019, from: https://energyathaas.wordpress.com/2015/05/04/is-the-future-of-electricity-generation-really-distributed/

Bossel, U. (2006). Does a Hydrogen Economy Make Sense? *Proceedings of the IEEE,* 94(10), 1826–1837. DOI: 10.1109/jproc.2006.883715.

Bowman, D. (2014). Geomagnetic Storms and the Power Grid. Retrieved June 24, 2019, from: https://www.spp.org/documents/23279/doug%20bowman%20-%20geomagneticstorms_grid_r3.pdf

Brattle Group. (2016). SB 350 Study: The Impacts of a Regional ISO-Operated Power Market on California: Analysis and Results. Retrieved June 19, 2019, from: https://www.ethree.com/wp-content/uploads/2017/02/Presentation-SenateBill350Study-Jul26_2016.pdf

Brayton Point Commerce Center. (n.d.). Retrieved April 13, 2019, from: http://www.braytonpointcommercecenter.com/

Breakthrough Energy. (n.d.). Retrieved March 30, 2019, from: http://www.b-t.energy/

Bronski, P., Creyts, J., Guccione, L., Madrazo, M., Mandel, J., Rader, B., . . . Tocco, H. (2014). *The Economics of Grid Defection: When and Where Distributed Solar Generation Plus Storage Competes with Traditional Utility Service.* Boulder, CO: Rocky Mountain Institute.

Brown, T., Bischof-Niemz, T., Blok, K., Breyer, C., Lund, H., and Mathiesen, B. (2018). Response to "Burden of Proof: A Comprehensive Review of the Feasibility of 100% Renewable-Electricity Systems." *Renewable and Sustainable Energy Reviews,* 92, 834–847. DOI: 10.1016/j.rser.2018.04.113.

Bruce, S., Temminghoff, M., Hayward, J., Schmidt, E., Munnings, C., Palfreyman, D., and Hartley, P. (2018). National Hydrogen Roadmap. Australia: CSIRO. Retrieved June 5, 2019, from: https://www.csiro.au/en/Do-business/Futures/Reports/Hydrogen-Roadmap

Budischak, C., Dewell, D., Thomson, H., Mach, L., Veron, D., and Kempton, W. (2013). Cost-Minimized Combinations of Wind Power, Solar Power and Electrochemical Storage, Powering the Grid up to 99.9% of the Time. *Journal of Power Sources,* 225, 60–74. DOI: 10.1016/j.jpowsour.2012.09.054.

Bureau of Economic Geology. Center for Energy Economics. (2013). A Primer on the Resource Adequacy Debate in Texas. Retrieved June 24, 2019, from: http://www.beg.utexas.edu/files/energyecon/think-corner/2013/A%20Primer%20on%20the%20Resource%20Adequacy%20debate%20in%20Texas%20122112.pdf

Burger, C., and Weinmann, J. (2016). European Utilities: Strategic Choices and Cultural Prerequisites for the Future. In Sioshansi, F., ed. *Future of Utilities—Utilities of the Future: How Technological Innovations in Distributed Energy Resources Will Reshape the Electrical Power Sector.* Cambridge, MA: Academic Press, pp. 303–321.

Burger, S., Schneider, I., Botterud, A., and Pérez-Arriaga, I. (2019). Fair, Equitable, and Efficient Tariffs in the Presence of Distributed Energy Sources. In Sioshansi, F., ed. *Consumer, Prosumer, Prosumager.* Cambridge, MA: Academic Press, pp. 155–185.

Burke, M., and Stephens, J. (2017). Energy Democracy: Goals and Policy Instruments for Societechnical Transitions. *Energy Research and Social Science,* 33, 35–48. DOI: 10.1016/j.erss.2017.09.024.

Bushnell, J., and Novan, K. (2018). *Setting with the Sun: The Impacts of Renewable Energy on Wholesale Power Markets.* Energy Institute WP 292. Berkeley, CA: Energy Institute at HAAS.

Byrd, S., Radcliff, T., Lee, S., Chada, B., Oiszewski, D., Matayoshi, Y., . . . Gosai, D. (2014). Solar Power and Energy Storage: Policy Factors vs. Improving Economics. Retrieved July 19, 2019, from: https://regmedia.co.uk/2014/08/06/morgan_stanley_energy_storage_blue_paper_2014.pdf

Byrne, J., Taminiau, J., Kurdgelashvili, L., and Kim, K. (2015). A Review of the Solar City Concept and Methods to Assess Rooftop Solar Electric Potential, with an Illustrative Application to the City of Seoul. *Renewable and Sustainable Energy Reviews,* 41, 830–844. DOI: 10.1016/j.rser.2014.08.023.

CAISO, PG&E, SCE, SDG&E with support from More than Smart. (2017). Coordination of Transmission and Distribution Operations in a High Distributed Energy Resource Electric Grid. Retrieved May 15, 2019, from: https://www.caiso.com/Documents/MoreThanSmartReport-CoordinatingTransmission_DistributionGridOperations.pdf

Caixin. (2008). Retrieved June 14, 2019, from: http://companies.caixin.com/2008-04-18/100050740.html

California Energy Commission. (2018). Adopted Building Standards. Retrieved June 14, 2019, from: http://www.energy.ca.gov/releases/2018_releases/2018-05-09_building_standards_adopted_nr.html

California ISO. (2016a). Fast Facts: What the Duck Curve Tells Us about Managing a Green Grid. Retrieved May 4, 2019, from: https://www.caiso.com/Documents/FlexibleResourcesHelpRenewables_FastFacts.pdf

California ISO. (2016b). Market Notice: Flexible Ramping Constraint Tariff Language Effective November 1, 2016. Retrieved May 24, 2019, from: http://www.caiso.com/Documents/FlexibleRampingConstraintTariffLanguageEffectiveNovember1_2016.html

California Public Utilities Commission. (2009). *General Information on Permitting Electric Transmission Projects at the California Public Utilities Commission.* Presentation by Transmission and Environmental Permitting Team. Slide 9. San Francisco, CA: California Public Utilities Commission.

California Public Utilities Commission. (n.d.). Resource Adequacy. Retrieved June 19, 2019, from: https://www.cpuc.ca.gov/RA/

Cameron, B. (2018). New Smart Meters Raise Privacy Questions. Retrieved July 23, 2019, from: https://www.poconorecord.com/news/20181118/new-smart-meters-raise-privacy-questions

Carbon Neutrality Coalition. (n.d.). Members of the Carbon Neutrality Coalition. Retrieved April 2, 2019, from: https://www.carbon-neutrality.global/members/

Carley, S., Konisky, D., and Ansolabehere, S. (2018). *Examining the Role of Nimbyism in Public Acceptance of Energy Infrastructure.* Panel Paper.

Castellanos, S., Sunter, D., and Kammen, D. (2017). Rooftop Solar Photovoltaic Potential in Cities: How Scalable Are Assessment Approaches? *Environmental Research Letters,* 12, 125005. DOI: 10.1088/1748-9326/aa7857.

Castillo, A. (2014). Risk Analysis and Management in Power Outage and Restoration: A Literature Survey. *Electric Power Systems Research,* 107, 9–15. DOI: 10.1016/j.epsr.2013.09.002.

Caughill, P. (2017). Hydrogen Power Storage Could Be an Important Part of a Fossil Fuel Free Future: Current Investments Look to Improve the Efficiency of the Technology. Retrieved June 5, 2019, from: https://futurism.com/hydrogen-power-storage-could-be-an-important-part-of-a-fossil-fuel-free-future

CBC. (2017). Laid Off Oil and Gas Workers Train for Alternative Energy Jobs as Wind Blows Alberta in New Direction. Retrieved June 22, 2019, from: https://www.cbc.ca/news/canada/calgary/alternative-energy-training-laid-off-oil-and-gas-1.4463217

CCCL. (2010). National Power Supply Enterprise Status Report. Retrieved May 23, 2019 from: http://www.competitionlaw.cn/info/1132/8590.htm

Center for Climate and Energy Solutions. (2019). Decoupling Policies. Retrieved July 14, 2019, from: https://www.c2es.org/document/decoupling-policies/

Centra Technology. (2011). Geomagnetic Storms, Contribution to the OECD Project—Future Global Shocks. Retrieved July 9, 2019, from: https://www.oecd.org/gov/risk/46891645.pdf

Chang, J., and Pfeifenberger, J. (2014). Well-Planned Electric Transmission Saves Customer Costs: Improved Transmission Planning Is Key to the Transition to a Carbon-Constrained Future. Retrieved July 24, 2019, from: https://brattlefiles.blob.core.windows.net/files/5813_well-planned_electric_transmission_saves_customers_costs_ppt.pdf

Chang, J., Pfeifenberger, J., Ruiz, P., and Van Horn, K. (2014). *Transmission to Capture Geographic Diversity of Renewables: Cost Savings Associated with Interconnecting Systems with High Renewables Penetration.* San Diego, CA: Brattle Group.

Chao, J. (2012). Parking Lot Science: Is Black Best? Retrieved June 14, 2019, from: https://newscenter.lbl.gov/2012/09/13/parking-lot-science/

Chediak, M., and Wells, K. (2013). Why the U.S. Power Grid's Days Are Numbered: Why the Electricity Grid's Days Are Numbered. Retrieved

July 19, 2019, from: https://www.bloomberg.com/news/articles/2013-08-22/why-the-u-dot-s-dot-power-grids-days-are-numbered#p4

Chen, J. (2018). Improving Market Design to Align with Public Policy. Retrieved June 28, 2019, from: https://nicholasinstitute.duke.edu/publications/improving-market-design-align-public-policy

Chen, M., Iyer, A., Shih, C., Kurdgelashvili, L., and Opila, R. (2017). A Critical Analysis on the Thin Crystalline Silicon PV Module of the Lightweight PV System. Retrieved June 14, 2019, from: https://ieeexplore.ieee.org/stamp/stamp.jsp?arnumber=8366671

Chen, O. (2017). Utilities Have Invested over $2.9 Billion in Distributed Energy Companies: DER Investment Is Facilitating the Transition to a Decentralized Energy System. Retrieved May 24, 2019, from: https://www.greentechmedia.com/articles/read/utilities-have-invested-over-2-9-billion-in-distributed-energy-companies#gs.djcs4s

Cheung, A. (2017). Cheung: Power Markets Need a Redesign—Here's Why. Retrieved June 14, 2019, from: https://about.bnef.com/blog/cheung-power-markets-need-redesign-heres/

Chew, A. (2015). Elon Musk Presents Tesla Powerwall and Powerpack Batteries. Retrieved May 1, 2019, from: https://www.youtube.com/watch?v=X4eY-0oKXmc

City of Fort Collins. (2015). Fort Collins Energy Policy. Retrieved May 24, 2019, from: https://www.fcgov.com/utilities/img/site_specific/uploads/Fort_Collins_2015_Energy_Policy_2.pdf

ClearSky. (n.d.). Retrieved June 14, 2019, from: http://www.clear-sky.com/

Cleveland, C., Castigliego, J., Cherne-Hendrick, M., Fox-Penner, P., Gopal, S., Hurley, L., . . . Zheng, K. (2019a). Carbon Free Boston Summary Report 2019. Boston Green Ribbon Commission. Retrieved May 31, 2019, from: https://www.greenribboncommission.org/wp-content/uploads/2019/01/Carbon-Free-Boston-Report-web.pdf

Cleveland, C., Cherne-Hendrick, M., Castigliego, J., Fox-Penner, P., Gopal, S., Hurley, L., . . . Zheng, K. (2019b). Carbon Free Boston Technical Summary 2019. Retrieved June 10, 2019, from: http://sites.bu.edu/cfb/files/2019/05/CFB_Technical_Summary_190514.pdf

Climate Home News. (2012). Stephen Hawking: Climate Disaster within 1000 Years. Retrieved November 26, 2018, from: http://www.climatechangenews.com/2012/01/06/stephen-hawking-warns-of-climate-disaster-ahead-of-70th-birthday/

Climate Institute. (2017). North American Supergrid. Transforming Electricity Transmission. Retrieved July 8, 2019, from: cleanandsecuregrid.org

Climate Ready Boston. (2016). Final Report, City of Boston. Retrieved July 8, 2019, from: https://www.boston.gov/departments/environment/climate-ready-boston

Coase, R. (1937). The Nature of the Firm. *Economica,* 4(16), 386–405. DOI: 10.1111/j.1468-0335.1937.tb00002.x.

Colthorpe, A. (2018). China's Biggest Flow Battery Project So Far Is Underway with Hundreds More Megawatts to Come. Retrieved June 5, 2019, from: https://www.energy-storage.news/news/chinas-biggest-flow-battery-project-so-far-is-underway-with-hundreds-more-m

Committee on Analytical Research Foundations for the Next-Generation Electric Grid. (2016). Analytic Research Foundations for the Next-Generation Electric Grid. Retrieved July 8, 2019, from: https://www.nap.edu/catalog/21919/analytic-research-foundations-for-the-next-generation-electric-grid

Cooper, J. (2016). The Innovation Platform Enables the Internet of Things. In Sioshansi, F., ed. *Future of Utilities—Utilities of the Future: How Technological Innovations in Distributed Energy Resources Will Reshape the Electrical Power Sector.* Cambridge, MA: Academic Press, pp. 91–108.

Corneli, S., Gimon, E., and Pierpont, B. (2019). *Wholesale Electricity Market Design for Rapid Decarbonization: Long-Term Markets, Working with Short-Term Energy Markets.* San Francisco, CA: Energy Innovation Policy & Technology LLC.

Corneli, S., and Kihm, S. (2015). Electric Industry Structure and Regulatory Responses in a High Distributed Energy Resources Future. FEUR Report No. 1. Retrieved July 1, 2019, from: https://escholarship.org/uc/item/2kf2n4kg

Council of European Energy Regulators. (2016). Key Support Elements of RES in Europe: Moving towards Market Integration. Retrieved May 30, 2019, from: https://www.ceer.eu/documents/104400/-/-/28b53e80-81cf-f7cd-bf9b-dfb46d471315

CPR News. (2018). Regulators OK Xcel's Early Shutdown of Pueblo Coal-Fired Generators. Retrieved June 22, 2019, from: https://www.cpr.org/news/story/regulators-ok-xcel-early-shutdown-of-pueblo-coal-fired-generators

CPUC. (2018). Retrieved May 29, 2019, from: http://docs.cpuc.ca.gov/PublishedDocs/Published/G000/M215/K380/215380424.PDF

Crawford, G., Crown, B., Johnston, S., Van Puyvelde, D., Watts, E., Brinsmead, T., . . . Bakker, T. (2017). Electricity Network Transformation Roadmap: Final Report. Retrieved July 23, 2019, from: https://www.energynetworks.com.au/sites/default/files/entr_final_report_april_2017.pdf

Crees, A. (n.d.). Bill Gates-Led Investors Back Two Startups Targeting Energy Storage. Retrieved October 25, 2019, from: https://www.chooseenergy.com/news/article/bill-gates-led-investors-back-two-startups-targeting-energy-storage/

Cross-Call, D., Goldenberg, C., Guccione, L., Gold, R., and O'Boyle, M. (2018). Navigating Utility Business Model Reform. Retrieved May 30, 2019, from: https://rmi.org/insight/navigating-utility-business-model-reform/

Cuadra, L., Salcedo-Sanz, S., Del Ser, J., Jiménez-Fernández, S., and Woo, Z. (2015). A Critical Review of Robustness in Power Grids Using Complex Networks Concepts. *Energies,* 8(9), 9211–9265. DOI: 10.3390/en8099211.

Curren, E. (2017). *The Solar Patriot: A Citizen's Guide to Helping America Win Clean Energy Independence.* Staunton, VA: New Sky Books.

Cutting the Cord. (1999). *The Economist,* October 7.

Dabiri, J. (2011). Potential Order-of-Magnitude Enhancement of Wind Farm Power Density via Counter-Rotating Vertical-Axis Wind Turbine Arrays. *Journal of Renewable and Sustainable Energy,* 3(4). DOI: 10.1063/1.3608170.

Dabiri, J., Greer, J., Koseff, J., Moin, P., and Peng, J. (2015). A New Approach to Wind Energy: Opportunities and Challenges. *AIP Conference Proceedings,* 1652(1), 51. DOI: 10.1063/1.4916168.

De Martini, P., and Kristov, L. (2015). Distribution Systems in a High Distributed Energy Resources Future. Future Electric Utility Regulation Report No. 2. Retrieved June 24, 2019, from: https://emp.lbl.gov/sites/default/files/lbnl-1003797.pdf

De Martini, P., Kristov, L., and Taft, J. (2019). *Operational Coordination across Bulk Power, Distribution and Customer Systems.* U.S. Department of Energy Electricity Advisory Board. Retrieved June 14, 2019, from: https://emp.lbl.gov/sites/default/files/lbnl-1003797.pdf

Decision Adopting Rules to Protect the Privacy and Security of the Electricity Usage Data of the Customers of Pacific Gas and Electric Company, Southern California Edison Company, and San Diego Gas & Electric Company, Public Utilities Commission of the St, 11-07-056, Public Utilities Commission of the State of California, July 28, 2011. Retrieved October 28, 2019, from: https://www.smartgrid.gov/files/Decision_Adopting_Rules_Protect_Privacy_d_Security_Electric2.pdf

Deign, J. (2019). Germany Looks to Put Thermal Storage Into Coal Plants: A New Pilot Will Replace Coal with Molten Salt to Create Giant Carnot

Batteries. Retrieved June 5, 2019, from: https://www.greentechmedia.com/articles/read/germany-thermal-storage-into-coal-plants

Deloitte. (n.d.). Utility 2.0 Winning Over the Next Generation of Utility Customers. Retrieved May 24, 2019, from: https://www2.deloitte.com/content/dam/Deloitte/us/Documents/energy-resources/us-e-r-utility-report.pdf

Department of Defense. (2015). National Security Implications of Climate-Related Risks and a Changing Climate. Retrieved June 14, 2019, from: http://archive.defense.gov/pubs/150724-congressional-report-on-national-implications-of-climate-change.pdf?source=govdelivery

Diesendorf, M., and Elliston, B. (2018). The Feasibility of 100% Renewable Electricity Systems: A Response to Critics. *Renewable and Sustainable Energy Reviews,* 93, 318–330. DOI: 10.1016/j.rser.2018.05.042.

Digiconomist. (n.d.). Bitcoin Energy Consumption Index. Retrieved March 29, 2019, from: https://digiconomist.net/bitcoin-energy-consumption

Dissanayaka, A., Wiebe, J., and Issacs, A. (2018). *Panhandle and South Texas Stability and System Strength Assessment.* Winnipeg, Canada: Electranix.

DNV GL Energy. (2014). *Integration of Renewable Energy in Europe.* Bonn, Germany: KEMA Consulting GmbH.

Dodds, P., Staffell, I., Hawkes, A., Li, F., Grünewald, P., McDowall, W., and Ekins, P. (2015). Hydrogen and Fuel Cell Technologies for Heating: A Review. *International Journal of Hydrogen Energy,* 40(5), 2065–2083. DOI: 10.1016/j.ijhydene.2014.11.059.

Doffman, Z. (2018). Network Effects: In 2019 IoT and 5G Will Push AI to the Very Edge. Retrieved May 15, 2019, from: https://www.forbes.com/sites/zakdoffman/2018/12/28/network-effects-in-2019-iot-and-5g-will-push-ai-to-the-very-edge/#56723b846bbe

Douglas Smith, L., Nayak, S., Karig, M., Kosnik, L., Konya, M., Lovett, K., . . . Luvai, H. (2014). Assessing Residential Customer Satisfaction for Large Electric Utilities. Retrieved May 24, 2019, from: http://www.umsl.edu/econ/Research/msl/workng/KosnikAmerenPaper.pdf

Downie, E. (2019a). China's Vision for a Global Grid: The Politics for Global Energy Interconnection. Retrieved July 25, 2019, from: https://reconnectingasia.csis.org/analysis/entries/global-energy-interconnection/

Downie, E. (2019b). *Powering the Globe: Lessons from Southeast Asia for China's Global Energy Interconnection Initiative.* Presentation at BNID, Boston, March 21.

Dunietz, J. (2017). Is the Power Grid Getting More Vulnerable to Cyber Attacks? Retrieved July 8, 2019, from: https://www.scientificamerican.com

/article/is-the-power-grid-getting-more-vulnerable-to-cyber-attacks/?redirect=1

Edison Electric Institute. (n.d.-a). EEI Report Finds Increased Transmission Investment. Retrieved October 27, 2019, from: https://www.eei.org/resourcesandmedia/newsroom/Pages/Press%20Releases/EEI%20Report%20Finds%20Increased%20Transmission%20Investment.aspx

Edison Electric Institute. (n.d.-b). Mutual Assistance. Retrieved November 8, 2018, from: http://www.eei.org/issuesandpolicy/electricreliability/mutualassistance/Pages/default.aspx

Editorial. (2019). On the Right Track. *Nature Energy,* 4, 169. DOI: 10.1038/s41560-019-0366-6.

Edmunds, C., Galloway, S., and Gill, S. (2017). Distributed Electricity Markets and Distribution Locational Marginal Prices: A Review. 52nd International Universities Power Engineering Conference (UPEC). Retrieved May 21, 2019, from: https://strathprints.strath.ac.uk/id/eprint/63483

EIPC. (n.d.). Retrieved June 14, 2019, from: https://www.eipconline.com/

Ekins, P., Strachan, N., Keppo, I., Usher, W., Skea, J., and Anandarajah, G. (2013). *The UK Energy System in 2050: Comparing Low-Carbon, Resilient Scenarios.* London: UKERC.

Electric Power Research Institute. (2018). U.S. National Electrification Assessment. Retrieved July 22, 2019, from: http://ipu.msu.edu/wp-content/uploads/2018/04/EPRI-Electrification-Report-2018.pdf

ElectricityPlans. (n.d.). Compare the Best Free Nights and Weekends Electricity Plans in Texas. Retrieved May 29, 2019, from: https://electricityplans.com/texas/compare/free-time-electricity-plans/

Ellis, A. (2016). *Improving Microgrid Cybersecurity.* Workshop Presentation on Microgrid Design. Sandia National Laboratories, November 24. Retrieved June 14, 2019, from: http://integratedgrid.com/wp-content/uploads/2017/01/5-Ellis-Improving-Microgrid-Cybersecurity.pdf

Enel. (2017). Annual Report 2017. Retrieved May 24, 2019, from: https://www.enel.com/content/dam/enel-com/governance_pdf/reports/annual-financial-report/2017/annual-report-2017.pdf

Energy Impact Partners. (n.d.). Retrieved June 14, 2019, from: https://www.energyimpactpartners.com/

Energy Networks Association. (2014). Electricity Prices and Network Costs. Retrieved June 14, 2019, from: https://www.energynetworks.com.au/sites/default/files/electricity-prices-and-network-costs_2.pdf

Energy Transition. (2014). The Re-Municipalization of the Hamburg Grid. Retrieved May 28, 2019, from: https://energytransition.org/2014/06/remunicipalization-of-hamburg-grid/

Energy Transitions Commission. (2017). The Future of Fossil Fuels: How to Steer Fossil Fuel Use in a Transition to a Low-Carbon Energy System. Retrieved June 14, 2019, from: https://www.copenhageneconomics.com/dyn/resources/Publication/publicationPDF/6/386/1485851778/copenhagen-economics-2017-the-future-of-fossil-fuels.pdf

EnergySage. (n.d.). Tesla Energy: Has Elon Musk Invented the First Clean Power Utility? Retrieved July 19, 2019, from: https://news.energysage.com/tesla-energy-has-elon-musk-invented-the-first-clean-power-utility/

ENTSO-E. (2015). The Electricity Highways: Preparing the Electricity Grid of the Future. Retrieved July 8, 2019, from: https://www.entsoe.eu/outlooks/ehighways-2050/

ENTSO-E. (2016). Real-Life Implementation of Electricity Projects of Common Interest—Best Practices. Retrieved June 14, 2019, from: https://docstore.entsoe.eu/Documents/SDC%20documents/AIM/entsoe_rl_impl_PCIs_web.pdf

ENTSO-E. (2018a). Insight Report Stakeholder Engagement. Retrieved July 23, 2019, from: https://tyndp.entsoe.eu/Documents/TYNDP%20documents/TYNDP2018/consultation/Communication/ENTSO_TYNDP_2018_StakeholderEngagement.pdf

ENTSO-E. (2018b). TYNDP 2018 Scenario Report. Retrieved July 24, 2019, from: https://docstore.entsoe.eu/Documents/TYNDP%20documents/TYNDP2018/Scenario_Report_2018_Final.pdf

ENTSO-E. (2019). Power Facts Europe 2019. Retrieved July 23, 2019, from: https://docstore.entsoe.eu/Documents/Publications/ENTSO-E%20general%20publications/ENTSO-E_PowerFacts_2019.pdf

ENTSO-E. (n.d.-a). System Development Reports. Retrieved July 19, 2019, from: https://www.entsoe.eu/publications/system-development-reports/

ENTSO-E. (n.d.-b). TYNDPs and Projects of Common Interests. Retrieved June 14, 2019, from: https://docstore.entsoe.eu/major-projects/ten-year-network-development-plan/TYNDP%20link%20with%20PCIs/Pages/default.aspx

EOS Energy Storage. (n.d.). Retrieved June 14, 2019, from: https://www.eosenergystorage.com

ERCOT. (n.d.-a). ERCOT Grid Information. Retrieved June 14, 2019, from: http://www.ercot.com/gridinfo/resource

ERCOT. (n.d.-b). ERCOT Wholesale Market Basics. Slide 8. Retrieved July 23, 2019, from: http://www.ercot.com/services/training/wholesale_presentations/Module%201%20-%20Market%20Overview%20-%20Sept%202007.ppt

EU GDPR.ORG. (n.d.). Retrieved February 14, 2019, from: https://eugdpr.org/the-regulation/

Eucalitto, G., Portillo, P., and Gander, S. (2018). *Governors Staying Ahead of the Transportation Innovation Curve: A Policy Roadmap for States.* Washington, DC: National Governors Association Center for Best Practices.

EUR-Lex. (2013). Document 32012R0347. Retrieved June 14, 2019, from: https://eur-lex.europa.eu/legal-content/EN/TXT/?qid=1446822144539&uri=CELEX:32013R0347#d1e93-65-1

Europe beyond Coal. (2019). Overview: National Coal Phase-Out Announcements in Europe. Retrieved June 22, 2019, from: https://beyond-coal.eu/wp-content/uploads/2019/02/Overview-of-national-coal-phase-out-announcements-Europe-Beyond-Coal-March-2019.pdf

European Commission. (2018). State Aid: Commission Approves Six Electricity Capacity Mechanisms to Ensure Security of Supply in Belgium, France, Germany, Greece, Italy and Poland—Factsheet. Retrieved July 6, 2019, from: https://europa.eu/rapid/press-release_MEMO-18-681_en.htm

Eurostat. (n.d.). Retrieved June 14, 2019, from: http://appsso.eurostat.ec.europa.eu/nui/submitViewTableAction.do

Executive Office of the President. (2013). Economic Benefits of Increasing Electric Grid Resilience to Weather Outages. Retrieved July 8, 2019, from: https://www.energy.gov/sites/prod/files/2013/08/f2/Grid%20Resiliency%20Report_FINAL.pdf

Fairley, P. (2004). The Unruly Power Grid. *IEEE Spectrum,* August 2004.

Fairley, P. (2019). China's Ambitious Plan to Build the World's Biggest Supergrid: A Massive Expansion Leads to the First Ultrahigh-Voltage AC-DC Power Grid. Retrieved July 23, 2019, from: https://spectrum.ieee.org/energy/the-smarter-grid/chinas-ambitious-plan-to-build-the-worlds-biggest-supergrid

Fares, R., and King, C. (2017). *Trends in Transmission, Distribution, and Administration Costs for U.S. Investor Owned Electric Utilities.* Austin, TX: The University of Texas at Austin. DOI: 10.1016/j.enpol.2017.02.036.

Farmer, M. (2018). Clean Energy Groups Urge FERC to Reconsider PJM Order. Retrieved June 14, 2019, from: https://www.nrdc.org/experts/miles-farmer/clean-energy-groups-urge-ferc-reconsider-pjm-order

Farmer, M., and Steinberger, K. (2017). “Baseload” in the Rearview Mirror of Today’s Electric Grid. Retrieved June 5, 2019, from: https://www.nrdc.org/experts/baseload-rearview-mirror-todays-electric-grid

Faruqui, A., and Aydin, M. (2017). Moving Forward with Electric Tariff Reform. *Regulation,* Fall, 42–48.

Faruqui, A., Grausz, L., and Bourbonnais, C. (2019). Transitioning to Modern Residential Rate Designs: Key Enabler of Renewable Energy Resources Integration. Retrieved May 21, 2019, from: https://www.fortnightly.com/fortnightly/2019/01/transitioning-modern-residential-rate-designs

Faruqui, A., and Palmer, J. (2011). Dynamic Pricing and Its Discontents. *Regulation,* 34(3), 16.

Fast Company (n.d.). Retrieved October 25, 2019, from: https://www.fastcompany.com/company/green-mountain-power

Federal Energy Regulatory Commission. (2016). Staff Report: Common Metrics Report. Retrieved May 23, 2019, from: https://www.ferc.gov/legal/staff-reports/2016/08-09-common-metrics.pdf?csrt=7790423430366596954

Federal Energy Regulatory Commission. (2018). Distributed Energy Resources Technical Considerations for the Bulk Power System. Retrieved June 5, 2019, from: https://www.ferc.gov/legal/staff-reports/2018/der-report.pdf

Federal Energy Regulatory Commission. (2019a). Order No. 1000—Transmission Planning and Cost Allocation. Retrieved June 14, 2019, from: https://www.ferc.gov/industries/electric/indus-act/trans-plan.asp

Federal Energy Regulatory Commission. (2019b). State of the Markets Reports 2018. Retrieved July 5, 2019, from: https://www.ferc.gov/market-oversight/reports-analyses/st-mkt-ovr/2018-A-3-report.pdf

Ferenstein, G. (2017). A Deeper Look at Silicon Valley’s Long-Term Politics. Retrieved July 23, 2019, from: https://www.brookings.edu/blog/techtank/2017/10/04/a-deeper-look-at-silicon-valleys-long-term-politics/

Fichera, J., and Klein, R. (2018). *Lowering Environmental and Capital Costs with Ratepayer-Backed Bonds.* New York: Saber Partners LLC.

Financial Tribune. (2019). New Nuclear Reactors to Come on Stream in Europe. Retrieved May 9, 2019, from: https://financialtribune.com/articles/energy/95960/new-nuclear-reactors-to-come-on-stream-in-europe

Florida Power and Light Company’s Amended Response to Staff’s Second Data Request No. 29 (2018). 20170215-EU, Florida Public Service Commission, April 23, 2018. Retrieved October 27, 2019, from: https://www.floridapsc.com/library/filings/2018/03152-2018/03152-2018.pdf

Forbes. (2018). How AI, IoT, and 5G Will Make a Difference in a Smarter World. Retrieved April 4, 2019, from: https://www.forbes.com/sites/intelai/2018/09/21/a-smarter-world-how-ai-the-iot-and-5g-will-make-all-the-difference/#76ea826230ab

Fox-Penner, P. (1997). *Electric Utility Restructuring: A Guide to the Competitive Era.* Public Utilities Reports.

Fox-Penner, P. (2001). Easing Gridlock on the Grid: Electricity Planning and Siting Compacts. *Electricity Journal,* 14(9), 11-30. DOI: 10.1016/S1040-6190(01)00242-1.

Fox-Penner, P. (2014). *Smart Power: Climate Change, the Smart Grid, and the Future of Electric Utilities.* Washington, DC: Island Press.

Fox-Penner, P., Gorman, W., and Hatch, J. (2018). Long-Term U.S Transportation Electricity Use Considering the Effect of Autonomous-Vehicles: Estimates and Policy Observations. *Energy Policy,* 122, 203–213. DOI: 10.1016/j.enpol.2018.07.033.

Fox-Penner, P., Harris, D., and Hesmondhalgh, S. (2013). A Trip to RIIO in Your Future? Great Britain's Latest Innovation in Grid Regulation. Retrieved July 14, 2019, from: https://www.fortnightly.com/fortnightly/2013/10/trip-riio-your-future

Fox-Penner, P., and Zarakas, W. (2013). *Analysis of Benefits: PSE&G's Energy Strong Program.* San Diego, CA: Brattle Group.

FRED. (n.d.). FRED Economic Data. Retrieved January 1, 2018, from: https://fred.stlouisfed.org/

Freeberg, E. (2013). *The Age of Edison.* New York: Penguin Books.

Frew, B., Becker, S., Dvorak, M., Andresen, G., and Jacobson, M. (2016). Flexibility Mechanisms and Pathways to a Highly Renewable US Electricity Future. *Energy,* 101, 65–78. DOI: 10.1016/j.energy.2016.01.079.

Frost & Sullivan. (2017). *Global Flow Battery Market, Forecast to 2023.* Mountain View, CA: Frost & Sullivan.

Frost & Sullivan. (2018). *Global Energy Storage Market Outlook, 2018.* Mountain View, CA: Frost & Sullivan.

Fürsch, M., Hagspiel, S., Jaggerman, C., Nagl, S., Lindenberger, D., and Troster, E. (2013). The Role of Grid Extensions in a Cost-Efficient Transformation of the European Electricity System until 2050. *Applied Energy,* 104, 642–652. DOI: 10.1016/j.apenergy.2012.11.050.

Gabaldón-Estevan, D., Peñalvo-López, E., and Solar, D. (2018). The Spanish Turn against Renewable Energy Development. *Sustainability,* 10(4), 1208. DOI: 10.3390/su10041208.

Gagnon, P., Margolis, R., Melius, J., Phillips, C., and Elmore, R. (2016). Rooftop Solar Photovoltaic Technical Potential in the United States: A Detailed Assessment. Technical Report NREL / TP-6A20-65298. Retrieved June 14, 2019, from: http://www.nrel.gov/docs/fy16osti/65298.pdf

Gagnon, P., Margolis, R., Melius, J., Phillips, C., and Elmore, R. (2018). Estimating Rooftop Solar Technical Potential across the US Using a Combination of GIS-Based Methods, Lidar Data, and Statistical Modeling. *Environmental Research Letters,* 13, 024027. DOI: 10.1088/1748-9326/aaa554.

Galloway, S. (2018). *The Four.* New York: Penguin.

Gao, J., Barzel, B., and Barabási, A. (2016). Universal Resilience Patterns in Complex Networks. *Nature,* 536, 238. DOI: 10.1038/nature16948.

Gardiner, M. (2014). Hydrogen for Energy Storage. Retrieved June 5, 2019, from: https://www.energy.gov/sites/prod/files/2014/08/f18/fcto_webinarslides_h2_storage_fc_technologies_081914.pdf

Gardner, C. (2011). We Are the 25%: Looking at Street Area Percentages and Surface Parking. Retrieved June 14, 2019, from: http://oldurbanist.blogspot.com/2011/12/we-are-25-looking-at-street-area.html

Gately, D. (1993). The Imperfect Price-Reversibility of World Oil Demand. *The Energy Journal,* 14, 163–182. DOI: 10.5547/ISSN0195-6574-EJ-Vol14-No4-11.

GE Power. (n.d.). Hydrogen Fueled Gas Turbines. Retrieved June 28, 2019, from: https://www.ge.com/power/gas/fuel-capability/hydrogen-fueled-gas-turbines

Geophysical Fluid Dynamics Laboratory. (2019). Global Warming and Hurricanes: An Overview of Current Research Results. Retrieved July 19, 2019, from: https://www.gfdl.noaa.gov/global-warming-and-hurricanes/

Gerarden, T., Newell, R., and Stavins, R. (2015). Deconstructing the Energy-Efficiency Gap: Conceptual Frameworks and Evidence. *American Economic Review,* 105, 183–186. DOI: 10.1257/aer.p20151012.

Gholami, A., Aminifar, F., and Shahidehpour, M. (2016). Front Lines against the Darkness: Enhancing the Resilience of the Electricity Grid through Microgrid Facilities. *IEEE Electrification Magazine,* 18–24. DOI: 10.1109/MELE.2015.2509879.

Ghosh, D. (2018). What You Need to Know about California's New Data Privacy Law. *Harvard Business Review.* Retrieved October 27, 2019, from: https://hbr.org/2018/07/what-you-need-to-know-about-californias-new-data-privacy-law

Gimon, E. (2016). Customer-Centric View of Electricity Service. In Sioshansi, F., ed. *Future of Utilities—Utilities of the Future: How Technological*

Innovations in Distributed Energy Resources Will Reshape the Electrical Power Sector. Cambridge, MA: Academic Press, pp. 75-90.

Gimon, E. (2017a). Flexibility, Not Resilience, Is the Key to Wholesale Electricity Market Reform. Retrieved July 21, 2019, from: https://www.greentechmedia.com/articles/read/flexibility-is-the-key-to-wholesale-electricity-market-reform#gs.djtx5t

Gimon, E. (2017b). New Financial Tools Proposed in Colorado Could Solve Coal Retirement Conundrum. Retrieved June 14, 2019, from: https://www.forbes.com/sites/energyinnovation/2017/04/19/new-financial-tools-proposed-in-colorado-could-solve-coal-retirement-conundrum/#71e6e44f11c5

Glachant, J.-M., Rossetto, N., and Vasconcelos, J. (2017). *Moving the Electricity Transmission System towards a Decarbonised and Integrated Europe: Missing Pillars and Roadblocks.* San Domenico di Fiesole (FI), Italy: European University Institute.

Glenk, G., and Reichelstein, S. (2019). Economics of Converting Renewable Power to Hydrogen. *Nature Energy,* 4, 216–222. DOI: 10.1038/s41560-019-0326-1.

Global CCS Institute. (2018). The Global Status of CCS. Retrieved June 5, 2019, from: https://www.globalccsinstitute.com/resources/global-status-report/download/

Global Energy Interconnection Development and Cooperation Organization. Meeting Global Power Demand with Clean and Green Alternatives, pp. 11–12. (n.d.). Retrieved June 14, 2019, from: https://www.geidco.org

Goggin, M., Gramlich, R., Shparber, S., and Silverstein, A. (2018). Customer Focused and Clean: Power Markets for the Future. Retrieved June 28, 2019, from: https://windsolaralliance.org/wp-content/uploads/2018/11/WSA_Market_Reform_report_online.pdf

Goldenberg, C., Dyson, M., and Masters, H. (2018). Demand Flexibility: The Key to Enabling a Low-Cost, Low-Carbon Grid. Retrieved June 5, 2019, from: https://rmi.org/wp-content/uploads/2018/02/Insight_Brief_Demand_Flexibility_2018.pdf

Golove, W., and Schipper, L. (1997). Restraining Carbon Emissions: Measuring Energy Use and Efficiency in the USA. *Energy Policy,* 25(7–9), 803–812. DOI: 10.1016/S0301-4215(97)00070-0.

Gonzalez-Longatt, F. (2015). Future Meshed HVDC Grids: Features, Challenges, and Opportunities. Retrieved July 25, 2019, from: https://www.slideshare.net/fglongatt/future-meshed-hvdc-grids-challenges-and-opportunities-29th-october-2015-portoviejo-ecuador

Governing. (2018). 2017 County Migration Rates, Population Estimates. Retrieved March 22, 2018, from: http://www.governing.com/gov-data/census/2017-county-migration-rates-population-estimates.html

Grahl-Madsen, L. (2010). *The Hydrogen Demonstration Society @ Lolland Island, Denmark.* Brussels, Belgium: IRD Fuel Cell Technology.

Gramlich, R., and Hogan, M. (2019). *Wholesale Electricity Market Design for Rapid Decarbonization: A Decentralized Markets Approach.* San Francisco, CA: Energy Innovation Policy & Technology LLC.

Graves, F., Carroll, R., and Haderlein, K. (2018). *State of Play in Retail Choice.* San Diego, CA: Brattle Group.

Graves, F., Ros, A., Sergici, S., Carroll, R., and Haderlein, K. (2018). Retail Choice: Ripe for Reform? Retrieved May 21, 2019, from: https://brattlefiles.blob.core.windows.net/files/14191_retail_choice_-_ripe_for_reform.pdf

Green, M., Hishikawa, Y., Dunlop, E., Levi, D., and Hohl-Ebinger, J. (2018). Solar Cell Efficiency Tables (Version 51). *Progress in Photovoltaics: Research and Applications,* 26(1), 3–12. DOI: 10.1002/pip.2978.

Green Mountain Power. (2017). Petition of Green Mountain Power Corporation for Approval of Temporary Limited Regulation Plan Pursuant 30 V.S.A. §§ 209, 218 and 218d. Retrieved May 30, 2019, from: https://greenmountainpower.com/wp-content/uploads/2018/01/2017-11-29-17-3232-PET-Final-Order.pdf

Greenberg, A. (2018). Stealthy Destructive Malware Infects Half a Million Routers. *Wired Magazine,* May 23.

Greenstone, M. (2016). India's Air-Conditioning and Climate Change Quandary. *New York Times,* October 26.

Griswold, A., and Karaian, J. (2018). It Took Amazon 14 Years to Make as Much in Net Profit as It Did Last Quarter. Retrieved May 29, 2019, from: https://qz.com/1196256/it-took-amazon-amzn-14-years-to-make-as-much-net-profit-as-it-did-in-the-fourth-quarter-of-2017/

Groebel, A. (2013). Role and Structure of the German Regulatory Authorities and the Role of BNetzA in Implementing the "Energiewende." Retrieved June 24, 2019, from: http://www.iei-la.org/admin/uploads/nopa/groebel.pdf

Groebel, A. (2017). Integrating Renewables in the Grid and the Market: Insights from the German Energiewende and Lessons Learnt. Workshop "Renewable Energy: The Future of Biofuels." Retrieved July 14, 2019, from: https://miamieuc.fiu.edu/events/general/2017/eu-jean-monnet-project-workshop

-on-renewable-energy-the-future-of-biofuels/2017-01-20-miami-integrating -renewables-in-the-grid-and-the-market-insights-from-the-german-energiew

Groebel, A. (2019). From Traditional to Clean Energy: Insights from the German Energiewende, Climate Risks and Regulation, Round Table. Retrieved July 14, 2019, from: http://chairgovreg.fondation-dauphine.fr/sites /chairgovreg.fondation-dauphine.fr/files/attachments/Groebel_OK.pdf

GSMA Intelligence. (n.d.). Definitive Data and Analysis for the Mobile Industry. Retrieved June 1, 2019, from: https://www.gsmaintelligence.com/

Gundlach, J., Minsk, R., and Kaufman, N. (2019). *Interactions between a Federal Carbon Tax and Other Climate Policies.* New York: Center on Global Energy Policy, Columbia SIPA.

Hadarau, S. (2016). *HVDC and Its Potential in Building the European Super-Grid.* Saarbrücken, Germany: Lap Lambert Academic Publishing.

Hadush, S., De Jonghe, C., and Belmans, R. (2015). The Implication of the European Inter-TSO Compensation Mechanism for Cross-Border Electricity Transmission Investments. *International Journal of Electrical Power and Energy Systems,* 73, 674–683. DOI: 10.1016/j.ijepes.2015.05.041.

Hanser, P., and Van Horn, K. (2016). The Repurposed Distribution Utility: Roadmaps to Getting There. In Sioshansi, F., ed. *Future of Utilities—Utilities of the Future: How Technological Innovations in Distributed Energy Resources Will Reshape the Electrical Power Sector.* Cambridge, MA: Academic Press, pp. 383–398.

Harris, D., Kolbe, A., Vilbert, M., and Villadsen, B. (2017). *Risk and Return for Regulated Industries.* Cambridge, MA: Academic Press.

Hart, D. (2018). Making "Beyond Lithium" a Reality: Fostering Innovation in Long-Duration Grid Storage. Retrieved June 5, 2019, from: https://itif.org /publications/2018/11/28/making-beyond-lithium-reality-fostering -innovation-long-duration-grid

Hart, D., Bonvillian, W., and Austin, N. (2019). Energy Storage for the Grid: Policy Options for Sustaining Innovation. MIT Energy Initiative Working Paper 2018–04. Retrieved June 5, 2019, from: http://energy.mit.edu/wp -content/uploads/2018/04/MITEI-WP-2018-04.pdf

Harvard Electricity Policy Group. (n.d.). William Hogan. Retrieved June 14, 2019, from: https://hepg.hks.harvard.edu/william-hogan

Harvey, A. (2017). Thermal Energy Storage for Concentrated Solar Power. Retrieved June 5, 2019, from: http://helioscsp.com/thermal-energy-storage -for-concentrated-solar-power/

Harvey, S., and Hogan, W. (2000). Issues in the Analysis of Market Power in California. Retrieved July 9, 2019, from: https://sites.hks.harvard.edu/fs/whogan/HHMktPwr_1027.pdf

Harvey, S., and Hogan, W. (2001). On the Exercise of Market Power through Strategic Withholding in California. Retrieved July 9, 2019, from: http://citeseerx.ist.psu.edu/viewdoc/download?doi=10.1.1.451.8360&rep=rep1&type=pdf

Hasan, A. (2016). Fact Sheet: Advancing Clean Energy Research and Development in the President's FY 2017 Budget. Retrieved June 5, 2019, from: https://obamawhitehouse.archives.gov/blog/2016/10/12/factsheet-advancing-clean-energy-research-and-development-presidents-fy-2017-budget

Hausker, K. (2018). *Technical and Economic Implications of the Clean Energy Transition.* Washington, DC: World Resources Institute.

Hawaii State Legislature. (2018). 2018 Archives: SB2939 SD2. Retrieved July 19, 2019, from: https://www.capitol.hawaii.gov/Archives/measure_indiv_Archives.aspx?billtype=SB&billnumber=2939&year=2018

Heard, B., Brook, B., Wigley, T., and Bradshaw, C. (2017). Burden of Proof: A Comprehensive Review of the Feasibility of 100% Renewable-Electricity Systems. *Renewable and Sustainable Energy Reviews,* 76, 1122–1133. DOI: 10.1016/j.rser.2017.03.114.

Heath, A. (2018). J. D. Power 2018 Electric Utility Residential Customer Survey. Retrieved May 24, 2019, from: https://www.smud.org/-/media/Documents/Corporate/About-Us/Board-Meetings-and-Agendas/2018/Aug/Strategic-Development-Committee—August-14-1-2018-JD-Power-Electric-Residential-Study-Board-Pres.ashx?la=en&hash=8BC06136089959F482DD7EE255E6E2CDB3B7FC41

Heidel, T., and Miller, C. (2017). Agile Fractal Systems: Reenvisioning Power System Architecture. Retrieved June 14, 2019, from: https://www.nae.edu/Publications/Bridge/176887/177000.aspx

Hern, A. (2018). Bitcoin's Energy Usage Is Huge—We Can't Afford to Ignore It. Retrieved June 4, 2019, from: https://www.theguardian.com/technology/2018/jan/17/bitcoin-electricity-usage-huge-climate-cryptocurrency

Hewson, B. (2018). *Ontario's Electricity Pricing and Rate Design.* Toronto, Canada: Ontario Energy Board.

Hezir, J., Knotek, M., Pablo, J., Kizer, A., Bushman, T., Arya, A., . . . Coan, J. (2019). Advancing the Landscape of Clean Energy Innovation. Breakthrough Energy. Retrieved June 5, 2019, from: http://www.b-t.energy/wp-content/

uploads/2019/02/Report_-Advancing-the-Landscape-of-Clean-Energy-Innovation_2019.pdf

Hillebrandt, K., Samadi, S., Fischedick, M., Eckstein, S., Höller, S., . . . Sellke, P. (2015). Pathways to Deep Decarbonization in Germany. DE 2015 Report. Retrieved July 8, 2019, from: http://deepdecarbonization.org/wp-content/uploads/2015/09/DDPP_DEU.pdf

Hirsch, A., Parag, Y., and Guerrero, J. (2018). Microgrids: A Review of Technologies, Key Drivers, and Outstanding Issues. *Renewable and Sustainable Energy Reviews,* 90, 402–411. DOI: 10.1016/j.rser.2018.03.040.

Hirsch, R. (1999). PURPA: The Spur to Competition and Utility Restructuring. *Electricity Journal,* 12(7), 60–72. DOI: 10.1016/S1040-6190(99)00060-3.

Hledik, R., Tsuchida, B., and Palfreyman, J. (2018). *Beyond Zero Net Energy?* Boston: Brattle Group.

HM Government. (2010). 2050 Pathways Analysis. Retrieved July 8, 2019, from: https://assets.publishing.service.gov.uk/government/uploads/system/uploads/attachment_data/file/68816/216-2050-pathways-analysis-report.pdf

Hoegh-Guldberg, O., Cai, R., Poloczanska, E., Brewer, P., Sundby, S., Hilmi, K., . . . Jung, S. (2018). The Ocean. In Climate Change 2014: Impacts, Adaptation, and Vulnerability. Part B: Regional Aspects. Contribution of Working Group II to the Fifth Assessment Report of the Intergovernmental Panel on Climate Change. Retrieved June 14, 2019, from: http://pure.iiasa.ac.at/15518

Hogan, M. (2018). Wholesale Market Design for a Low-Carbon Power System. Retrieved June 28, 2019, from: https://www.raponline.org/wp-content/uploads/2018/04/rap_hogan_wholesale_market_design_2018_feb_28.pdf

Hogan, W. (2016). Electricity Market Design: Political Economy and the Clean Energy Transition. Retrieved April 22, 2019, from: https://sites.hks.harvard.edu/fs/whogan/Hogan_IHS_110916.pdf

Hood, G. (2018). Coal-Fired Past or Green-Powered Future? Pueblo Looks for a New Economic Leg Up. Retrieved June 14, 2019, from: https://www.cpr.org/news/story/coal-fired-past-or-green-powered-future-pueblo-looks-for-a-new-economic-leg-up

Hopkins, A., Horowitz, A., Knight, P., Takahashi, K., Comings, T., Kreycik, P., . . . Koo, J. (2017). Northeastern Regional Assessment of Strategic Electrification. Retrieved July 8, 2019, from: https://neep.org/sites/default/files/Strategic%20Electrification%20Regional%20Assessment.pdf

Hughes, T. (1993). *Networks of Power: Electrification in Western Society, 1880–1930.* Reprint edition. Baltimore: Johns Hopkins University Press.

Iannelli, J. (2017). Why Didn't FPL Do More to Prepare for Irma? *Miami New Times,* July 19.

Institute for Local Self-Reliance. (n.d.). Retrieved June 14, 2019, from: https://ilsr.org/

Institute of Medicine. (2007). *Rising above the Gathering Storm: Energizing and Employing America for a Brighter Economic Future.* Washington, DC: The National Academies Press.

International Energy Agency. (2015a). Energy Efficiency Market Report. Retrieved June 15, 2019, from: https://www.iea.org/publications/freepublications/publication/MediumTermEnergyefficiencyMarketReport2015.pdf

International Energy Agency. (2015b). Key World Energy Statistics 2015. Retrieved June 15, 2019, from: http://www.iea.org/publications/freepublications/publication/KeyWorld_Statistics_2015.pdf

International Energy Agency. (2015c). Monthly Energy Review March 2015. Retrieved June 15, 2019, from: https://www.eia.gov/totalenergy/data/monthly/archive/00351503.pdf

International Energy Agency. (2015d). Technology Roadmap: Hydrogen and Fuel Cells. Retrieved June 5, 2019, from: https://www.iea.org/publications/freepublications/publication/TechnologyRoadmapHydrogenandFuelCells.pdf

International Energy Agency. (2017a). Energy Technology Perspectives 2017: Catalysing Energy Technology Transformations. Retrieved June 15, 2019, from: https://www.iea.org/publications/freepublications/publication/EnergyTechnologyPerspectives2017ExecutiveSummaryEnglishversion.pdf

International Energy Agency. (2017b). World Energy Outlook 2017. Retrieved June 14, 2019, from: https://www.iea.org/weo2017/

International Energy Agency. (2018). Global Energy and CO_2 Status Report: The Latest Trends in Energy and Emissions in 2018. Retrieved April 1, 2019, from: https://www.iea.org/geco/data/

International Energy Agency. (2019). Is Pumped Storage Hydropower Capacity Forecast to Expand More Quickly than Stationary Battery Storage? Retrieved June 5, 2019, from: https://www.iea.org/newsroom/news/2019/march/will-pumped-storage-hydropower-capacity-expand-more-quickly-than-stationary-b.html

International Energy Agency. (n.d.-a). Retrieved May 25, 2019, from: https://www.iea.org/statistics/?country=USA&year=2016&category=Electricity&indicator=TPESbySource&mode=chart&dataTable=ELECTRICITYANDHEAT

International Energy Agency. (n.d.-b). Statistics: Global Energy Data at your Fingertips. Retrieved March 25, 2019, from: www.iea.org/statistics

International Hydropower Association. (2019). 2019 Hydropower Status Report. Retrieved June 5, 2019, from: https://www.hydropower.org/status2019

International Rivers. (2019). Environmental Impacts of Dams. Retrieved July 9, 2019, from: https://www.internationalrivers.org/environmental-impacts-of-dams

Inventys. (n.d.). Retrieved April 2, 2019, from: http://inventysinc.com/technology/

IPCC. (2005). Intergovernmental Panel on Climate Change Special Report on Carbon Dioxide Capture and Storage. Retrieved June 5, 2019, from: https://www.ipcc.ch/report/carbon-dioxide-capture-and-storage/

IRENA. (2017). Electricity Storage and Renewables: Cost and Markets to 2030. Retrieved June 14, 2019, from: https://www.irena.org/-/media/Files/IRENA/Agency/Publication/2017/Oct/IRENA_Electricity_Storage_Costs_2017.pdf

Ishugah T., Li, Y., Wang, R., and Kiplagat, J. (2014). Advances in Wind Energy Resource Exploitation in Urban Environment: A Review. *Renewable and Sustainable Energy Reviews,* 37, 613–626. DOI: 10.1016/j.reser.2014.05.053.

ISO New England. (2018a). Proposed 2019 Operating and Capital Budgets. Retrieved May 23, 2019 from: https://www.iso-ne.com/static-assets/documents/2018/08/4_isone_2019_proposed_op_cap_budget.pdf

ISO New England. (2018b). Tariff Section III.13.1.4. Retrieved March 29, 2019, from: https://www.iso-ne.com/static-assets/documents/regulatory/tariff/sect_3/mr1_sec_13_14.pdf

ISO New England. (n.d.). New England's Electricity Use. Retrieved February 8, 2019, from: https://www.iso-ne.com/about/key-stats/electricity-use

Jacobson, M., Camerson, M., Hennessy, E., Petkov, I., Meyer, C., Gambhir, T., . . . Delucchi, M. (2018a). 100% Clean and Renewable Wind, Water, and Sunlight (WWS) All-Sector Energy Roadmaps for 53 Towns and Cities in North America. *Sustainable Cities and Society,* 42, 22–37. DOI: 10.1016/j.scs.2018.06.031.

Jacobson, M., and Delucchi, M. (2011). Providing All Global Energy with Wind, Water, and Solar Power, Part I: Technologies, Energy Resources, Quantities and Areas of Infrastructure, and Materials. *Energy Policy,* 39, 1154–1169. DOI: 10.1016/j.enpol.2010.11.040.

Jacobson, M., Delucchi, M., Bazouin, G., Bauer, Z., Heavey, C., Fisher, E., . . . Yeskoo, T. (2015). 100% Clean and Renewable Wind, Water, and Sunlight (WWS) All-Sector Energy Roadmaps for the 50 United States. *Energy and Environmental Science,* 8, 2093. DOI: 10.1039/c5ee01283j.

Jacobson, M., Delucchi, M., Cameron, M., and Mathiesen, B. (2018b). Matching Demand with Supply at Low Cost in 139 Countries among 20 World Regions with 100% Intermittent Wind, Water, and Sunlight (WWS) for All Purposes. *Renewable Energy,* 123, 236–248. DOI: 10.1016/j.renene.2018.02.009.

Jakubowski, A. (2018). Phasing Out Coal in the French Energy Sector. Retrieved July 23, 2019, from: https://www.etui.org/content/download/33822/322399/file/6+A+Jakubowski+-+Phasing+out+coal+in+the+French+energy+sector.pdf

Jenkins, C., Cook, P., Ennis-King, J., Undershultz, J., Boreham, C., Dance, T., . . . Urosevic, M. (2012). Safe Storage and Effective Monitoring of CO_2 in Depleted Gas Fields. *Proceedings of the National Academy of Sciences of the United States of America,* E35–E41. DOI: 10.1073/pnas.1107255108.

Jenkins, J., and Thernstrom, S. (2017). Deep Decarbonization of the Electric Power Sector: Insights from Recent Literature. Retrieved June 14, 2019, from: https://www.innovationreform.org/wp-content/uploads/2018/02/EIRP-Deep-Decarb-Lit-Review-Jenkins-Thernstrom-March-2017.pdf

Johnson, N. (2018). Lessons from Boulder's Bad Breakup. *Grist,* January 19.

Johnson, S. (2018). The Political Education of Silicon Valley. *Wired Magazine,* July 24.

Johnstone, P., and Hielscher, S. (2017). *Phasing Out Coal, Sustaining Coal Communities? Living with Technological Decline in Sustainability Pathways.* Falmer, UK: Science Policy Research Unit (SPRU), School of Business Management and Economics, University of Sussex.

Joskow, P. (2019). Challenges for Wholesale Electricity Markets with Intermittent Renewable Generation at Scale: The US Experience. *Oxford Review of Economic Policy,* 35(2), 291–331. DOI: 10.1093/oxrep/grz001.

Joskow, P., and Schmalensee, R. (1983). *Markets for Power: An Analysis of Electric Utility Deregulation.* Cambridge, MA: The MIT Press.

Kahn, A. (1988). *The Economics of Regulation: Principles and Institutions.* Cambridge, MA: The MIT Press.

Kamal, M., and Saraswat, S. (2014). Emerging Trends in Tall Building Design: Environmental Sustainability through Renewable Energy Technologies. *Civil Engineering and Architecture,* 2, 116–120. DOI: 10.13189/cea.2014.020302.

Kammen, D., and Sunter, D. (2016). City-Integrated Renewable Energy for Urban Sustainability. *Science,* 352(6288), 922–928. DOI: 10.1126/science.aad9302.

Kann, S. (2016). How the Grid Was Won: Three Scenarios for the Distributed Grid in 2030. Retrieved June 6, 2019, from: https://www.greentechmedia.com/articles/read/how-the-grid-was-won#gs.gqj5wg

Kantamneni, A., Winkler, R., Gauchia, L., and Pearce, J. (2016). Emerging Economic Viability of Grid Defection in a Northern Climate Using Solar Hybrid Systems. *Energy Policy,* 95, 378–389. DOI: 10.1016/j.enpol.2016.05.013.

Kapperman, J. (2010). Geomagnetic Storms and Their Impacts on the U.S. Power Grid (Meta-R-319). Retrieved July 9, 2019, from: https://www.ferc.gov/industries/electric/indus-act/reliability/cybersecurity/ferc_meta-r-319.pdf

Karteris, M., Slini, T., and Papadopoulos, A. (2013). Urban Solar Energy Potential in Greece: A Statistical Calculation Model of Suitable Built Roof Areas for Photovoltaics. *Energy and Buildings,* 62, 459–468. DOI: 10.1016/j.enbuild.2013.03.033.

Kaufmann, R., Gopal, S., Tang, X., Raciti, S., Lyons, P., Geron, N., and Craig, F. (2013). Revisiting the Weather Effect on Energy Consumption: Implications for the Impact of Climate Change. *Energy Policy,* 62, 1377–1384. DOI: 10.1016/j.enpol.2013.07.056.

Kaye, K. (2015). Hurricane Wilma: Ten Years Later, We Reap Benefits. *Sun Sentinel.* Retrieved October 24, 2019, from: https://www.sun-sentinel.com/local/broward/fl-hurricane-wilma-10-years-later-20151017-story.html

Kellner, T. (2016). Neural Networks and Dynamite: AI Engineer Peter Kirk Talks about His Fascination with Coal Power Plants. Retrieved October 27, 2019, from: https://www.ge.com/reports/neural-networks-and-dynamite-ai-engineer-peter-kirk-talks-about-his-fascination-with-coal-power-plants/

Kelly, M. (2014). Two Years after Hurricane Sandy, Recognition of Princeton's Microgrid Still Surges. Retrieved July 19, 2019, from: Princeton University, Office of Communications: https://www.princeton.edu/news/2014/10/23/two-years-after-hurricane-sandy-recognition-princetons-microgrid-still-surges

Kempner, M., and Ondieki, A. (2018). After Wrangling over Georgia Nuclear Plant, Cost Concerns Remain. Retrieved June 5, 2019, from: https://www.ajc.com/business/after-wrangling-over-georgia-nuclear-plant-cost-concerns-remain/9iGHX9Ugo7QPkli9LoqGbM/

Kenny, T. (2019). How to Invest in Utility Stocks. Retrieved June 6, 2019, from: https://www.thebalance.com/how-to-invest-in-utility-stocks-416833

Khan, L. (2018). The New Brandeis Movement: America's Antimonopoly Debate. *Journal of European Competition Law & Practice.*

Kiesling, L., Munger, M., and Theisen, A. (2018). From Airbnb to Solar: Toward a Transaction Cost Model of a Retail Electricity Distribution Platform. Retrieved May 21, 2019, from: https://www.researchgate.net/publication/326995058_From_Airbnb_to_Solar_Toward_A_Transaction_Cost_Model_of_a_Retail_Electricity_Distribution_Platform

Kihm, S. (2018). From a Regulation Mindset to an Entrepreneurial Orientation? Can (Should) Utilities Make the Switch? Disruption and Innovation in the Electric Utility Industry. Wisconsin Public Utility Institute. Retrieved October 27, 2019, from: https://wpui.wisc.edu/wp-content/uploads/sites/746/2018/12/Kihm-Disruption-and-Innovation-sent-to-WPUI-9-14-18.pdf

Knieps, G. (2016). The Evolution of Smart Grids Begs Disaggregated Nodal Pricing. In Sioshansi, F., ed. *Future of Utilities—Utilities of the Future: How Technological Innovations in Distributed Energy Resources Will Reshape the Electrical Power Sector.* Cambridge, MA: Academic Press, pp. 267–280.

Knight, R. (2011). City of Winter Park Our Municipalization Story. Retrieved June 28, 2019, from: https://www.southdaytona.org/egov/documents/1302183733_26702.pdf

Kopnecki, D. (2018). Tesla Shares Soar on Surprise Third-Quarter Profit That Beats Wall Street Expectations. Retrieved January 26, 2019, from: https://www.cnbc.com/2018/10/24/tesla-earnings-q3-2018.html

Kossin, J., Hall, T., Knutson, T., Kunkel, K., Trapp, R., Waliser, D., and Wehner, M. (2017). Extreme Storms. In Wuebbles, D., Fahey, D., Hibbard, D., Dokken, D., Stewart, B., and Maycock T., eds. *Climate Science Special Report: Fourth National Climate Assessment.* Volume 1. Washington, DC: U.S. Global Change Research Program, pp. 257–276.

Koza, F. (2017). Industry Webinar: Project 2013–03 Geomagnetic Disturbance Mitigation. Retrieved July 19, 2019, from: www.nerc.com/pa/Stand/WebinarLibrary/Project_2013_03_Webinar_2017_07_27_Slides.pdf

Kraftwerk Forschung. (n.d.). Hydrogen Gas Turbines. Retrieved May 6, 2019, from: https://kraftwerkforschung.info/en/hydrogen-gas-turbines/

Kron, W. (2016). *Floods in the Atacama Desert.* Munich, Germany: Munich RE.

Kuffner, A. (2018). Power over Solar: R.I. Seeks to Strike a Development Balance. Retrieved June 14, 2019, from: https://www.providencejournal.com/news/20180808/power-over-solar-ri-seeks-to-strike-development-balance

Kwok, G., and Haley, B. (2018). *Exploring Pathways to Deep Decarbonization for the Portland General Electric Service Territory.* Portland, OR: Portland General Electric.

Lacey, S. (2014). Resiliency: How Superstorm Sandy Changed America's Grid. Retrieved June 3, 2019, from: https://www.greentechmedia.com/articles/featured/resiliency-how-superstorm-sandy-changed-americas-grid#gs.g4dxcu. U.S. Energy Information Administration.

Laitner, J., Nadel, S., Elliott, R., Sachs, H., and Khan, A. (2012). The Long-Term Energy Efficiency Potential: What the Evidence Suggests. Report Number E121. Retrieved June 14, 2019, from: https://aceee.org/sites/default/files/publications/researchreports/e121.pdf

Lamonica, M. (2011). A Moore's Law for Computers and Energy Efficiency. Retrieved June 14, 2019, from: https://www.cnet.com/news/a-moores-law-for-computers-and-energy-efficiency/

Lanier, J. (2011). *You Are Not a Gadget.* First Vintage Books.

Larsen, P. (2016). A Method to Estimate the Costs and Benefits of Undergrounding Electricity Transmission and Distribution Lines. Retrieved July 8, 2019, from: https://emp.lbl.gov/sites/all/files/lbnl-1006394_pre-publication.pdf

Lazar, J. (2014). Performance-Based Regulations for EU Distribution System Operators. Retrieved May 21, 2019, from: RAP: https://www.raponline.org/knowledge-center/performance-based-regulation-for-eu-distribution-system-operators/

Lazard. (2017). *Lazard's Levelized Cost of Storage Analysis—Version 3.0.* Hamilton, Bermuda: Lazard.

Lean Energy U.S. (n.d.). CCA by State. Retrieved May 28, 2019, from: https://leanenergyus.org/cca-by-state/

Lechtenböhmer, S., Nilsson, L., Åhman, M., and Schneider, C. (2016). Decarbonising the Energy Intensive Basic Materials Industry through Electrification—Implications for Future EU Electricity Demand. *Energy,* 115, 1623–1631. DOI: 10.1016/j.energy.2016.07.110.

Ledec, G., and Quintero, J. (2003). Good Dams and Bad Dams: Environmental Criteria for Site Selection of Hydroelectric Projects. Retrieved June 5, 2019, from: http://siteresources.worldbank.org/LACEXT/Resources/258553-1123250606139/Good_and_Bad_Dams_WP16.pdf

Leeuwen, C. van, and Mulder, M. (2018). Power-To-Gas in Electricity Markets Dominated by Renewables. *Applied Energy,* 232, 258–272. DOI: 10.1016/j.apenergy.2018.09.217.

Lehmann, H., and Peter, S. (2003). *Assessment of Roof Facade Potentials for Solar Use in Europe.* Aachen, Germany: Institute for Sustainable Solutions and Innovations.

Leisch, J., and Cochran J. (2015). Greening the Grid. Retrieved July 8, 2019, from: https://www.nrel.gov/docs/fy15osti/63033.pdf

Levenda, A., Mahmoudi, D., and Sussman, G. (2015). The Neoliberal Politics of "Smart": Electricity Consumption, Household Monitoring, and the Enterprise Form. *Canadian Journal of Communication,* 40(4). DOI: 10.22230/cjc.2015v40n4a2928.

Littell, D., Kadoch, C., Baker, P., Bharvirkar, R., Dupuy, M., Hausauer, B., . . . Xuan, W. (2017). Next-Generation Performance Based Regulation Emphasizing Utility Performance to Unleash Power Sector Innovation. Technical Report of National Renewable Energy Laboratory, NREL / TP-6A50-68512. Retrieved July 14, 2019, from: https://www.nrel.gov/docs/fy17osti/68512.pdf

Littlechild, S. (2018). The Regulation of Retail Competition in US Residential Electricity Markets. Retrieved May 21, 2019, from: https://www.eprg.group.cam.ac.uk/wp-content/uploads/2018/03/S.-Littlechild_28-Feb-2018.pdf

Loftus, P., Cohen, A., Long, J., and Jenkins, J. (2014). A Critical Review of Global Decarbonization Scenarios: What Do They Tell Us about Feasibility? *Wiley Interdisciplinary Reviews: Climate Change,* 6(1), 23. DOI: 10.1002/wcc.324.

Lovins, A. (2011). *Reinventing Fire: Bold Business Solutions for the New Energy Era.* White River Junction, VT: Chelsea Green.

Lovins, A., and Lovins, H. (1982). *Brittle Power.* Andover, MA: Brick House.

Lowrey, D. (2017). *RRA Financial Focus Berkshire Hathaway Energy.* New York: S&P Global Market Intelligence.

Lubershane, A. (2019). What to Do about DERs. Retrieved June 5, 2019, from: https://medium.com/@alubershane/what-to-do-about-ders-2957d087fca

Lyseng, B., Niet, T., Keller, V., Palmer-Wilson, K., Robertson, B., Rowe, A., and Wild, P. (2018). System-Level Power-To-Gas Energy Storage for High Penetrations of Variable Renewables. *International Journal of Hydrogen Energy,* 43(4), 1966–1979. DOI: 10.1016/j.ijhydene.2017.11.162.

MacDonald, A., Clack, C., Alexander, A., Dunbar, A., Wilczak, J., and Xie, Y. (2016). Future Cost-Competitive Electricity Systems and Their Impact on US CO_2 Emissions. *Nature Climate Change,* 6, 526–531. DOI: 10.1038/nclimate2921.

Magill, B. (2015). Defecting from the Power Grid? Unlikely, Analysts Says. Retrieved July 19, 2019, from: https://www.climatecentral.org/news/defecting-from-the-power-grid-18891

Mahone, A., Kahn-Lang, J., Li, V., Ryan, N., Subin, Z., Allen, D., . . . Price, S. (2018). Deep Decarbonization in a High Renewables Future: Updated Results from the California PATHWAYS Model. Retrieved July 22, 2019, from: https://www.ethree.com/wp-content/uploads/2018/06/Deep_Decarbonization_in_a_High_Renewables_Future_CEC-500-2018-012-1.pdf

Mai, T., Jadun, P., Logan, J., McMillan, C., Muratori, M., Steinberg, D., . . . Nelson, B. (2018). NREL Electrification Futures Study: Scenarios of Electric Technology Adoptions and Power Consumption for the United States. Retrieved July 8, 2019, from: https://www.nrel.gov/docs/fy18osti/71500.pdf

Mainzer, K., Fath, K., McKenna, R., Stengel, J., Fichtner, W., and Schultmann, F. (2014). A High-Resolution Determination of the Technical Potential for Residential-Roof-Mounted Photovoltaic Systems in Germany. *Solar Energy,* 105, 715–731. DOI: 10.1016/j.solener.2014.04.015.

Majumdar, A., and Deutch, J. (2018). Research Opportunities for CO_2 Utilization and Negative Emissions at the Gigatonne Scale. *Joule,* 2(5), 805–809. DOI: 10.1016/joule.2018.04.018.

Maloney, P. (2017). Navigant Sees $221B Energy as a Service Market by 2026. Retrieved May 15, 2019, from: https://www.utilitydive.com/news/navigant-sees-221b-energy-as-a-service-market-by-2026/448093/

Manjoo, F. (2017). Silicon Valley's Politics: Liberal, with One Big Exception. *New York Times,* September 6.

Manwaring, M. (2012). Understanding Pumped Storage Hydropower. Retrieved June 5, 2019, from: https://www.ntc.blm.gov/krc/uploads/712/12%20-%20Understanding%20Pumped%20Storage%20Hydro%20-%20Manwaring.pdf

Mapdwell. (n.d.). Retrieved February 26, 2018, from: https://www.mapdwell.com/en/solar/company

Margolis, R., Gagnon, P., Melius, J., Phillips, C., and Elmore, R. (2017). Using GIS-Based Methods and Lidar Data to Estimate Rooftop Solar Technical Potential in US Cities. *Environmental Research Letters,* 12, 074013. DOI: 10.1088/1748-9326/aa7225.

Marnay, C. (2016). Microgrids: Finally Finding Their Place. In Sioshansi, F., ed. *Future of Utilities—Utilities of the Future: How Technological Innovations in Distributed Energy Resources Will Reshape the Electrical Power Sector.* Cambridge, MA: Academic Press, pp. 51–70.

Martin, M. (2016). Overview of the Agile, Fractal Grid. Retrieved July 8, 2019, from: http://www.electric.coop/wp-content/uploads/2016/07/Achieving_a_Resilient_and_Agile_Grid.pdf

Maximilian, A., Baylis, P., and Hausman, C. (2017). Climate Change Is Projected to Have Severe Impacts on the Frequency and Intensity of Peak

Electricity Demand across the United States. *Proceedings of the National Academy of Sciences of the United States of America,* 114, 1886–1891. DOI: 10.1073/pnas.1613193114.

McCalley, J., Bushnell, J., Krishnan, V., and Cano, S. (2012). Transmission Design at the National Level: Benefits, Risks and Possible Paths Forward. Power Systems Engineering Research Center. Retrieved May 4, 2019, from: https://pserc.wisc.edu/documents/publications/papers/fgwhitepapers/McCalley_PSERC_White_Paper_Transmission_Overlay_May_2012.pdf

McKibben, B. (2015). Power to the People: Why the Rise of Green Energy Makes Utility Companies Nervous. *New Yorker,* June 29.

Meeus, L., and Glachant, J.-M. (2018). *Electricity Network Regulation in the EU: The Challenges Ahead for Transmission and Distribution.* Cheltenham, UK: Edward Elgar Publishing.

Metz, A. (2018). European Utilities Have Increased Their Activity in Energy Cloud Platforms. Retrieved May 24, 2019, from: https://energypost.eu/european-utilities-have-increased-their-activity-in-energy-cloud-platforms/

Meyer, A., and Pac, G. (2013). Environmental Performance of State-Owned and Privatized Eastern European Energy Utilities. *Energy Economics,* 36, 205–214. DOI: 10.1016/j.eneco.2012.08.019.

Microgrid Institute. (n.d.). Retrieved June 14, 2019, from: http://www.microgridinstitute.org/

Miller, C., Martin, M., Pinney, D., and Walker, G. (2014). *Achieving a Resilient and Agile Grid.* Arlington, VA: National Rural Electric Cooperative Association (NRECA).

Miller, L., Brunsell, N., Mechem, D., Gans, F., Monaghan, A., Vautard, R., and Kleidon, A. (2015). Two Methods for Estimating Limits to Large-Scale Wind Power Generation. *Proceedings of the National Academy of Sciences of the United States of America,* 112(36), 11169–11174. DOI: 10.1073/pnas.1408251112.

Miller, L., Smil, V., Wagner, G., and Keith, D. (2016). Establishing Practical Estimates for City-Integrated Solar PVs and Wind. Science eLetter, July 18. Retrieved October 24, 2019, from: https://keith.seas.harvard.edu/publications/establishing-practical-estimates-city-integrated-solar-pv-and-wind

Miller, N. (2018). *Inertia, Frequency, and Stability.* Presentation at NAGF and ESIG Frequency Response and Energy Storage Workshop. Washington, DC: HickoryLedge.

Milligan, M., Frew, B., Bloom, A., Ela, E., Botterud, A., Townsend, A., and Levin, T. (2016). Wholesale Electricity Market Design with Increasing Levels of Renewable Generation: Revenue Sufficiency and Long-Term Reliability. *Electricity Journal,* 29(2), 26–38. DOI: 10.1016/j.tej.2016.02.005.

Mills, A., and Wiser, R. (2010). *Implications of Wide-Area Geographic Diversity for Short-Term Variability of Solar Power. Office of Energy Efficiency and Renewable Energy, Solar Energy Technologies Program.* Washington, DC: U.S. Department of Energy.

Ming, Z., Olson, A., Jiang, H., Mogadali, M., and Schlag, N. (2019). Resource Adequacy in the Pacific Northwest. Retrieved July 22, 2019, from: https://www.ethree.com/wp-content/uploads/2019/03/E3_Resource_Adequacy_in_the_Pacific-Northwest_March_2019.pdf

Mission Innovation. (n.d.). Retrieved May 25, 2019, from: http://mission-innovation.net/about-mi/overview/

Mission Support Center. (2016). *Cyber Threat and Vulnerability Analysis of the U.S. Electric Sector.* Idaho Falls: Mission Support Center, Idaho National Laboratory.

Mithraratne, N. (2009). Roof-Top Wind Turbines for Microgeneration in Urban Houses in New Zealand. *Energy and Buildings,* 41(10), 1013–1018. DOI: 10.1016/j.enbuild.2019.05.003.

Monitoring Analytics LLC. (2019). Energy Market. 2019 Quarterly State of the Market Report for PJM: January through March, 99. Retrieved July 23, 2019, from: https://www.monitoringanalytics.com/reports/PJM_State_of_the_Market/2019/2019q1-som-pjm-sec3.pdf

Morgan, K. (2014). Two Years after Hurricane Sandy, Recognition of Princeton's Microgrid Still Surges. Princeton University Office of Communications, October 23. Retrieved June 14, 2019, from: https://www.princeton.edu/news/2014/10/23/two-years-after-hurricane-sandy-recognition-princetons-microgrid-still-surges

Morgan Stanley. (2017). What Cheap, Clean Energy Means for Global Utilities. Retrieved November 26, 2018, from: https://www.morganstanley.com/ideas/solar-wind-renewable-energy-utilities

Mufson, S. (2011). Before Solyndra, a Long History of Failed Government Energy Projects. *Washington Post,* November 12.

Mukhopadhyay, S., and Hastak, M. (2016). Public Utility Commissions to Foster Resilience Investment in Power Grid Infrastructure. *Procedia—Social and Behavioral Sciences,* 218, 5–12. DOI: 10.1016/j.sbspro.2016.04.005.

Munich RE. (n.d.). NatCatSERVICE. Number of Relevant Natural Loss Events Worldwide 2013–2018. Retrieved June 14, 2019, from: https://natcatservice

.munichre.com/events/1?-lter=eyJ5ZWFyRnJvbSI6MTk4MCwieWVhclRvIjo yMDE3fQ%3D%3D&type

Nadel, S. (2016). Pathways to Cutting Energy Use and Carbon Emissions in Half. Retrieved June 14, 2019, from: https://aceee.org/sites/default/files /pathways-cutting-energy-use.pdf

Nagabhushan, D., and Thompson, J. (2019). Carbon Capture and Storage in the United States Power Sector: The Impact of 45Q Federal Tax Credits. Clean Air Task Force. Retrieved June 5, 2019, from: https://www.catf.us/wp-content /uploads/2019/02/CATF_CCS_United_States_Power_Sector.pdf

National Academies of Sciences, Engineering, and Medicine. (2017). *Enhancing the Resilience of the Nation's Electricity System.* Washington, DC: The National Academies Press.

National Conference of State Legislatures. (2019). State Renewable Portfolio Standards and Goals. Retrieved May 29, 2019, from: http://www.ncsl.org /research/energy/renewable-portfolio-standards.aspx

National Electrical Safety Code. (2017). IEE C2-2017. Retrieved July 8, 2019, from: https://standards.ieee.org/standard/C2-2017.html

National Rural Electric Cooperative Association. (2015). The 51st State: A Cooperative Path to a Sustainable Future. Retrieved May 24, 2019, from: https://www.cooperative.com/value-of-membership/Documents/51st-State -Report-Phase-I.pdf

National Rural Electric Cooperative Association. (2017). The Role of the Consumer-Centric Utility. National Rural Electric Cooperative Association. Retrieved May 24, 2019, from: https://sepapower.org/resource/51st-state -ideas-role-consumer-centric-utility/

National Rural Electric Cooperative Association. (n.d.). Retrieved November 15, 2019, from: https://www.cooperative.com/nreca/Pages/default.aspx

National Statistics Digest of UK Energy Statistics (DUKES). (2012). DUKES Chapter 5: Statistics on Electricity from Generation through Sales. Retrieved July 8, 2019, from: https://www.gov.uk/government/statistics/electricity -chapter-5-digest-of-united-kingdom-energy-statistics-dukes

Navigant. (2017). Energy as a Service. Retrieved July 8, 2019, from: https://www .navigantresearch.com/reports/energy-as-a-service

Navigant. (2018). *Scoring with the Energy Cloud Playbook: Examples of Disruption and Innovation in the Electric Industry.* Chicago: Navigant.

Nedler, C. (2013). Can the Utility Industry Survive the Energy Transition? A New Paper from the Edison Electric Institute Raises Numerous Doubts.

Retrieved June 14, 2019, from: https://www.greentechmedia.com/articles/read/can-the-utility-industry-survive-the-energy-transition#gs.58lvn1

Nelsen, A. (2016). Portugal Runs for Four Days Straight on Renewable Energy Alone. *The Guardian,* May 18.

Nelson, M., Ramamurthy, A., Czerwinski, M., Light, M., and Shellenberger, M. (2017). The Power to Decarbonize: Characterizing the Impact of Hydroelectricity, Nuclear, Solar, and Wind on the Carbon Intensity of Energy. Retrieved June 5, 2019, from: https://static1.squarespace.com/static/56a45d683b0be33df885def6/t/5a02016eec212dc32217e28f/1510080893757/Power+to+Decarbonize+%283%29.pdf

Nelson, T., and MacNeill, J. (2016). Role of Utility and Pricing in the Transition. In Sioshansi, F., ed. *Future of Utilities—Utilities of the Future: How Technological Innovations in Distributed Energy Resources Will Reshape the Electrical Power Sector.* Cambridge, MA: Academic Press, pp. 109–128.

NERC. (2018). Grid Security Exercise GridEx IV. Lessons Learned. White Paper. Retrieved July 8, 2019, from: https://www.nerc.com/pa/CI/CIPOutreach/GridEX/GridEx%20IV%20Public%20Lessons%20Learned%20Report.pdf

New Buildings Institute. (2018). 2018 Getting to Zero Status Update. Retrieved July 18, 2019, from: https://newbuildings.org/resource/2018-getting-zero-status-update/

New Energy Finance. (2009). Bridging the Valley of Death: Addressing the Scarcity of Seed and Scale-Up Capital for the Next Generation Clean Energy Technologies. Retrieved June 5, 2019, from: https://www.cleanegroup.org/wp-content/uploads/CESA-NEF-scale-up-capital-clean-energy-dec09.pdf

New York State Department of Public Service. (2018). 2017 Electric Reliability Performance Report. Retrieved July 19, 2019, from: http://www3.dps.ny.gov/W/PSCWeb.nsf/96f0fec0b45a3c6485257688006a701a/d82a200687d96d3985257687006f39ca/$FILE/16972359.pdf/2017%20Electric%20Reliability%20Performance%20Report.pdf

Newell, S., Pfeifenberger, J., Chang, J., and Spees, K. (2017). How Wholesale Power Markets and State Environmental Policies Can Work Together. Retrieved June 14, 2019, from: https://www.utilitydive.com/news/how-wholesale-power-markets-and-state-environmental-policies-can-work-toget/446715/

Newell, S., Spees, K., Yang, Y., Metzler, E., and Pedtke, J. (2018). *Opportunities to More Efficiently Meet Seasonal Capacity Needs in PJM.* San Diego, CA: Brattle Group.

Newsham, J. (2015). Five Things You Should Know about Jerrold Oppenheim, Theo MacGregor. *Boston Globe,* April 19.

NOAA National Centers for Environmental Information. (2019). Billion-Dollar Weather and Climate Disasters: Overview. Retrieved July 19, 2019, from: https://www.ncdc.noaa.gov/billions/overview

NOAA Office for Coastal Management. (n.d.). Fast Facts: Hurricane Costs. Retrieved July 19, 2019, from: https://coast.noaa.gov/states/fast-facts /hurricane-costs.html

Noon, C. (2019). The Hydrogen Generation: These Gas Turbines Can Run on the Most Abundant Element in the Universe. Retrieved June 5, 2019, from: https://www.ge.com/reports/hydrogen-generation-gas-turbines-can-run -abundant-element-universe/

Nordhaus, W. (2013). *The Climate Casino.* New Haven, CT: Yale University Press.

North American Supergrid. (2017). Retrieved March 1, 2019, from: Cleanand-securegrid.org/2017/11/28/info

Northeast Energy Efficiency Partnership (NEEP). (2017). Northeastern Regional Assessment of Strategic Electrification. Retrieved June 14, 2019, from: https://neep.org/sites/default/files/Strategic%20Electrification%20 Regional%20Assessment.pdf

NorthWestern Energy. (2018). 2018 Annual Report. Retrieved June 5, 2019, from: http://www.northwesternenergy.com/our-company/investor-relations /annual-reports

Nuclear Energy Institute. (2019). Nuclear by the Numbers. Retrieved March 29, 2019, from: https://www.nei.org/resources/fact-sheets/nuclear-by-the-numbers

NuScale. (2019). NuScale's SMR Design Clears Phases 2 and 3 of Nuclear Regulatory Commission's Review Process. Retrieved July 23, 2019, from: https://newsroom.nuscalepower.com/press-release/company/nuscales-smr -design-clears-phases-2-and-3-nuclear-regulatory-commissions-revie

Ofgem. (2010). RIIO—A New Way to Regulate Energy Networks. Factsheet 93. Retrieved July 14, 2019, from: https://www.ofgem.gov.uk/ofgem-publications /64031/re-wiringbritainfspdf

Ofgem. (2017). RIIO-ED1 Annual Report. Retrieved July 14, 2019, from: https://www.ofgem.gov.uk/system/files/docs/2017/12/riio-ed1_annual _report_2016-17.pdf

Ofgem. (2018a). Review of the RIIO Framework and RIIO-1 Performance. Retrieved July 14, 2019, from: https://www.ofgem.gov.uk/system/files/docs /2018/03/cepa_review_of_the_riio_framework_and _riio-1_performance.pdf

Ofgem. (2018b). RIIO-2 Business Plans Draft Guidance Document. Retrieved July 14, 2019, from: https://www.ofgem.gov.uk/system/files/docs/2018/12/riio-2_business_plans_-_updated_guidance_december_2018_vs_4.pdf

Onyx Solar. (n.d.). Retrieved March 29, 2019, from: www.onyxsolar.com

Order Adopting Regulatory Policy Framework and Implementation Plan. (2015). 14M-0101, New York Public Service Commission, February 26, 2015. Retrieved October 27, 2019, from: http://documents.dps.ny.gov/public/Common/ViewDoc.aspx?DocRefId=%7B0B599D87-445B-4197-9815-24C27623A6A0%7D

Order Authorizing Acquisition and Disposition of Jurisdictional Facilities. (2018). EC18-32-000, Federal Energy Regulatory Commission, April 3, 2018. Retrieved October 27, 2019, from: https://www.ferc.gov/CalendarFiles/20180403165704-EC18-32-000.pdf

Order Resetting Retail Energy Markets and Establishing Further Process. (2016). 15-M-0127, New York Public Service Commission, February 23, 2016. Retrieved October 27, 2019, from: http://www3.dps.ny.gov/W/PSCWeb.nsf/ArticlesByTitle/A6FFDA3D233FF24185257F68006F6D78?OpenDocument

Oregon Department of Energy. (2017). Oregonians' Guide to Siting and Oversight of Energy Facilities. Retrieved July 23, 2019, from: https://www.oregon.gov/energy/facilities-safety/facilities/Documents/Fact-Sheets/EFSC-Public-Guide.pdf

Orvis, R., and Aggarwal, S. (2018). Refining Competitive Electricity Market Rules to Unlock Flexibility. *Electricity Journal,* 31(5), 31–37. DOI: 10.1016/j.tej.2018.05.012.

Orvis, R., and O'Boyle, M. (2018). It's Time to Refine How We Talk about Wholesale Markets. Retrieved June 14, 2019, from: https://www.greentechmedia.com/articles/read/its-time-to-refine-how-we-talk-about-wholesale-markets#gs.d5hb5x

Osborn, D., and Waight, M. (2014). Conceptual Interregional HVDC Network. MISO. Retrieved July 8, 2019, from: https://www.puc.nh.gov/electric/Wholesale%20Investigation/IR%2015-124%20Comments%20D%20Osborn%204-27-15.PDF

Ouyang, M., and Dueñas-Osorio, L. (2014). Multi-Dimensional Hurricane Resilience Assessment of Electric Power Systems. *Structural Safety,* 48, 15–24. DOI: 10.1016/j.strusafe.2014.01.001.

Pacific Gas and Electric Company. (n.d.). Electric Reliability Reports. Retrieved November 8, 2018, from: https://www.pge.com/en_US/residential/outages

/planning-and-preparedness/safety-and-preparedness/grid-reliability /electric-reliability-reports/electric-reliability-reports.page

Palast, G., Oppenheim, J., and MacGregor, T. (2003). *Democracy and Regulation: How the Public Can Govern Essential Services.* Sterling, VA: Pluto Press.

Palmer, K., Butraw, D., and Keyes, A. (2017). Carbon Trading for Integrating Carbon Adders into Wholesale Electricity Markets. Retrieved July 8, 2019, from: https://www.rff.org/publications/reports/lessons-from-integrated-resource-planning-and-carbon-trading-for-integrating-carbon-adders-into-wholesale-electricity-markets/

Panteli, M., and Mancarella, P. (2015). Influence of Extreme Weather and Climate Change on the Resilience of Power Systems: Impacts and Possible Mitigation Strategies. *Electric Power Systems Research,* 127, 259–270. DOI: 10.1016/j.epsr.2015.06.012.

Parsons, J. (2008). *The Value of Long-Term Contracts for New Investments in Generation.* Cambridge, MA: MIT Center for Energy and Environmental Policy Research.

Patel, S. (2019). ENGIE to Exit 20 Countries, Refine Transition Growth Strategy. Retrieved June 14, 2019, from: https://www.powermag.com/engie-to-exit-20-countries-refine-transition-growth-strategy/

Pawar, R., Bromhal, G., Carey, J., Foxall, W., Korre, A., Ringrose, P., . . . White, J. (2015). Recent Advances in Risk Assessment and Risk Management of Geologic CO_2 Storage. *International Journal of Greenhouse Gas Control,* 40, 292–311. DOI: 10.1016/j.ijggc.2015.06.014.

Pearlstein, S. (2019). CVS Bought Your Local Drugstore, Mail-Order Pharmacy and Health Insurer. What's Next, Your Hospital? Retrieved May 28, 2019, from: https://www.washingtonpost.com/business/cvs-bought-your-local-drugstore-mail-order-pharmacy-and-health-insurer-whats-next-your-hospital/2019/01/31/4946dcda-1f2c-11e9-9145-3f74070bbdb9_story.html?utm_term=.4168ed40d59b

Pechman, C. (2016). Modernizing the Electric Distribution Utility to Support the Clean Energy Economy. Retrieved May 30, 2019, from: https://www.energy.gov/sites/prod/files/2017/01/f34/Modernizing%20the%20Electric%20Distribution%20Utility%20to%20Support%20the%20Clean%20Energy%20Economy_0.pdf

Pechman, C. (2017). Determining the Scope of the Electric Distribution Utility of the Future: Prepared for SEPA's 51st State Initiative. Retrieved October 24, 2019, from: https://sepapower.org/resource/51st-state-ideas-determining-scope-electric-distribution-utility-future/

Penn, I. (2018). How Zinc Batteries Could Change Energy Storage. *New York Times,* September 26.

Pentland, W. (2013). Why The Utility Death Spiral Myth Needs to Die. Retrieved July 19, 2019, from: https://www.forbes.com/sites/williampentland/2013/12/02/why-the-utility-death-spiral-myth-needs-to-die/#41c4574f768d

Pepall, L., and Richards, D. (2019). Big-Tech and the Resurgence of Antitrust. Retrieved May 28, 2019, from: https://econofact.org/big-tech-and-the-resurgence-of-antitrust

Pereira, G., Pereira da Silva, P., and Soule, D. (2019). Designing Markets for Innovative Electricity Services in the EU: The Roles of Policy, Technology, and Utility Capabilities. In Sioshansi, F., ed. *Consumer, Prosumer, Prosumager.* Cambridge, MA: Academic Press, pp. 355–382.

Pfeifenberger, J., Chang, J., Aydin, O., and Oates, D. (2016). *The Role of RTO/ISO Markets in Facilitating Renewable Generation Development.* San Diego, CA: Brattle Group.

Pfeifenberger, J., Chang, J., and Sheilendranath, A. (2015). Toward More Effective Transmission Planning: Addressing the Costs and Risks of an Insufficiently Flexible Electricity Grid. Retrieved August 1, 2019, from: https://brattlefiles.blob.core.windows.net/files/5950_toward_more_effective_transmission_planning_addressing_the_costs_and_risks_of_an_insufficiently_flexible_electricity_grid.pdf

Pfeifenberger, J., Chang, J., Sheilendranath, A., Hagerty, J., Levin, S., and Wren, J. (2019). Cost Savings Offered in Competition in Electric Transmission: Experience to Date and the Potential for Additional Customer Value. Retrieved July 23, 2019, from: https://brattlefiles.blob.core.windows.net/files/16726_cost_savings_offered_by_competition_in_electric_transmission.pdf

Phelan, D. (2015). *Protecting Customers: Data Privacy across Utility Sectors.* Silver Spring, MD: National Regulatory Research Institute.

Phillips, A. (2015). Germany Just Got 78 Percent of its Electricity from Renewable Sources. Retrieved June 5, 2019, from: https://thinkprogress.org/germany-just-got-78-percent-of-its-electricity-from-renewable-sources-ac4a323c840c/

PJM. (2017). 2017 RTEP Process Scope and Input Assumptions. White Paper. Retrieved June 14, 2019, from: https://www.pjm.com/-/media/library/reports-notices/2017-rtep/20170731-rtep-input-assumptions-and-scope-whitepaper.ashx?la=en

PJM. (2019). 2018 / 2019 RPM Base Residual Auction Results. Retrieved June 28, 2019, from: https://www.pjm.com/~/media/markets-ops/rpm/rpm-auction-info/2018-2019-base-residual-auction-report.ashx

PJM. (n.d.-a). 2019 / 2020 RPM Third Incremental Auction Results. Retrieved May 24, 2019, from: https://learn.pjm.com/-/media/markets-ops/rpm/rpm-auction-info/2019-2020/2019-2020-third-incremental-auction-report.ashx?la=en

PJM. (n.d.-b). Capacity Market. Retrieved April 20, 2019, from: https://learn.pjm.com/three-priorities/buying-and-selling-energy/capacity-markets.aspx

Plumer, B. (2019). The New Climate Battleground. *New York Times,* June 28, pp. B1, B4.

Pounds, M., and Fleshler, D. (2017). FPL Spent Billions to Protect System, But Why Did Irma Kill Power Anyway? *Sun Sentinel.* Retrieved June 14, 2019, from: https://www.sun-sentinel.com/business/fl-bz-fpl-irma-performance-20170915-story.html

PowerSouth Energy Cooperative. (2017). Compressed Air Energy Storage: McIntosh Power Plant, McIntosh, Alabama. Retrieved June 5, 2019, from: http://www.powersouth.com/wp-content/uploads/2017/07/CAES-Brochure-FINAL.pdf

Pramaggiore, A., and Jensen, V. (2017). Building the Utility Platform. Retrieved May 21, 2019, from: https://www.fortnightly.com/fortnightly/2017/07/building-utility-platform

PSEG. (2018). PSE&G Unveils Next Phase of "Energy Strong" Investments. Retrieved July 19, 2019, from: http://investor.pseg.com/press-release/featured/pseg-unveils-next-phase-energy-strong-investments

Public Power for Your Community. (n.d.). Benefits of Public Power. Retrieved May 28, 2019, from: https://www.publicpower.org/system/files/documents/municipalization-benefits_of_public_power.pdf

Purchala, K. (2018). The EU's Electricity Market: The Good, the Bad, and the Ugly. *Politico,* October 18.

PwC. (2015). A Different Energy Future: Where Energy Transformation Is Leading Us. Retrieved June 6, 2019, from: https://www.pwc.com/ca/en/power-utilities/publications/pwc-global-power-and-utilities-survey-2015-05-en.pdf

PwC. (2017). Global State of Information Security Survey 2017. Retrieved July 8, 2019, from: https://www.pwc.com/gsiss2017

Pye, S., Anandarajah, G., Fais, B., McGlade, C., and Strachan, N. (2015). Pathways to Deep Decarbonization in the United Kingdom. UK 2015 Report.

Retrieved July 8, 2019, from: http://deepdecarbonization.org/wp-content/uploads/2015/09/DDPP_GBR.pdf

Pyper, J. (2015). A Culture Shift Gains Momentum in the Century-Old Utility Industry: How Leading Utilities Are Reforming Their Businesses—And Where There's Still More Work to Do. Retrieved May 29, 2019, from: https://www.greentechmedia.com/articles/read/a-culture-shift-takes-gains-traction-in-the-utility-industry#gs.ff2mth

Pyper, J. (2018a). It's Official: All New California Homes Must Incorporate Solar. Retrieved June 14, 2019, from: https://www.greentechmedia.com/articles/read/solar-mandate-all-new-california-homes#gs.QHHMFB9r

Pyper, J. (2018b). Xcel to Replace 2 Colorado Coal Units with Renewables and Storage. Retrieved June 14, 2019, from: https://www.greentechmedia.com/articles/read/xcel-retire-coal-renewable-energy-storage#gs.c2zcp4

Rai, V., and Zarnikau, J. (2016). Retail Competition, Advanced Metering Investments, and Product Differentiation: Evidence from Texas. In Sioshansi, J., ed. *Future of Utilities—Utilities of the Future: How Technological Innovations in Distributed Energy Resources Will Reshape the Electric Power Sector.* Cambridge, MA: Academic Press, pp. 153–173.

Renewable Energy Buyers Alliance. (n.d.). Retrieved April 22, 2019, from: https://businessrenewables.org

Renewable Energy Policy Network for the 21st Century. (2017). Renewables 2017 Global Status Report. Retrieved June 14, 2019, from: http://www.ren21.net/gsr-2017/

Retière, N., Muratore, G., Kariniotakis, G., Michiorri, A., Frankhauser, P., Caputo, J., . . . Poirson, A. (2017). Fractal Grid—Towards the Future Smart Grid. Retrieved July 8, 2019, from: https://hal.archives-ouvertes.fr/hal-01518413/document

Reyna, J., and Chester, M. (2015). The Growth of Urban Building Stock: Unintended Lock-in and Embedded Environmental Effects. *Journal of Industrial Ecology,* 19(4), 524–537. DOI: 10.1111/jiec.12211.

Rhame, J. (2018). It's Time for Retirement Savers to Consider Utilities as More than Just a Dividend Investment. Retrieved May 24, 2019, from: https://www.marketwatch.com/story/retirement-savers-should-consider-utilities-as-more-than-just-a-dividend-investment-2018-10-01

Rissman, J., and Marcacci, S. (2019). How Clean Energy R&D Policy Can Help Meet Decarbonization Goals. Retrieved June 5, 2019, from: https://www.forbes.com/sites/energyinnovation/2019/02/28/how-clean-energy-rd-policy-can-help-meet-decarbonization-goals/#255227682229

Roach, J. (2015). For Storing Electricity, Utilities Are Turning to Pumped Hydro. Retrieved June 5, 2019, from: https://e360.yale.edu/features/for_storing_electricity_utilities_are_turning_to_pumped_hydro

Roberts, D. (2013). Solar Panels Could Destroy U.S. Utilities, According to U.S. Utilities. Retrieved July 19, 2019, from: https://grist.org/climate-energy/solar-panels-could-destroy-u-s-utilities-according-to-u-s-utilities/

Roberts, D. (2018a). A Tiny, Beleaguered Government Agency Seeks an Energy Holy Grail: Long-Term Energy Storage. Retrieved June 5, 2019, from: https://www.vox.com/energy-and-environment/2018/9/20/17877850/arpa-e-long-term-energy-storage-days

Roberts, D. (2018b). That Natural Gas Power Plant with No Carbon Emissions or Air Pollution? It Works. Retrieved June 5, 2019, from: https://www.vox.com/energy-and-environment/2018/6/1/17416444/net-power-natural-gas-carbon-air-pollution-allam-cycle

Roberts, D. (2018c). This Company May Have Solved One of the Hardest Problems in Clean Energy. Retrieved June 5, 2019, from: https://www.vox.com/energy-and-environment/2018/2/16/16926950/hydrogen-fuel-technology-economy-hytech-storage

Rocky Mountain Institute. (2015). *The Economics of Load Defection: How Grid-Connected Solar Plus Battery Systems Will Compete with Traditional Electric Service, Why It Matters, and Possible Paths Forward.* Boulder, CO: Rocky Mountain Institute.

Rogers, J., and Williams, S. (2015). *Lighting the World: Transforming Our Energy Future by Bringing Electricity to Everyone.* New York: St. Martin's Press.

Romm, J. (2004). *The Hype about Hydrogen: Fact and Fiction in the Race to Save the Climate.* Washington, DC: Island Press.

Romm, J. (2009). Ignore the Media Hype and Keep Googling, Think Progress. Retrieved March 29, 2019, from: https://thinkprogress.org/ignore-the-media-hype-and-keep-googling-the-energy-impact-of-web-searches-is-very-low-d98b38acfefa/

Romm, J. (2010). Debunking the Myth of the Internet as Energy Hog, Again: How Information Technology Is Good for Climate. Retrieved June 4, 2019, from: https://thinkprogress.org/debunking-the-myth-of-the-internet-as-energy-hog-again-how-information-technology-is-good-for-15bcb63e6333/

Ros, A. (2017). An Econometric Assessment of Electricity Demand in the United States Using Utility-Specific Panel Data and the Impact of Retail Competition on Prices. *The Energy Journal,* 38(4). DOI: 10.5547/01956574.38.4.aros.

Ross, C., and Guhathakurta, S. (2017). Autonomous Vehicles and Energy Impacts: A Scenario Analysis. *Energy Procedia,* 143, 47–52. DOI: 10.1016/j.egypro.2017.12.646.

Rote, M. (2019). Costa Rica Has Run on 100% Renewable Energy for 299 Days. Retrieved June 5, 2019, from: https://www.under30experiences.com/blog/costa-rica-has-run-on-100-renewable-energy-for-299-days

Rouse, G., and Kelly, J. (2011). Electricity Reliability: Problem, Progress and Policy Solutions. Retrieved July 8, 2019, from: http://www.galvinpower.org/sites/default/files/Electricity_Reliability_031611.pdf

Ryerson, J. (2019). Is Blockchain Overhyped? *New York Times,* February 15, p. 15.

S&C Electric Company. (2017). *Microgrid Cybersecurity: Protecting and Building the Grid of the Future.* White Paper. Chicago: S&C Electric Company.

S&P Global Platts. (2018). World Electric Power Plants Database, March 2018. Retrieved April 22, 2019, from: https://www.spglobal.com/platts/en/products-services/electric-power/world-electric-power-plants-database

S&P Global Platts. (2019). Top 250 Global Energy Company Rankings: 2019 Top 250 Companies. Retrieved March 29, 2019, from: https://top250.platts.com/Top250Rankings

Safaei, H., and Keith, D. (2015). How Much Bulk Energy Storage Is Needed to Decarbonize Electricity? *Energy and Environmental Science,* 8, 3409–3417. DOI: 10.1039/C5EE01452B.

Salisbury, S. (2010). Hurricane Wilma Five Years Later: Storm Taught Hard Lesson. Retrieved July 8, 2019, from: https://www.palmbeachpost.com/weather/hurricanes/hurricane-wilma-five-years-later-storm-taught-hard-lesson/wxXi109ERademERqe2fNuO/

Sandalow, D. (2012). Hurricane Sandy and Our Energy Infrastructure. Retrieved July 8, 2019, from: https://www.energy.gov/articles/hurricane-sandy-and-our-energy-infrastructure

Sandia National Laboratories. (2006). Solar FAQs. Retrieved January 18, 2019, from: http://www.sandia.gov/~jytsao/Solar%20FAQs.pdf

Sanger, D. (2018). Russian Hackers Appear to Shift Focus to U.S. Power Grid. *New York Times,* July 17.

Santos, T., Gomes, N., Freire, S., Brito, M., Santos, L., and Tenedório, J. (2014). Applications of Solar Mapping in the Urban Environment. *Applied Geography,* 51, 48–57. DOI: 10.1016/j.apgeog.2014.03.008.

Sarralde, J., Quinn, D., Wiesmann, D., and Steemers, K. (2015). Solar Energy and Urban Morphology: Scenarios for Increasing the Renewable Energy

Potential of Neighbourhoods in London. *Renewable Energy,* 73, 10–17. DOI: 10.1016/j.renene.2014.06.028.

Sartor, O. (2018). Implementing Coal Transition: Insights from Case Studies of Major Coal-Consuming Economies. A Summary Report of the Coal Transitions Project Based on Inputs Developed under the Coal Transitions Research Project. Retrieved July 23, 2019, from: https://coaltransitions.files.wordpress.com/2018/09/coal_synthesis_final.pdf

Scheibe, A. (2018). *Utilization of Scenarios in European Electricity Policy: The Ten-Year Network Development Plan.* Oxford: The Oxford Institute for Energy Studies.

Schipper, L., and Grubb, M. (2000). On the Rebound? Feedback between Energy Intensities and Energy Uses in IEA Countries. *Energy Policy,* 28(6–7), 367–388. DOI: 10.1016/S0301-4215(00)00018-5.

Schmidt, O., Hawkes, A., Gambhir, A., and Staffell, I. (2017). The Future Cost of Electrical Energy Storage Based on Experience Rates. *Nature Energy,* 2, 17110. DOI: 10.1038/nenergy.2017.110.

Schumacher, E. (2010). *Small Is Beautiful.* Reprint Edition. New York: Harper Perennial.

Schwartz, L., Wei, M., Morrow, W., Deason, J., Schiller, S., Leventis, G., . . . Teng, J. (2017). Electricity End Uses, Energy Efficiency, and Distributed Energy Resources Baseline. Retrieved June 14, 2019, from: http://eta-publications.lbl.gov/sites/default/files/lbnl-1006983.pdf

Scott, I., and Bernell, D. (2015). Planning for the Future of the Electric Power Sector through Regional Collaboratives. *Electricity Journal,* 28(1). DOI: 10.1016/j.tej.2014.12.002.

Seel, J., Mills, A., Wiser, R., Deb, S., Asokkumar, A., Hassanzadeh, M., and Aarbali, A. (2018). Impacts of High Variable Renewable Energy Futures on Wholesale Electricity Prices, and on Electric-Sector Decision Making. Retrieved June 28, 2019, from: http://eta-publications.lbl.gov/sites/default/files/report_pdf_0.pdf

SEMARNAT-INECC. (2016). Mexico's Climate Change Mid-Century Strategy. Retrieved July 8, 2019, from: https://unfccc.int/files/focus/long-term_strategies/application/pdf/mexico_mcs_final_cop22nov16_red.pdf

Sepulveda, N., Jenkins, F., and Sisternes, R. (2018). The Role of Firm Low-Carbon Electricity Resources in Deep Decarbonization of Power Generation. *Joule,* 2, 2403–2420. DOI: 10/1016/j.joule.2018.08.006.

Shaner, M., Davis, S., Lewis, N., and Caldeira, K. (2018). Geophysical Constraints on the Reliability of Solar and Wind Power in the United States. *Energy and Environmental Science,* (4), 914–925. DOI: 10.1039/c7ee03029k.

Shehabi, A., Smith, S., Horner, N., Azevedo, I., Brown, R., Koomey, J., . . . and Lintner, W. (2016). United States Data Center Energy Usage Report. Retrieved June 14, 2019, from: https://www.osti.gov/servlets/purl/1372902/

Shellenberger, M. (2017). The Nuclear Option: Renewables Can't Save the Planet—But Uranium Can. *Foreign Affairs,* 96(5), 159–165.

Shukla, A., Sudhakar, K., and Baredar, P. (2017). Recent Advancement in BIPV Product Technologies: A Review. *Energy and Buildings,* 140, 188–195. DOI: 10.1016/j.enbuild.2017.02.015.

Siemens. (2014). Fact Sheet: High-Voltage Direct Current Transmission (HVDC). Retrieved May 21, 2019, from: https://www.siemens.com/press/pool/de/feature/2013/energy/2013-08-x-win/factsheet-hvdc-e.pdf

Siemens. (2015). Kick-Off for World's Largest Electrolysis System in Mainz. Retrieved July 1, 2019, from: https://press.siemens.com/global/en/feature/kick-worlds-largest-electrolysis-system-mainz

Siemens Gamesa. (n.d.). Electric Thermal Energy Storage: GWh Scale and for Different Applications. Retrieved June 5, 2019, from: https://www.siemensgamesa.com/en-int/products-and-services/hybrid-and-storage/thermal-energy-storage-with-etes

Sierra Club. (n.d.). Energy Facilities Siting. Retrieved June 28, 2019, from: https://www.sierraclub.org/policy/energy/energy-facilities

Silverman, J. (2019). Big Tech Is Watching. *New York Times,* January 20, p. 10.

Silverstein, A., Gramlich, R., and Goggin, M. (2018). A Customer-Focused Framework for Electric System Resilience. Retrieved May 21, 2019, from: https://gridprogress.files.wordpress.com/2018/05/customer-focused-resilience-final-050118.pdf

Simeone, C. (2017a). Part 1: Cost-of-Service Retired More Coal. Retrieved May 21, 2019, from: https://kleinmanenergy.upenn.edu/blog/2017/12/12/cost-service-retired-more-coal

Simeone, C. (2017b). Part 2: Future Coal Retirements and a NOPR Disconnect. Retrieved May 21, 2019, from: https://kleinmanenergy.upenn.edu/blog/2017/12/13/part-2-future-coal-retirements-and-nopr-disconnect

Simeone, C. (2017c). Part 3: Utilities Continue Coal Retreat, Advance on Gas and Renewables. Retrieved from: https://kleinmanenergy.upenn.edu/blog/2017/12/13/part-3-utilities-continue-coal-retreat-advance-gas-and-renewables

Simmons, D. (2018). 2019 Outlook: Utilities, amid Rising Volatility and Questions about Peaking Growth, Safe-Haven Utilities Are Poised for a

Leadership Role. Retrieved July 16, 2019, from: https://www.fidelity.com/viewpoints/investing-ideas/2019-outlook-utilities

Simon, C. (2018). Global Power for Global Powers. Retrieved August 1, 2019, from: https://news.harvard.edu/gazette/story/2018/04/harvard-talk-outlines-plan-for-global-energy-sharing/

Simonov, E. (2018). The Risks of a Global Supergrid. Retrieved August 1, 2019, from: https://www.thethirdpole.net/en/2018/07/24/the-risks-of-a-global-supergrid/

Sivaram, V. (2018a). *Digital Decarbonization: Promoting Digital Innovations to Advance Clean Energy Systems.* New York: Council on Foreign Relations.

Sivaram, V. (2018b). *Taming the Sun: Innovations to Harness Solar Energy and Power the Planet.* Cambridge, MA: The MIT Press.

SkyscraperPage.com. (n.d.). Cities and Buildings. Retrieved July 18, 2019, from: https://skyscraperpage.com/cities/?10=1

Slashdot. (2017). Underwater Pumped-Storage Hydroelectric Project Completes Its First Practical Test. Retrieved April 2, 2019, from: https://hardware.slashdot.org/story/17/03/05/1758231/underwater-pumped-storage-hydroelectric-project-completes-its-first-practical-test

Slaughter, A. (2015). *Electricity Storage: Technologies, Impacts, and Prospects.* Houston, TX: Deloitte Center for Energy Solutions.

Smart Energy Consumer Collaborative. (2018). Consumer Platform of the Future. Retrieved May 24, 2019, from: https://smartenergycc.org/consumer-platform-of-the-future-report/

SmartGrid Consumer Collaborative. (2017a). Consumer Pulse and Market Segmentation Study—Wave 6. Retrieved May 24, 2019, from: http://smartenergycc.org/wp-content/uploads/2017/05/SGCCs-Consumer-Pulse-and-Market-Segmentation-Study-Wave-6-Executive-Summary.pdf

SmartGrid Consumer Collaborative. (2017b). Spotlight on Millennials. Retrieved May 24, 2019, from: http://smartenergycc.org/wp-content/uploads/2017/08/SGCC-Spotlight-on-Millennials-Report-Executive-Summary-8-8-17.pdf

Smil, V. (2008). *Energy in Nature and Society.* Cambridge, MA: The MIT Press.

Smil, V. (2010). *Energy Transitions: History, Requirements, Prospects.* Santa Barbara, CA: Praeger Press.

Smith, G. (2015). Bill Gates Is Doubling His Billion-Dollar Bet on Renewables. Retrieved June 5, 2019, from: http://fortune.com/2015/06/26/bill-gates-renewables-investment-solar-depleted-uranium-battery-storage/

Smith, R., and MacGill, I. (2016). The Future of Utility Customers and the Utility Customer of the Future. In Sioshansi, F., ed. *Future of Utilities—Utilities of the Future: How Technological Innovations in Distributed Energy Resources Will Reshape the Electric Power Sector.* Cambridge, MA: Academic Press 2016, pp. 343–362.

Solar District Heating. (n.d.). About SDH. Retrieved March 29, 2019, from: https://www.solar-district-heating.eu/en/about-sdh/

Solar Energy Industries Association. (2017). Solar Market Insight Report. Retrieved June 14, 2019, from: https://www.seia.org/research-resources/solar-market-insight-report-2017-year-review

Solar Energy Industries Association. (2019). U.S. Solar Market Insight. Retrieved March 29, 2019, from: https://www.seia.org/us-solar-market-insight

Space Weather Prediction Center. (n.d.). *Geomagnetic Storms and the US Power Grid.* Silver Spring, MD: NOAA.

Spector, J. (2016). Vancouver Leapfrogs Energy Efficiency, Adopts Zero-Emissions Building Plan. Retrieved October 25, 2019, from: https://www.greentechmedia.com/articles/read/vancouver-leapfrogs-energy-efficiency-adopts-zero-emissions-building-plan

Spector, J. (2018). PG&E Proposes World's Biggest Batteries to Replace South Bay Gas Plants. Retrieved June 5, 2019, from: https://www.greentechmedia.com/articles/read/pge-proposes-worlds-biggest-batteries-to-replace-south-bay-gas-plants#gs.h0rm9n

Spees, K., and Chang, J. (2017). *A Dynamic Clean Energy Market.* San Diego, CA: Brattle Group.

Spees, K., Newell, S., Pfeifenberger, J., and Chang, J. (2018). *Market Design 3.0: A Vision for the Clean Electricity Grid of the Future.* San Diego, CA: Brattle Group.

St. John, J. (2018a). 5 Predictions for the Global Energy Storage Market in 2019. Retrieved March 29, 2019, from: https://www.greentechmedia.com/articles/read/five-predictions-for-the-global-energy-storage-market-in-2019#gs.h07whf

St. John, J. (2018b). Illinois Decision Opens the Path to Shared Utility-Customer Microgrids. Retrieved July 8, 2019, from: https://www.greentechmedia.com/articles/read/illinois-decision-opens-the-path-to-shared-utility-customer-microgrids#gs.bzh7z6

St. John, J. (2018c). The Shifting Makeup of the Fast-Growing US Energy Storage Market. Retrieved March 29, 2019, from: https://www.greentechmedia.com/articles/read/tracking-the-shifting-makeup-of-the-us-energy-storage-market#gs.h08igw

Starn, J. (2018). Power Worth Less than Zero Spreads as Green Energy Floods the Grid. Retrieved July 9, 2019, from: https://www.bloomberg.com/news/articles/2018-08-06/negative-prices-in-power-market-as-wind-solar-cut-electricity

State of Colorado. (2010). "Clean Air, Clean Jobs Act" House Bill 10–1365. Retrieved June 22, 2019, from: https://www.csg.org/sslfiles/dockets/2012cycle/32Abills/1232a01cocoalgasutility.pdf

Statistics Canada. (2018). Electric Power, Electric Utilities and Industry, Annual Supply and Disposition, Table 25-10-0021-01. Retrieved July 8, 2019, from: https://www150.statcan.gc.ca/t1/tbl1/en/tv.action?pid=2510002101

Stern, R. (2012). Solar Eclipsed: Why the Sun Won't Power Phoenix. *Phoenix New Times,* December 27.

Stone, A. (2016). Britain Pioneered Performance-Based Utility Regulation. How Has It Worked Thus Far? Exploring the RIIO Model and Its Potential Impact beyond the UK. Retrieved July 14, 2019, from: https://www.greentechmedia.com/articles/read/britain-was-a-leader-in-performance-based-utility-regulation-how-has-it-wo

Strauss, L. (2018). Why Utility Stocks Are Worth a Second Look. Retrieved July 16, 2019, from: https://www.barrons.com/articles/why-utility-stocks-are-worth-a-second-look-1531344310?mod=article_signInButton?mod=article_signInButton?mod=article_signInButton?mod=article_signInButton

Strbac, G., Mancarella, P., and Pudjianto, D. (2009). Advanced Architecture and Control Concepts for MORE MICROGRIDS: DH1. Microgrid Evolution Roadmap in the EU. Retrieved July 16, 2019, from: http://www.microgrids.eu/documents/676.pdf

Strubell, E., Ganesh, A., and McCallum, A. (2019). *Energy and Policy Considerations for Deep Learning in NLP.* 57th Annual Meeting of the Association for Computational Linguistics (ACL), Florence, Italy.

Stutz, B., Le Pierres, N., Kuznik, F., Johannes, K., Del Barrio, E., Bédécarrats, J.-P., . . . Minh, D. (2016). Storage of Thermal Solar Energy. *Comptes Rendus Physique,* 18(7–8), 401–414. DOI: 10.1016/j.crhy.2017.09.008.

Sun, C., Hahn, A., and Liu, C. (2018). Cyber Security of a Power Grid: State-of-the-Art. *International Journal of Electrical Power and Energy Systems,* 99, 45–56. DOI: 10.1016/j.ijepes.2017.12.020.

Sunrun. (2018). Affordable, Clean, Reliable Energy. Retrieved July 14, 2019, from: https://www.sunrun.com/sites/default/files/affordable-clean-reliable-energy.pdf

Sunter, D., Dabiri, J., and Kammen, D. (2016). Confirming Practical Estimates for City-Integrated Photovoltaic and Wind Power Densities. Science eLetter,

July 18. Retrieved October 24, 2019, from: https://science.sciencemag.org/content/352/6288/922/tab-e-letters

Sweeney, J. (2002). *The California Electricity Crisis.* Stanford, CA: Hoover Institution Press.

Szulecki, K. (2018). Conceptualizing Energy Democracy. *Environmental Politics,* 27(1), 21–41. DOI: 10.1080/09644016.2017.1387294.

Tabors, R. (2016). Valuing Distributed Energy Resources (DER) via Distribution Locational Marginal Prices (DLMP). Retrieved May 21, 2019, from: https://www.energy.gov/sites/prod/files/2016/06/f32/4_Transactive%20Energy%20Panel%20-%20Richard%20Tabors%2C%20MIT%20Energy%20Initiative.pdf

Tabors, R., Caramanis, M., Ntakou, E., Parker, G., VanAlstyne, M., Centolella, P., and Hornby, R. (2017). Distributed Energy Resources: New Markets and New Products. *Proceedings of the 50th Hawaii International Conference on System Sciences.* Retrieved May 21, 2019, from: https://papers.ssrn.com/sol3/papers.cfm?abstract_id=2964982

Taft, J. (2016). *Comparative Architecture Analysis: Using Laminar Structure to Unify Multiple Grid Architectures.* Washington, DC: Grid Modernization Laboratory Consortium, U.S. Department of Energy.

Taylor, G., Ledgerwood, S., Broehm, R., and Fox-Penner, P. (2015). *Market Power and Market Manipulation in Energy Markets: From California Crisis to the Present.* Reston, VA: Public Utilities Reports, Inc.

Taylor, J., Dhople, S., and Callaway, D. (2016). Power Systems without Fuel. *Renewable and Sustainable Energy Reviews,* 57, 1322–1336. DOI: 10.1016/j.rser.2015.12.083.

Temple, J. (2017). Potential Carbon Capture Game Changer Nears Completion. Retrieved June 5, 2019, from: https://www.technologyreview.com/s/608755/potential-carbon-capture-game-changer-nears-completion/

Temple, J. (2018). At This Rate, It's Going to Take Nearly 400 Years to Transform the Energy System. Retrieved July 1, 2019, from: https://www.technologyreview.com/s/610457/at-this-rate-its-going-to-take-nearly-400-years-to-transform-the-energy-system/

Tepper, Jonathan. (n.d.). Publications. Retrieved October 27, 2019, from: http://jonathan-tepper.com/publications/

The Atlantic. (n.d.). Welcome Solar: Ushering in the Age of a New Energy. Retrieved November 26, 2018, from: https://www.theatlantic.com/sponsored/thomson-reuters-why-2025-matters/solar-power/210/

The White House Washington. (2016). United States Mid-Century Strategy for Deep Decarbonization. Retrieved July 8, 2019, from: https://unfccc.int/files/focus/long-term_strategies/application/pdf/mid_century_strategy_report-final_red.pdf

The World Bank. (2015). World Bank Development Indicators 2015, Table 3.6. Retrieved June 14, 2019, from: http://wdi.worldbank.org/table/3.6

Thorp, C. (2019). State Bill for Coal-Fired Power Plant Communities One Step Closer to Law. Retrieved May 8, 2019, from: https://www.craigdailypress.com/news/state-bill-for-coal-fired-power-plant-communities-one-step-closer-to-law/

Tierney, S. (2016). The Value of "DER" to "D": The Role of Distributed Energy Resources in Supporting Local Electric Distribution System Reliability. Retrieved May 21, 2019, from: https://www.analysisgroup.com/globalassets/content/news_and_events/news/value_of_der_to-_d.pdf

Tierney, S. (2017). About That National Conversation on Resilience of the Electric Grid. Retrieved July 8, 2018, from: https://www.utilitydive.com/news/about-that-national-conversation-on-resilience-of-the-electric-grid-the-ur/512545/

Tierney, S. (2018). *Resource Adequacy and Wholesale Market Structure for a Future Low-Carbon Power System in California.* White Paper. Boston: Analysis Group.

Tokarska, K., Gillett, N., Weaver, A., Arora, V., and Eby, M. (2016). The Climate Response to Five Trillion Tonnes of Carbon. *Nature Climate Change,* 6, 851–855. DOI: 10.1038/nclimate3036.

Ton, D., and Wang, W.-T. (2015). A More Resilient Grid: The U.S. Department of Energy Joins with Stakeholders in an R&D Plan. *IEEE Power and Energy Magazine,* 13(3), 26–34. DOI: 10.1109/MPE.2015.2397337.

Tong, J., and Wellinghoff, J. (2014). Rooftop Parity: Solar for Everyone, including Utilities. Retrieved July 23, 2019, from: https://www.fortnightly.com/fortnightly/2014/08/rooftop-parity

Townsend, A., and Havercroft, I. (2019). The LCFS and CCS Protocol: An Overview for Policymakers and Project Developers. Retrieved June 5, 2019, from: https://www.globalccsinstitute.com/wp-content/uploads/2019/05/LCFS-and-CCS-Protocol_digital_version-2.pdf

Trabish, H. (2012). How Electricity Gets Bought and Sold in California. Retrieved July 23, 2019, from: https://www.greentechmedia.com/articles/read/how-electricity-gets-bought-and-sold-in-california#gs.r89hei

Trabish, H. (2014). Jon Wellinghoff: Utilities Should Not Operate the Distribution Grid. Retrieved May 24, 2019, from: https://www.utilitydive.com/news/jon-wellinghoff-utilities-should-not-operate-the-distribution-grid/298286/

Trabish, H. (2017). Illinois Energy Reform Set to Shape New Solar Business Models for Utilities. Retrieved July 14, 2019, from: https://www.utilitydive.com/news/illinois-energy-reform-set-to-shape-new-solar-business-models-for-utilities/504590/

Trabish, H. (2018). Business Models: What Utilities Can Learn from Amazon and Netflix about the Future of Ratemaking. Retrieved May 15, 2019, from: https://www.utilitydive.com/news/business-models-what-utilities-can-learn-from-amazon-and-netflix-about-the/530415/

Trabish, H. (2019). An Emerging Push for Time of Use Rates Sparks New Debates about Customer and Grid Impacts. Retrieved July 14, 2019, from: https://www.utilitydive.com/news/an-emerging-push-for-time-of-use-rates-sparks-new-debates-about-customer-an/545009/

TransWest Express LLC. (n.d.). Schedule and Timeline. Retrieved July 19, 2019, from: http://www.transwestexpress.net/about/timeline.shtml

Tsuchida, B., Sergici, S., Mudge, B., Gorman, W., Fox-Penner, P., and Schoene, J. (2015). *Comparative Generation Costs of Utility-Scale and Residential Solar PV in Xcel Energy Colorado's Service Area.* San Diego, CA: Brattle Group.

Tucker, A. (2016). Boston Startup Will Help GE Make Coal-Fired Power Plants Cleaner with Software. Retrieved May 15, 2019, from: https://www.ge.com/reports/boston-startup-will-help-ge-make-coal-fired-power-plants-cleaner-with-software

UNFCCC. (2015). Paris Agreement. Retrieved June 14, 2019, from: https://unfccc.int/process/conferences/pastconferences/paris-climate-change-conference-november-2015/paris-agreement

United Nations. (2014). Department of Economic and Social Affairs, Population Division. World Urbanization Prospects: The 2014 Revision, Highlights. ST / ESA / SER.A / 352. Retrieved June 14, 2019, from: http://esa.un.org/unpd/wup/Highlights/WUP2014-Highlights.pdf

United Nations Framework Convention on Climate Change. (2017). Enhancing Financing for the Research, Development and Demonstration of Climate Technologies. UNFCCC, Technology Executive Committee Working Paper. Retrieved June 5, 2019, from: https://unfccc.int/ttclear/docs/TEC_RDD%20finance_FINAL.pdf

U.S. Department of Energy. (2017). 2016 Wind Technologies Market Report. Retrieved June 19, 2019, from: https://www.energy.gov/sites/prod/files/2017/08/f35/2016_Wind_Technologies_Market_Report_0.pdf

U.S. Department of Energy. (2018). 2017 Hydropower Market Report. Retrieved June 5, 2019, from: https://www.energy.gov/eere/water/downloads/2017-hydropower-market-report

U.S. Department of Energy's (DOE) Office of Energy Efficiency and Renewable Energy. (n.d.-a). DOE Technical Targets for Hydrogen Production from Electrolysis. Retrieved May 4, 2019, from: https://www.energy.gov/eere/fuelcells/doe-technical-targets-hydrogen-production-electrolysis

U.S. Department of Energy's (DOE) Office of Energy Efficiency and Renewable Energy. (n.d.-b). Fuel Cell Technologies Office. Retrieved May 2, 2019, from: https://www.energy.gov/eere/fuelcells/fuel-cell-technologies-office

U.S. Department of Energy's (DOE) Office of Fossil Energy and the Oak Ridge National Laboratory. (2017). Accelerating Breakthrough Innovation in Carbon Capture, Utilization, and Storage. Retrieved June 5, 2019, from: https://www.energy.gov/fe/downloads/accelerating-breakthrough-innovation-carbon-capture-utilization-and-storage

U.S. Department of the Interior Bureau of Reclamation. (2019). Grand Coulee Dam Statistics and Facts. Retrieved March 25, 2019, from: https://www.usbr.gov/pn/grandcoulee/pubs/factsheet.pdf

U.S. Department of Transportation. (2014). Beyond Traffic 2045: Trends and Choices, Draft Report. Retrieved June 14, 2019, from: https://cms.dot.gov/sites/dot.gov/files/docs/Draft_Beyond_Traffic_Framework.pdf

U.S. Energy Information Administration. (2012). Annual Energy Review. Retrieved July 18, 2019, from: https://www.eia.gov/totalenergy/data/annual/showtext.php?t=ptb0802a

U.S. Energy Information Administration. (2014). Energy Transportation Use 2014. Retrieved June 14, 2019, from: http://www.eia.gov/Energyexplained/?page=us_energy_transportation

U.S. Energy Information Administration. (2017). Form EIA-860 Annual Electric Generator Report. Retrieved June, 14, 2019, from: www.eia.gov/electricity/data/eia860

U.S. Energy Information Administration. (2018). Annual Energy Outlook 2018 with Projections to 2050. Retrieved March 21, 2018, from: https://www.eia.gov/outlooks/aeo/pdf/AEO2018.pdf

U.S. Energy Information Administration. (2019a). Annual Energy Outlook 2019. Retrieved June 14, 2019, from: https://www.eia.gov/outlooks/aeo/data

/browser/#/?id=8-AEO2019®ion=0-0&cases=ref2019&start=2017&end =2050&f=A&linechart=ref2019-d111618a.6-8-AEO2019~ref2019-d111618a .75-8-AEO2019&ctype=linechart&sourcekey=0

U.S. Energy Information Administration. (2019b). Preliminary Monthly Electric Generator Inventory (Based on Form EIA-860M as a Supplement to Form EIA-860). Retrieved April 22, 2019, from: https://www.eia.gove /electricity/data/eia860M

U.S. Environmental Protection Agency. (n.d.). Community Choice Aggregation. Retrieved May 28, 2019, from: https://www.epa.gov/greenpower /community-choice-aggregation

U.S. Global Change Research Program. (2018). Fourth National Climate Assessment. Retrieved July 14, 2019, from: https://nca2018.globalchange.gov/

Utah Office of Energy Development. (2013). *Guide to Permitting Electric Transmission Lines in Utah.* Salt Lake City: Utah Office of Energy Development.

Van Hertem, D., Gomis-Bellmunt, O., and Liang, J. (2016). *HVDC Grids: For Offshore and Supergrid of the Future.* Hoboken, NJ: Wiley-IEEE Press.

Van Nuffel, L., Rademaekers, K., Yearwood, J., Graichen, V., Lopez, M., Gonzalez, A., . . . Marias, F. (2017). European Energy Industry Investments. European Parliament. Directorate-General for Internal Policies. Retrieved July 24, 2019, from: http://www.europarl.europa.eu/RegData/etudes/STUD /2017/595356/IPOL_STU(2017)595356_EN.pdf

Varadarajan, U., Posner, D., and Fisher, J. (2018). *Harnessing Financial Tools to Transform the Electric Sector.* Oakland, CA: Sierra Club.

Vaughan, A. (2016). New Battery Power-Storage Plants Scheduled to Keep UK Lights On. *The Guardian,* December 9.

Véliz, K., Kauffman, R., Cleveland, C., and Stoner, A. (2017). The Effect of Climate Change on Electricity Expenditures in Massachusetts. *Energy Policy,* 106(C), 1–11. DOI: 10.1016/j.enpol.2017.03.016.

Viribright. (n.d.). Comparing LED vs CFL vs Incandescent Light Bulbs. Retrieved July 18, 2019, from: http://www.viribright.com/lumen-output -comparing-led-vs-cfl-vs-incandescent-wattage/

Viscusi, W., Vernon, J., and Harrington, Jr., J. (1998). *Economics of Regulation and Antitrust.* 2nd ed. Cambridge, MA: The MIT Press.

Vlerick Business School. (n.d.). Outlook on the European DSO Landscape 2020. Retrieved June 14, 2019, from: https://home.kpmg/content/dam/kpmg /pdf/2016/05/Energy-Outlook-DSO-2020.pdf

Vries, A. de (2018). Bitcoin's Growing Energy Problem, *Joule,* 2, 801–809. DOI: 10.1016/j.joule.2018.04.016.

Vrins, J., Artze, H., Shandross, R., and Lawrence, M. (2015). From Grid to Cloud: A Network of Networks—In Search of an Orchestrator. *Fortnightly Magazine,* October.

Wadud, Z., MacKenzie, D., and Leiby, P. (2016). Help or Hindrance? The Travel, Energy and Carbon Impacts of Highly Automated Vehicles. *Transportation Research Part A: Policy and Practice,* 86, 1–18. DOI: 10.1016/j.tra.2015.12.001.

Wakabayashi, D. (2018). California Passes Sweeping Law to Protect Online Privacy. *New York Times,* June 28.

Walker, R. (2017). Artificial Intelligence in Business: Balancing Risk and Reward. Retrieved May 21, 2019, from: https://www.pega.com/insights/resources/artificial-intelligence-business-balancing-risk-and-reward

Wallace-Wells, D. (2019). *The Uninhabitable Earth.* New York: Tim Duggan Books.

Walton, R. (2018). Xcel, Boulder Agree on Separation Details in March towards Municipal Utility. Retrieved July 1, 2019, from: https://www.utilitydive.com/news/xcel-boulder-agree-on-separation-details-in-march-towards-municipal-utilit/541323/

Wang, J., Lu, K., Ma, L., Wang, J., Dooner, M., Miao, S., . . . Wang, D. (2017). Overview of Compressed Air Energy Storage and Technology Development. *Energies,* 10, 991. DOI: 10.1016/j.egypro.2014.12.423.

Wang, N., Phelan, P., Harris, C., Langevin, J., Nelson, B., and Sawyer, K. (2018). Past Visions, Current Trends, and Future Context: A Review of Building Energy, Carbon, and Sustainability. *Renewable and Sustainable Energy Reviews,* 82, 976–993. DOI: 10.1016/j.rser.2017.04.114.

Watts, N., Amann, M., Ayeb-Karlsson, S., Belesova, K., Bouley, T., Boykoff, M., . . . Costello, A. (2018). The Lancet Countdown on Health and Climate Change: From 25 Years of Inaction to a Global Transformation for Public Health. *The Lancet,* 391(10120), 581–630. DOI: 10.1016/S0140-6736(17)32464-9.

Weaver, J. (2017). World's Largest Battery: 200mw / 800mwh Vanadium Flow Battery—Site Work Ongoing. Retrieved June 5, 2019, from: https://electrek.co/2017/12/21/worlds-largest-battery-200mw-800mwh-vanadium-flow-battery-rongke-power/

Webber, M. (2016). *Thirst for Power: Energy, Water and Human Survival.* New Haven, CT: Yale University Press.

Weigert, K. (2017). *Grid Security Is National Security: Cyber Threats to Energy Infrastructure and Cities.* Working Paper Number 2017–01. Chicago: The Chicago Council on Global Affairs.

Weiser, S. (2018). Groups Request PUC Reconsider Approval of Xcel's Fuel Switching Plan. Retrieved July 23, 2019, from: https://pagetwo

.completecolorado.com/2018/10/02/groups-request-puc-reconsider-approval-of-xcels-fuel-switching-plan/

Wellinghoff, J., and Cusick, K. (2017). Alternative Transmission Solutions: An Analysis of the Emerging Business Opportunity for Advanced Transmission Technologies and FERC-Driven Requirements on Transmission Planning and Selection. Retrieved August 1, 2019, from: http://grid-8990.kxcdn.com/wp-content/uploads/2017/11/Alternative-Transmission-Solutions-pfeifen

Wellinghoff, J., and Tong, J. (2015). Wellinghoff and Tong: A Common Confusion over Net Metering Is Undermining Utilities and the Grid. Retrieved May 30, 2019, from: https://www.utilitydive.com/news/wellinghoff-and-tong-a-common-confusion-over-net-metering-is-undermining-u/355388/

Wesoff, E. (2015). NextEra on Storage: "Post 2020, There May Never Be Another Peaker Built in the US." Retrieved October 25, 2019, from: https://www.greentechmedia.com/articles/read/nextera-on-storage-post-2020-there-may-never-be-another-peaker-built-in-t

Weston, D. (2018). Hornsdale Battery Has "Significant Impact" on Market. Retrieved June 28, 2019, from: https://www.windpowermonthly.com/article/1520406/hornsdale-battery-significant-impact-market

Wikipedia. (n.d.-a). Hurricane Harvey. Retrieved July 19, 2019, from: https://en.wikipedia.org/wiki/Hurricane_Harvey

Wikipedia. (n.d.-b). List of Major Power Outages. Retrieved May 8, 2019, from: https://en.wikipedia.org/wiki/List_of_major_power_outages

Wikiquote. (n.d.). Dwight D. Eisenhower. Retrieved March 2, 2019, from: https://en.wikiquote.org/wiki/Dwight_D._Eisenhower

Williams, J., Haley, B., Kahrl, F., Moore, J., Jones, A., Torn, M., and McJeon, H. (2014). Pathways to Deep Decarbonization in the United States. The U.S. Report of the Deep Decarbonization Pathways Project of the Sustainable Development Solutions Network and the Institute for Sustainable Development and International Relations. Retrieved June 14, 2019, from: https://usddpp.org/downloads/2014-technical-report.pdf

Williams, M. (2018). *Powering Culture Change: How Redding Utility Looked Inward to Better Serve Its Community.* Sacramento: California Municipal Utilities Association, pp. 28–30.

Williamson, O. (1975). *Markets and Hierarchies.* New York: Free Press.

Williamson, O. (1979). Transaction–Cost Economics: The Governance of Contractual Relations. *Journal of Law and Economics,* 22(2), 233–262. DOI: 10.1086/466942.

Wilson, J. (2019). Electric Companies Overspend by Billions, Driving Up Utility Bills, Report Finds. *USA Today,* February 18, B1.

Wimberly, J. (2018). *The Glass Half Full for Utility Customer Service.* EcoPinion Consumer Survey Report No. 32. Los Angeles: DEFG & Russell Research.

Wimberly, J., and Treadway, N. (2018). 2018 Regulator Survey on Customer Service Metrics. Retrieved May 30, 2019, from: http://defgllc.com/publication/2018-regulator-survey-on-customer-service-metrics/

Wolf, G. (2018). Supergrids Are Possible: A History Mingled with Facts on the Hybrid DC Breaker. Retrieved July 23, 2019, from: https://www.tdworld.com/hvdc/supergrids-are-possible

Wood, E. (2018). Princeton University's Microgrid: How to Partner, Not Part from the Grid. Retrieved July 8, 2019, from: https://microgridknowledge.com/princeton-universitys-microgrid-partner-part-central-grid/

Wood, E. (2019). Hawaii Traverses New Ground with Microgrid Tariff. Retrieved July 8, 2019, from: https://microgridknowledge.com/microgrid-tariff-hawaii/

Wood, L., Hemphill, R., Howat, J., Cavanagh, R., and Borenstein, S. (2016). Recovery of Utility Costs: Utility, Consumer, Environmental and Economist Perspectives, LBNL-1005742 Report No. 5. Retrieved July 8, 2019, from: https://emp.lbl.gov/sites/all/files/lbnl-1005742_1.pdf

World Bank Group. (2018). State and Trends of Carbon Pricing 2018. Retrieved July 17, 2019, from: https://openknowledge.worldbank.org/bitstream/handle/10986/29687/9781464812927.pdf?sequence=5&isAllowed=y

World Economic Forum. (2015). *Intelligent Assets: Unlocking the Circular Economy Potential.* Cologny, Switzerland: World Economic Forum.

Wu, T. (2018). *The Curse of Bigness: Antitrust in the New Gilded Age.* New York: Columbia Global Reports.

Xcel. (2014). Overhead vs Underground, Xcel Colorado Information Sheet, 14-05-042. Retrieved July 23, 2019, from: https://www.xcelenergy.com/staticfiles/xe/Corporate/Corporate%20PDFs/OverheadVsUnderground_FactSheet.pdf

Yates, D., Quan Luna, B., Rasmussen, R., Bratcher, D., Garre, L., Chen, F., . . . Friis-Hansen, P. (2014). Stormy Weather: Assessing Climate Change Hazards to Electric Power Infrastructure: A Sandy Case Study. *IEEE Power and Energy Magazine,* 12(5), 66–75. DOI: 10.1109/MPE.2014.2331901.

Zarakas, W., Sergici, S., Bishop, H., Zahniser-Word, J., and Fox-Penner, P. (2014). Utility Investment in Resiliency: Balancing Benefit with Cost in an

Uncertain Environment. *Electricity Journal,* 27(5), 31–39. DOI: 10.1016/j.tej.2014.05.005.

Zehong, L. (2017). China's Future Power Grids. Retrieved July 18, 2019, from: https://stanford.app.box.com/s/dxiyh6b196wg7y1aoje7b7u47nfg3uyi

Zoback, M., and Gorelick, S. (2012). Earthquake Triggering and Large-Scale Geologic Storage of Carbon Dioxide. *Proceedings of the National Academy of the Sciences of the United States of America,* 109(26), 10164–10168. DOI: 10.1073/pnas.1202473109.

Zummo, P. (2018). The Value of the Grid. Retrieved May 24, 2019, from: https://www.publicpower.org/system/files/documents/Value%20of%20the%20Grid_1.pdf

Acknowledgments

During the years it has taken me to write this book, I have accumulated a pathetically large collection of debts to colleagues and friends. This work knits an endless stream of questions, favors, slide decks, and documents into something I hope is a partial reward for their kindness.

I must first thank the three organizations that allowed me to take on this project. At Boston University, the Institute for Sustainable Energy (ISE) has provided wonderful support, starting with Gloria Waters and Jacquie Ashmore at the top and including Jenny Hatch (for early leadership and Chapter 13), David Jermain, Cutler Cleveland, Michael Walsh, Justin Ren, Bai Li, Sophia Xuehai Xiong, Peishan Wang, Karla Kim, Robert Perry, Laura Hurley, Lam Tan Tjien, Guillermo Perriera, and Ali Ammar. I also thank the ISE's funders, including the Hewlett Foundation, where Matt Baker gets special thanks for his encouragement, as well as Dan Adler at the Energy Foundation and Bloomberg Philanthropies, all of whom generously supported the work of the ISE over the last three years.

I also received critical encouragement and support from my colleagues at Energy Impact Partners (EIP), starting at the top with Hans Kobler (and earlier, Steve Hellman). Without Hans's infectious enthusiasm, I am sure I would have abandoned ship along the way. I am also very grateful to EIP colleagues Shayle Kann, Andy Lubershane, Kevin Fitzgerald, Cassie Bowe, Michael Donnelly, Sameer Reddy, Lindsay Luger, Evan Pittman, Madison Freeman, and Vienna Poiesz—the nicest as well as sharpest strategists in the industry today, bar none.

Finally, my career-long colleagues at the Brattle Group have been a source of unparalleled expertise and industry wisdom as well as friendship. I thank all of my Regulatory and Energy Markets colleagues there, including Judy Chang, Sam Newell, Kathleen Spees, Frank Graves, Ahmad Faruqui, Ryan Hledik, Sanem Sergici, Phillip Hanser, Alexis Maniatis, Bill Zarakas, Dean Murphy, Kathleen Spees, Jurgen Weiss, Mike Hagerty, Pablo Ruiz, Roger Lueken, Dan Jang, Romkaew Broehm, Gary Taylor, Ira Shavel, Bob Mudge, Jose Garcia, Serena Hesmondhalgh, Carlos Lapuerta, and most of all, my forever assistant Marianne Gray. Special thanks to Sanem and Hannes for very thoughtful edits of Appendix A.

Over almost two years of writing, many generous souls helped along the way. For the first two chapters, I thank Amory Lovins, Steve Nadel, Joe Romm, Mikhail Chester, Chris Hoehne, Chris MacCahill, and Henry Kelly. For help with Chapters 4 and 5, I am grateful to Jim Fama, James Mandel, Erfan Ibrahim, Christopher Sprague, and Rob Lee. For Chapters 6–8, I thank Bob Rowe, Bill Thompson, Sonya Aggarwal, Hal Harvey, Rob Gramlich, Michael Goggins, Karl Hausker, Mateo Jaramillo, Dan Berwick, Mike Boots, Shirley Mengrong Cheng, Yuan Ren, Frank Wang, Paul Joskow, Michael Caramanis, David Hart, Scott Willensky, Jim Hoecker, and *Smart Power*'s heroine Heidi Bishop, now at the World Resources Institute.

Chapters 9–12 were aided enormously by input from many colleagues, including Sue Kelly, Richard Kaufmann, Severin Borenstein, John Rhodes, Terry Sobolewski, Patricia DiOrio, Jon Wellinghoff, Audrey Zibelman, Jan Vrins, Perry Sioshansi, Jamie Wimberly, Patty Durand, Lisa Wood, Josh Wong, Carl Pechman, Mary Powell, and Lynn Keisling. Finally, Chapters 13 and 14 owe thanks to Dan Ford, Sheldon Simon, Douglas Simmons, Jay Horine, Jeffrey Holschuh, Stephen Byrd, and Mike Lapides.

I must also thank the commercial team behind this work, starting with my agent, Albert LaFarge. What other man of letters can toggle between extensive commentary on Shakespeare, the NPR News Hour, and Pink Floyd lyrics, all within a single blistering tennis game? Albert wisely brought me to Jeff Dean and Janice Audet at Harvard University Press—editors possessed of unrelenting courteousness and professionalism. Half-anticipating the worst, I experienced the best, and I thank them and their colleagues at the Press. Alexandra Kokkevi, my graphic artist, also deserves thanks.

There are three heroic souls without whom this book simply would not exist. Ryan Hopping patiently packed and moved a dozen boxes of files *twice* over the course of the book and then typed an endless series of drafts, all while juggling his many other duties. Olena Pechak smartly untangled the literature surrounding the very toughest issues in the book, while Nicole Mikkelson's impressive smarts and organizational skills took the book over the finish line. I am grateful beyond words to all three of them.

Above all, I thank my wife, Susan Vitka, for her encouragement, love, and sorely tested patience. I would promise never to do this again, but I know she wouldn't believe me.

Index